Cognitive and Neurophysiological Models of Brain Asymmetry

Cognitive and Neurophysiological Models of Brain Asymmetry

Editors

Sebastian Ocklenburg
Onur Güntürkün

MDPI • Basel • Beijing • Wuhan • Barcelona • Belgrade • Manchester • Tokyo • Cluj • Tianjin

Editors

Sebastian Ocklenburg
Psychology
MSH Medical School
Hamburg
Germany

Onur Güntürkün
Psychology
Ruhr University Bochum
Bochum
Germany

Editorial Office
MDPI
St. Alban-Anlage 66
4052 Basel, Switzerland

This is a reprint of articles from the Special Issue published online in the open access journal *Symmetry* (ISSN 2073-8994) (available at: www.mdpi.com/journal/symmetry/special_issues/Cognitive_Neurophysiological_Models_Brain_Asymmetry).

For citation purposes, cite each article independently as indicated on the article page online and as indicated below:

LastName, A.A.; LastName, B.B.; LastName, C.C. Article Title. *Journal Name* **Year**, *Volume Number*, Page Range.

ISBN 978-3-0365-4250-8 (Hbk)
ISBN 978-3-0365-4249-2 (PDF)

Contents

About the Editors

Sebastian Ocklenburg

Sebastian Ocklenburg is professor for research methods in psychology at the Institute for Cognitive and Affective Neuroscience at MSH Medical School in Hamburg, Germany. His research focuses on hemispheric asymmetries and left-handedness in humans, with a special interest in ontogenetic processes.

Onur Güntürkün

Onur Güntürkün is professor of biopsychology at the Institute of Cognitive Neuroscience at Ruhr University, Bochum. His research focuses on learning, cognition, and hemispheric asymmetries in birds, humans and other species.

Editorial

Cognitive and Neurophysiological Models of Brain Asymmetry

Sebastian Ocklenburg [1,2,3,*] and Onur Güntürkün [3]

[1] Department of Psychology, Medical School Hamburg, 20457 Hamburg, Germany
[2] ICAN Institute for Cognitive and Affective Neuroscience, Medical School Hamburg, 20457 Hamburg, Germany
[3] Faculty of Psychology, Institute of Cognitive Neuroscience, Biopsychology, Ruhr University Bochum, 44801 Bochum, Germany; onur.guentuerkuen@ruhr-uni-bochum.de
* Correspondence: sebastian.ocklenburg@medicalschool-hamburg.de

Abstract: Asymmetry is an inherent characteristic of brain organization in both humans and other vertebrate species, and is evident at the behavioral, neurophysiological, and structural levels. Brain asymmetry underlies the organization of several cognitive systems, such as emotion, communication, and spatial processing. Despite this ubiquity of asymmetries in the vertebrate brain, we are only beginning to understand the complex neuronal mechanisms underlying the interaction between hemispheric asymmetries and cognitive systems. Unfortunately, despite the vast number of empirical studies on brain asymmetries, theoretical models that aim to provide mechanistic explanations of hemispheric asymmetries are sparse in the field. Therefore, this Special Issue aims to highlight empirically based mechanistic models of brain asymmetry. Overall, six theoretical and four empirical articles were published in the Special Issue, covering a wide range of topics, from human handedness to auditory laterality in bats. Two key challenges for theoretical models of brain asymmetry are the integration of increasingly complex molecular data into testable models, and the creation of theoretical models that are robust and testable across different species.

Keywords: neuroscience; brain; asymmetry; laterality; functional hemispheric asymmetries; structural hemispheric asymmetries; theoretical models

Citation: Ocklenburg, S.; Güntürkün, O. Cognitive and Neurophysiological Models of Brain Asymmetry. *Symmetry* **2022**, *14*, 971. https://doi.org/10.3390/sym14050971

Received: 5 May 2022
Accepted: 6 May 2022
Published: 9 May 2022

Publisher's Note: MDPI stays neutral with regard to jurisdictional claims in published maps and institutional affiliations.

1. Introduction

Research on symmetry and asymmetry in the nervous system is a central part of neuroscience [1–4]. Over the last decade, tremendous progress has been made in research on brain asymmetries, due to large-scale consortium or databank studies [5]. For example, large-scale databank studies have investigated the genetics of structural brain asymmetries [6] and handedness [7–10], as well as the influence of early life factors on handedness [11], and the role of epigenetic processes in handedness ontogenesis [12]. Unfortunately, despite the vast number of data-driven studies on brain asymmetries, more recent publications featuring theoretical models that aim to provide mechanistic explanations of hemispheric asymmetries are sparse in the field. This is in line with the larger development in psychology and neuroscience, which has been called the "theory crisis" [13]. Following earlier works [14], Eronen and Bringmann (2021) argue that the theoretical foundations of psychology are shaky, and that, instead of gathering more and more data, the field needs to shift to developing better theories, which, in turn, inform empirical research.

Research on symmetry and asymmetry in the brain has always been a field driven by influential theories. A few examples are the McManus dextral/chance (DC) model of handedness and language dominance [15], the pathological left-handedness model [16], the right-hemisphere and valence model of emotional lateralization [17], and the Geschwind–Galaburda–Behan model [18–20]. However, since approximately the year 2000, the number of new, influential theories about symmetry and asymmetry in the brain has declined considerably. Moreover, many of the older theories were formulated before the widespread use of modern genetic and neuroimaging techniques, and were often not in line with newer empirical findings obtained with such techniques [21]. Unfortunately, not all authors updated their theories to reflect newer empirical findings, leading to a lack of theoretical models that aim to provide mechanistic explanations of hemispheric asymmetries. Therefore, it was the aim of the present Special Issue to highlight empirically based mechanistic models of brain asymmetry. To this end, we invited several experts in the field of hemispheric asymmetries to contribute theoretical or empirical papers on models of brain asymmetries. Both submissions covering research in humans and research in non-human model species were welcome. Overall, ten excellent articles were submitted to the Special Issue, covering a broad range of theoretical models on hemispheric asymmetries in a wide range of species. Six articles were review papers and four presented new empirical research. In the following sections, we will shortly discuss each of these contributions to the Special Issue.

2. Theoretical Articles

As mentioned above, the McManus DC model of handedness and language dominance [15] was one of the most influential models of handedness ontogenesis. In his contribution to the Special Issue, entitled "Cerebral Polymorphisms for Lateralisation: Modelling the Genetic and Phenotypic Architectures of Multiple Functional Modules", McManus presents an update and extension of this model with two central changes [22]. Building upon the 2014 polygenic revision of the DC model [21], McManus presents an extended polygenic DC model informed by recent studies on the genetics of handedness and, in particular, the role of cilia. Moreover, the model is further extended to include cerebral polymorphisms that are based on a multitude of functional modules for different lateralized cognitive systems, such as language, praxis, and visuo-spatial functioning. This idea is in line with recent findings showing that multidimensional phenotypes improve the genetic analysis of laterality traits [23]. It is very encouraging to observe how a leading model in the field is adjusted to fit with recent empirical findings, even decades after it was first published.

Similarly to McManus, Paracchini focusses on human handedness in her contribution, entitled "Recent Advances in Handedness Genetics" [24]. She highlights several recent advances in the understanding of the genetics of handedness, based on databank studies with large sample sizes, but also highlights the importance of phenotyping, i.e., which handedness measure is used in a study.

The contribution by Guy Vingerhoets, Robin Gerrits, and Helena Verhelst, entitled "Atypical Brain Asymmetry in Human Situs Inversus: Gut Feeling or Real Evidence?" [25], is also related to the ontogenesis of handedness and other forms of hemispheric asymmetries in humans. Whether or not individuals with situs inversus also show inverted hemispheric asymmetries has been discussed for decades [26,27]. While the data pattern found in previous studies in humans has been inconclusive, Vingerhoets et al. (2021) suggest a model that assumes that cilia play a critical role in determining whether someone with situs inversus shows reversed hemispheric asymmetries or not. They suggest that greater attention needs to be paid to the subtypes of situs inversus, and that situs inversus with a ciliary etiology is related to reversed hemispheric asymmetries, while situs inversus with a non-ciliary etiology is related to typical hemispheric asymmetries.

In addition to these theoretical articles that were mainly focused on research in human subjects, three theoretical articles integrated findings from human subjects with comparative research. In her contribution, "It Is Not Just in the Genes", Martina Manns integrates human and animal research, with a focus on birds, to create a new multi-level model for asymmetry formation [28]. The model focuses on the cellular processes that determine hemispheric asymmetries during embryonic patterning, neural differentiation, and refinement of neural circuits.

In their contribution, entitled "Structural Brain Asymmetries for Language: A Comparative Approach across Primates", Yannick Becker and Adrien Meguerditchian also take a strongly comparative look at hemispheric asymmetries, but focus on primates instead of birds [29]. They highlight how in several non-human primate species, human-like structural brain asymmetries can be found in brain areas that are homologous to key language regions in the human brain. This finding proposes a challenge for models of human language lateralization, as it suggests that these structural asymmetries in language areas did not develop for language per se, as they are also present in non-linguistic primates. Becker and Meguerditchian suggest that gestural communication may be a key factor here. Intriguingly, this idea is very much in line with the "from hand to mouth" theory about the origins of language, proposed by Mike Corballis 20 years ago [30]. Mike Corballis also contributed a theoretical paper to this Special Issue [31]. Entitled "Asymmetry research in human subjects and in non-human species—How Asymmetries Evolved: Hearts, Brains, and Molecules", this article gives a cross-species overview of the evolution of asymmetries in the body and brain, and their potential molecular basis.

3. Empirical Articles

In addition to these six theoretical articles, the Special Issue includes four empirical articles. Pamela Villar González and co-workers presented a study on dichotic listening in Silbo Gomero, a form of whistled Spanish, entitled "Lateralization of Auditory Processing of Silbo Gomero" [32]. Whistled languages are highly interesting in the context of theoretical models of language lateralization [33]. While whistled languages typically use the full lexical and syntactic properties of the spoken languages they are derived from, their acoustic properties differ from the acoustic properties of spoken languages. While the left hemisphere typically shows dominance for processing spoken languages, the right hemisphere is dominant for processing spectral cues, pitch, and melodic lines [34], all of which are central for understanding whistled languages. Testing the assumptions of theoretical models of language lateralization using both spoken and whistled stimuli may allow us to better disentangle which lateralized processes are relevant for language.

The contribution by Stuart Washington and co-workers also belongs to the acoustic domain, entitled "Hemispheric and Sex Differences in Mustached Bat Primary Auditory Cortex Revealed by Neural Responses to Slow Frequency Modulations" [35]. This study convincingly shows how unusual model species can yield very informative results in research on hemispheric asymmetries. Washington et al., (2021) investigated hemispheric asymmetries in the primary auditory cortex of mustached bats (*Pteronotus parnellii*). Similarly to humans, these bats show leftward asymmetry for complex social vocalizations. Washington et al., (2021) demonstrated that this asymmetry is driven by spectro-temporal processing differences, which, to some extent, mirrors the findings in humans. This work highlights that using a broader range of model species in laterality research than those typically used (e.g., rats, mice, pigeons, and chickens) could be very beneficial to test laterality models in an evolutionary context.

The contribution by Gisela Kaplan and Lesley J. Rogers, entitled "Brain Size Associated with Foot Preferences in Australian Parrots", shows another important empirical technique to test laterality models in an evolutionary context [36]. In this study, the authors did not analyze data from one species, but assessed foot preferences and brain masses in 25 psittacine species from Australia. Importantly, they found that birds with larger brain masses showed stronger foot preferences. We expect to observe more multi-species studies aimed at testing evolutionary models of laterality in the future. Clearly, a theoretical model is stronger if its predictions hold true across different species.

Finally, the contribution by Hao Cheng and co-workers, entitled "A Simulation on Relation between Power Distribution of Low-Frequency Field Potentials and Conducting Direction of Rhythm Generator Flowing through 3D Asymmetrical Brain Tissue", reported an EEG simulation, taking into account brain asymmetries [37]. Their work may be helpful for testing theoretical models of EEG asymmetries.

4. Conclusions

Taken together, the ten articles included in the present Special Issue, "Cognitive and neurophysiological models of brain asymmetry", give several insights into theoretical models of hemispheric asymmetries in 2022. Clearly, one of the key challenges identified in several articles is integrating the increasingly complex findings of molecular genetic and epigenetic studies in humans and non-human animal species into testable theoretical models. Before the wide availability of molecular research methods, models were typically based on statistical distributions of phenotypes. Molecular research has clearly shown that many of these models were oversimplified, and that the field needs to adjust. Particularly, cilia function needs to be integrated into theoretical models about the ontogenesis of hemispheric asymmetries. In addition, we are convinced that the next decade will observe stronger cross-species integration in theoretical models of hemispheric asymmetries, particularly in the context of evolutionary models. Along with this, more research in non-typical model species, to test specific aspects of theoretical models, will emerge. We hope that the theoretical and empirical articles presented in this Special Issue will lead to empirical studies testing these models in various contexts.

Author Contributions: Conceptualization, S.O. and O.G.; writing—original draft preparation, S.O. and O.G.; writing—review and editing, S.O. and O.G. All authors have read and agreed to the published version of the manuscript.

Funding: This research received no external funding.

Institutional Review Board Statement: Not applicable.

Informed Consent Statement: Not applicable.

Data Availability Statement: Not applicable.

Conflicts of Interest: The authors declare no conflict of interest.

References

1. Mundorf, A.; Peterburs, J.; Ocklenburg, S. Asymmetry in the Central Nervous System: A Clinical Neuroscience Perspective. *Front. Syst. Neurosci.* **2021**, *15*, 733898. [CrossRef]
2. Ocklenburg, S.; Berretz, G.; Packheiser, J.; Friedrich, P. Laterality 2020: Entering the next decade. *Laterality* **2021**, *26*, 265–297. [CrossRef]
3. Vallortigara, G.; Rogers, L.J. A function for the bicameral mind. *Cortex* **2020**, *124*, 274–285. [CrossRef]
4. Wiper, M.L. Evolutionary and mechanistic drivers of laterality: A review and new synthesis. *Laterality* **2017**, *22*, 740–770. [CrossRef]
5. Kong, X.-Z.; Postema, M.C.; Guadalupe, T.; de Kovel, C.; Boedhoe, P.S.W.; Hoogman, M.; Mathias, S.R.; Rooij, D.; Schijven, D.; Glahn, D.C.; et al. Mapping brain asymmetry in health and disease through the ENIGMA consortium. *Hum. Brain Mapp.* **2022**, *43*, 167–181. [CrossRef] [PubMed]
6. Sha, Z.; Schijven, D.; Carrion-Castillo, A.; Joliot, M.; Mazoyer, B.; Fisher, S.E.; Crivello, F.; Francks, C. The genetic architecture of structural left-right asymmetry of the human brain. *Nat. Hum. Behav.* **2021**, *5*, 1226–1239. [CrossRef] [PubMed]
7. Cuellar-Partida, G.; Tung, J.Y.; Eriksson, N.; Albrecht, E.; Aliev, F.; Andreassen, O.A.; Barroso, I.; Beckmann, J.S.; Boks, M.P.; Boomsma, D.I.; et al. Genome-wide association study identifies 48 common genetic variants associated with handedness. *Nat. Hum. Behav.* **2021**, *5*, 59–70. [CrossRef] [PubMed]
8. de Kovel, C.G.F.; Francks, C. The molecular genetics of hand preference revisited. *Sci. Rep.* **2019**, *9*, 5986. [CrossRef] [PubMed]
9. Sha, Z.; Pepe, A.; Schijven, D.; Carrión-Castillo, A.; Roe, J.M.; Westerhausen, R.; Joliot, M.; Fisher, S.E.; Crivello, F.; Francks, C. Handedness and its genetic influences are associated with structural asymmetries of the cerebral cortex in 31,864 individuals. *Proc. Natl. Acad. Sci. USA* **2021**, *118*, e2113095118. [CrossRef] [PubMed]
10. Wiberg, A.; Ng, M.; Al Omran, Y.; Alfaro-Almagro, F.; McCarthy, P.; Marchini, J.; Bennett, D.; Smith, S.; Douaud, G.; Furniss, D. Handedness, language areas and neuropsychiatric diseases: Insights from brain imaging and genetics. *Brain* **2019**, *142*, 2938–2947. [CrossRef] [PubMed]
11. de Kovel, C.G.F.; Carrión-Castillo, A.; Francks, C. A large-scale population study of early life factors influencing left-handedness. *Sci. Rep.* **2019**, *9*, 584. [CrossRef] [PubMed]
12. Odintsova, V.V.; Suderman, M.; Hagenbeek, F.A.; Caramaschi, D.; Hottenga, J.-J.; Pool, R.; Heijmans, B.T.; Hoen, P.A.C.; van Meurs, J.; Isaacs, A.; et al. DNA methylation in peripheral tissues and left-handedness. *Sci. Rep.* **2022**, *12*, 5606. [CrossRef] [PubMed]

13. Eronen, M.I.; Bringmann, L.F. The Theory Crisis in Psychology: How to Move Forward. *Perspect. Psychol. Sci.* **2021**, *16*, 779–788. [CrossRef] [PubMed]
14. Meehl, P.E. Theory-Testing in Psychology and Physics: A Methodological Paradox. *Philos. Sci.* **1967**, *34*, 103–115. [CrossRef]
15. McManus, I.C. Handedness, language dominance and aphasia: A genetic model. *Psychol. Med. Monogr. Suppl.* **1985**, *8*, 1–40. [CrossRef] [PubMed]
16. Satz, P.; Orsini, D.L.; Saslow, E.; Henry, R. The pathological left-handedness syndrome. *Brain Cogn.* **1985**, *4*, 27–46. [CrossRef]
17. Palomero-Gallagher, N.; Amunts, K. A short review on emotion processing: A lateralized network of neuronal networks. *Brain Struct. Funct.* **2022**, *227*, 673–684. [CrossRef]
18. Geschwind, N.; Galaburda, A.M. Cerebral lateralization. Biological mechanisms, associations, and pathology: I. A hypothesis and a program for research. *Arch. Neurol.* **1985**, *42*, 428–459. [CrossRef]
19. Geschwind, N.; Galaburda, A.M. Cerebral lateralization. Biological mechanisms, associations, and pathology: II. A hypothesis and a program for research. *Arch. Neurol.* **1985**, *42*, 521–552. [CrossRef]
20. Geschwind, N.; Galaburda, A.M. Cerebral lateralization. Biological mechanisms, associations, and pathology: III. A hypothesis and a program for research. *Arch. Neurol.* **1985**, *42*, 634–654. [CrossRef]
21. Armour, J.A.L.; Davison, A.; McManus, I.C. Genome-wide association study of handedness excludes simple genetic models. *Heredity* **2014**, *112*, 221–225. [CrossRef] [PubMed]
22. McManus, C. Cerebral Polymorphisms for Lateralisation: Modelling the Genetic and Phenotypic Architectures of Multiple Functional Modules. *Symmetry* **2022**, *14*, 814. [CrossRef]
23. Schmitz, J.; Zheng, M.; Lui, K.F.H.; McBride, C.; Ho, C.S.-H.; Paracchini, S. Quantitative multidimensional phenotypes improve genetic analysis of laterality traits. *Transl. Psychiatry* **2022**, *12*, 68. [CrossRef] [PubMed]
24. Paracchini, S. Recent Advances in Handedness Genetics. *Symmetry* **2021**, *13*, 1792. [CrossRef]
25. Vingerhoets, G.; Gerrits, R.; Verhelst, H. Atypical Brain Asymmetry in Human Situs Inversus: Gut Feeling or Real Evidence? *Symmetry* **2021**, *13*, 695. [CrossRef]
26. Brown, N.A.; Hoyle, C.I.; McCarthy, A.; Wolpert, L. The development of asymmetry: The sidedness of drug-induced limb abnormalities is reversed in situs inversus mice. *Development* **1989**, *107*, 637–642. [CrossRef]
27. Morgan, M.J. The asymmetrical genetic determination of laterality: Flatfish, frogs and human handedness. *Ciba Found Symp.* **1991**, *162*, 234–247. [CrossRef]
28. Manns, M. It Is Not Just in the Genes. *Symmetry* **2021**, *13*, 1815. [CrossRef]
29. Becker, Y.; Meguerditchian, A. Structural Brain Asymmetries for Language: A Comparative Approach across Primates. *Symmetry* **2022**, *14*, 876. [CrossRef]
30. Corballis, M.C. *From Hand to Mouth. The Origins of Language*; Princeton University Press: Princeton, NJ, USA, 2002.
31. Corballis, M.C. How Asymmetries Evolved: Hearts, Brains, and Molecules. *Symmetry* **2021**, *13*, 914. [CrossRef]
32. Villar González, P.; Güntürkün, O.; Ocklenburg, S. Lateralization of Auditory Processing of Silbo Gomero. *Symmetry* **2020**, *12*, 1183. [CrossRef]
33. Güntürkün, O.; Güntürkün, M.; Hahn, C. Whistled Turkish alters language asymmetries. *Curr. Biol.* **2015**, *25*, R706–R708. [CrossRef] [PubMed]
34. Scott, S.K.; McGettigan, C. Do temporal processes underlie left hemisphere dominance in speech perception? *Brain Lang.* **2013**, *127*, 36–45. [CrossRef] [PubMed]
35. Washington, S.D.; Pritchett, D.L.; Keliris, G.A.; Kanwal, J.S. Hemispheric and Sex Differences in Mustached Bat Primary Auditory Cortex Revealed by Neural Responses to Slow Frequency Modulations. *Symmetry* **2021**, *13*, 1037. [CrossRef]
36. Kaplan, G.; Rogers, L.J. Brain Size Associated with Foot Preferences in Australian Parrots. *Symmetry* **2021**, *13*, 867. [CrossRef]
37. Cheng, H.; Ge, M.; Belkacem, A.N.; Fu, X.; Xie, C.; Song, Z.; Chen, S.; Chen, C. A Simulation on Relation between Power Distribution of Low-Frequency Field Potentials and Conducting Direction of Rhythm Generator Flowing through 3D Asymmetrical Brain Tissue. *Symmetry* **2021**, *13*, 900. [CrossRef]

Review

Cerebral Polymorphisms for Lateralisation: Modelling the Genetic and Phenotypic Architectures of Multiple Functional Modules

Chris McManus 

Research Department of Clinical, Educational and Health Psychology and Research Department of Medical Education, University College London, London WC1E 6BT, UK; i.mcmanus@ucl.ac.uk

Abstract: Recent fMRI and fTCD studies have found that functional modules for aspects of language, praxis, and visuo-spatial functioning, while typically left, left and right hemispheric respectively, frequently show atypical lateralisation. Studies with increasing numbers of modules and participants are finding increasing numbers of module combinations, which here are termed *cerebral polymorphisms*—qualitatively different lateral organisations of cognitive functions. Polymorphisms are more frequent in left-handers than right-handers, but it is far from the case that right-handers all show the lateral organisation of modules described in introductory textbooks. In computational terms, this paper extends the original, monogenic McManus DC (dextral-chance) model of handedness and language dominance to multiple functional modules, and to a polygenic DC model compatible with the molecular genetics of handedness, and with the biology of visceral asymmetries found in primary ciliary dyskinesia. Distributions of cerebral polymorphisms are calculated for families and twins, and consequences and implications of cerebral polymorphisms are explored for explaining aphasia due to cerebral damage, as well as possible talents and deficits arising from atypical inter and intra-hemispheric modular connections. The model is set in the broader context of the testing of psychological theories, of issues of laterality measurement, of mutation-selection balance, and the evolution of brain and visceral asymmetries.

Keywords: cerebral polymorphisms; handedness; cerebral dominance; DC model; genetics; polygenic model; brain asymmetry; visceral asymmetry; bilateral language; functional modules

Citation: McManus, C. Cerebral Polymorphisms for Lateralisation: Modelling the Genetic and Phenotypic Architectures of Multiple Functional Modules. *Symmetry* **2022**, *14*, 814. https://doi.org/10.3390/sym14040814

Academic Editors: Sebastian Ocklenburg and Onur Güntürkün

Received: 4 November 2021
Accepted: 16 February 2022
Published: 14 April 2022

1. Introduction

People are mostly right-handed or left-handed, and since handedness is determined by the brain, handedness is a lateralised *cerebral polymorphism*, people having qualitatively different brain organisations. Language in most people is controlled by the left cerebral hemisphere, as Dax [1] and Broca [2] realised in the nineteenth century [3]. Two decades after Broca, left-handers were wrongly thought to mirror right-handers, with Sir William Gowers in 1887 stating that "speech-processes go on chiefly in the left hemisphere in right-handed persons, in the right hemisphere in left-handed persons" [4] (pp. 131–132). It took a further six decades for it to be accepted that most left-handers, like right-handers, actually have left-hemisphere language [5] (p. 331). Modern estimates suggest that about 30% of left-handers have language in their right hemisphere as do about 5% of right-handers, although estimates vary [6]. Since handedness can be right or left and language dominance can be right or left, there are four lateral combinations for this cerebral polymorphism.

The terminology of polymorphisms can be confusing, and in this paper I will refer to individual functional processes such as *language dominance*, *visuo-spatial processing* or *handedness* as *modules* [7], with individual modules lateralised to the right or left side of the brain; in particular, handedness will always be treated as a module. Different neural organisations have been referred to as combinations of *multiple modular traits* [8], *phenotypes of brain functional organisation* [9], or what we began to call *cerebral polymorphisms* [10–12].

Cerebral polymorphism is a useful portmanteau term for the variability found in lateralised cerebral organisation, with specific details relating to particular functional modules.

A major interest of *cerebral polymorphisms for lateralisation* comes from there being *qualitative variation* between individuals in their brain organisation, which may relate to specific skills, talents, deficits and responses to damage. Whereas most studies of individual differences in brain functioning consider continuous measures, cerebral polymorphisms explicitly consider behaviours and functions that show categorically different behaviours. In an everyday, colloquial sense, when people say of a person that "their brain seems to be organised differently to mine", there may be a deeper element of truth, with qualitative differences perhaps explaining different responses to damage, as well as talents and deficits. The latter idea is far from new, dating back at least to Orton's theorising on dyslexia [13,14], with recent work confirming that dyslexia is more prevalent in left-handers [15].

2. The Prevalence of Cerebral Polymorphisms

Cerebral polymorphisms have not been well described in the literature, and as Gerrits et al. say, "Little is known about the relationships between lateralised functions, in part because there is a paucity of studies measuring multiple functional asymmetries in the same individuals" [16] (p. 14061), with most studies "exploring asymmetries of a single cognitive function at a time" [17].

2.1. Three-Module Studies

Vingerhoets et al. tabulated eight studies of multiple modules [9] (p. 7), fTCD being used in four studies, fMRI used in two studies, and dichotic listening and lesions in one study each. Seven of the eight studies looked at only three modules (including handedness).

A much earlier study from 1983, by McGlone and Davidson, used dichotic listening and tachistoscopic methods, albeit both somewhat unreliable methods, to study *language* and *visuo-spatial processing* in relation to *handedness*, and found all eight possible combinations of the three modules [18,19]. Those and other data were modelled in McManus' 1985 monograph on the DC genetic model of handedness and lateralisation, with calculations provided for the proportions of the three modules [20]. Those and other calculations will be expanded later in this paper.

A very important, but relatively ignored, earlier study looking at three modules was Bryden, Hécaen and De Agostini's 1983 analysis of 270 patients with unilateral brain lesions [21]. They analysed three modules, *language* (indicated by aphasia), *visuo-spatial analysis* (indicated by agnosia), and *handedness*, and concluded that the lateralisation of language and visuo-spatial analysis showed *statistical independence*, a term to be discussed later.

Although most three-module studies have looked at *handedness* and a typically left-hemisphere function (*language*) and a typically right-hemisphere function (*visuo-spatial ability*), the central interest of the 2021 study of Kroliczak et al. was praxis-assessed using fMRI in 125 participants, the three modules being *tool use pantomime*, *word generation* and *handedness*, all typically in the left hemisphere [22]. In total, 66% of 125 participants had both praxis and language in the left hemisphere, 22% had atypical praxis, 2% had atypical language, and 10% had atypical language and praxis. Atypical praxis dominance was more frequent in left-handers than right-handers

In 2018, Beking used fTCD to study two typical right hemisphere modules (*mental rotation*, MR and a *chimeric faces task*, CF), as well as *word generation*, WG, a left-hemisphere task in 55 participants [23]. Handedness was not reported. Seven of the eight possible combinations of modules were found, with only 32 (58%) having typical lateralisation (RRL for MR, CF and WG), 2 being entirely mirrored (LLR), and considering just MR and CF, 6 had LR for MR and CF, and 1 had RL [23] (See Appendix Table A3, p. 66 of reference 23).

2.2. Four or More Modules

Four modules. Studies with four or more modules have only appeared in the past decade or so, with that of Badzekova-Trajkov et al. being seminal. Using fMRI, 87 participants were assessed on four modules (*frontal lobe speech production*, *temporal lobe face processing*, *parietal lobe spatial processing* and *handedness*), and found 12 of the 16 possible lateralised combinations [24]. The lateralised parietal lobe (landmark) task was also independent of other modules, in particular not being correlated with handedness.

Five modules. Three recent studies have increased the number of modules being analysed, and are also characterised by upweighting the numbers of atypical individuals (typically left-handers). In 2020, Gerrits et al. used fMRI in a sample of 63 left-handers to study five modules (*language*, *praxis*, *spatial attention*, *face recognition*, and *emotional prosody*), and found only 27 participants with the 'typical' pattern of language and praxis on the left, and the other modules on the right, with 19 of the 32 possible lateralised patterns being found [16].

Six modules. The important 2020 study of Emma Karlsson used fMRI in 67 participants who were over-selected for left-handedness and other predictors of atypical dominance, to assess lateralisation of six different modules (*verbal fluency*, *face recognition*, *perception of neutral bodies*, *emotional prosody*, *emotional vocalisation*, and *handedness*), finding 30 of the 64 possible lateral combinations [25] (p. 114).

Seven modules. In 2019 and 2021, Woodhead et al. used fTCD to study seven modules, including six different language tasks (list generation, phonological decision, semantic decision, sentence generation, sentence comprehension and syntactic decision) as well as handedness [26,27]. The results were expressed as correlations of raw laterality indices, so the number of independent combinations is unclear, but the data are clearly multivariate and do not fit a 'unitary theory'.

This brief review makes clear that as the number of modules increases, so the number of cerebral polymorphic combinations increases. Larger numbers of modules do though mean that not all combinations are described, which probably reflects some combinations being intrinsically rarer and relatively small sample sizes making it harder to find some modules than others. As an example, with six modules, and $2^6 = 64$ potential combinations, a study with N = 67 suggested it was statistically extremely unlikely to encounter them all.

Some studies have attempted to find a typology for cerebral polymorphisms; for instance, Karlsson [25], following earlier researchers, used terms such as 'Traditional', 'Reversed' and 'Right Hemispheric', but only 17, 4 and 2 individuals, respectively, fitted into those categories, with the remaining 44 participants in 'Other patterns', with 25 different types [25].

Despite the variability described in the various studies, there is also little doubt that the population as a whole shows a modal pattern of lateral organisation, which is by far the most frequent, and is described in introductory textbooks for a typical right-hander, with language in the left hemisphere, visuo-spatial processing in the right, etc. As Gerrits et al. conclude, " ... while typical organization is generally maintained, it is subject to more variation than is often assumed" (p. 14057), which "raise[s] a host of questions to be addressed by future research" [16] (p. 14063). Addressing those questions is what the rest of the present paper will attempt to do. In particular, it will assess how polymorphisms originate, and why some polymorphic forms are more prevalent than others, and the extent to which they can be explained and their proportions predicted using genetic models originating in the study of handedness. There are many difficult questions, not least the potential number of modules and their possible inter-combinations, and the consequences of different lateral organisations of the inter-combinations. For most purposes, the modelling will be kept simple, so that its broad approach can be seen and a ground plan set out, rather than the details being explored, which can be left to future work.

2.3. Dynamic Shifts in Functional Lateralization

The modelling in the present study assumes that functional lateralisation is fixed or static within individuals. Studies of dynamic variation in functional lateralisation are rare, an important exception being in female participants in relation to estradiol and progesterone levels during the menstrual cycle [28,29]. It may be that more sensitive methods such as double biofeedback [30] or bimanual control of an avatar [31] will, in the future, be able to assess dynamic changes more effectively. For present purposes, the assumption of fixed asymmetries will probably be sufficient.

3. Patterns of Cerebral Organisation and Terminologies

Theoretical thinking about how modules may be organised has gone through several stages, and the literature contains multiple terms used for describing different *patterns of cerebral polymorphisms*, some of which are summarised below.

- **Cerebral dominance**. An idea dating back to the nineteenth century is of strong *cerebral dominance*, with a *major (or leading) hemisphere*, and *a minor hemisphere*, language in the left hemisphere directly *causing* visuo-spatial function to be in the right hemisphere.
- **Hemispheric dominance/functional lateralisation.** The tendency for a module or function to be predominantly organised in one hemisphere both in individuals and in the population.
- **Typical or traditional.** The typical or traditional pattern [25] ("typical functional segregation" [17]) is that described in neuropsychology textbooks for a typical right-hander, with left-hemisphere language, right-hemisphere visuo-spatial processing, etc. It is useful in modelling polymorphisms to describe each module as being *typical* or *atypical*, rather than right or left (where typical depends on the type of module, e.g., verbal on the left or visuo-spatial on the right).
- **Complementarity.** Bryden et al. distinguished two very different meanings of complementarity, *causal* and *statistical* [21]. *Statistical complementarity* refers to the "the normal state of affairs", in the sense of the mode in statistics, in the normative sense of language in the left hemisphere and visuo-spatial processing in the right hemisphere, i.e., the typical or traditional pattern [21]. Bryden et al. also describe *causal complementarity*, "implying that one hemisphere carries out a particular set of functions because the complementary functions are located to the other hemisphere". They suggest the latter "seems to have become part of the lore of contemporary neuropsychology, especially as viewed by the popular press". They continue that the strong idea "that the right hemisphere acquires its functions only in response to the specialization of the left hemisphere" [32,33],"cannot be correct at the level of the individual, although the lateralisation of language may well have preceded that of visuospatial processes in the population" [21].
- **Statistical independence**. Statistical independence is used by Bryden et al. in their 1983 analyses of aphasia, agnosia and handedness after unilateral lesions, where they use chi-square tests to show that there is only a small association between the presence of aphasia and the presence of agnosia after a unilateral lesion [21]. Causal complementarity should result in a strong negative association between aphasia and agnosia, the presence of one after a unilateral lesion meaning that the other should be absent. Bryden (1990) refers to *statistical complementarity*, whereby random allocation of modules to the right or left will sometimes result in a complementary pattern whereby language is on the left and visuo-spatial function is on the right, without any underlying causal process [34].
- **Reversed complementarity or mirror reversal**. This is used to describe individuals with the reverse of the typical pattern of complementarity (e.g., where the typical complementary pattern is LRRRR, with one verbal and four non-verbal tasks, then the reversed or mirror pattern is RLLLL). Karlsson found 4 such individuals out of 67, compared with 17 who were complementary (LRRRR), and the remaining 46 individuals showed 22 other combinations, excluding handedness [25] (p. 114). Additionally, referred to as 'mirror-reversed' [16], and by analogy with *situs inversus totalis* has been called *mens inversus totalis* [9,17] (p. 8).
- **Crossed laterality**. This is an old term for describing individuals who are right-handed but have language in the right-hemisphere, occurring particularly in cases of

'crossed aphasia', which seemed to be the opposite of what Broca had described, with right-handers being aphasic after right-hemisphere damage [35,36].

- **The unitary theory**. 'Unitary theory', as described by Woodhead et al., applied specifically to the cerebral lateralisation of language, the authors stating that "at the population level, we may ask [when] all language tasks show a similar degree of lateralisation, and at the individual level, [when] people show consistent differences in laterality profiles across tasks" [26] (p. 17). The authors suggest that although "the majority of people appear to have language laterality driven by a single process affecting all types of task", there is a "a minority showing fractionation of language asymmetry", particularly in left-handers [26,27].
- **Crowding.** The functional crowding hypothesis dates back to Lansdell in 1969 [37], Levy in 1969 [38] and Teuber in 1974 [39], reviewed by Groen et al. [40]. "[C]ompetition for neural resources would result in a functional deficit if multiple functions rely on the same hemisphere", and has also been called the "cognitive laterality profile" hypothesis [41], "load imbalance" [42], and the "parallel processing" [43,44] account. Crowding predicts that individuals with multiple modules in the same hemisphere should underperform compared with those with modules spread across the hemispheres [40]. *Atypical functional segregation* is said to be characterised by functional crowding, in contrast to typical functional organisation and reversed complementarity (*reversed functional segregation*) [9], although studies have found very limited evidence of any functional deficit with crowding [45]. The term crowding has also been used as a simple description of two modules being in the same hemisphere when they are usually in separate hemispheres, with no implication of functional disadvantage [46]. *Pseudo-crowding* has been used to refer to the case where modules are adjacent because they overlap in their functions which may benefit both [22,47].

4. Cerebellar Asymmetries

The majority of this article concerns 'cerebral polymorphisms' in the narrow sense of the right and left cerebral hemispheres. However, 'Cerebral' also has a broader meaning, originating in the Latin *cerebrum*, the entire brain, with the Oxford English Dictionary defining 'cerebral' as "Pertaining or relating to the brain, or to the cerebrum", which is the sense in, say, 'cerebrovascular disease. 'Cerebral polymorphisms' can therefore include the fore-, mid- and hind-brain, including the cerebellum. Early fMRI and other imaging techniques ignored the posterior fossa, looking only at supra-tentorial structures, but later studies revealed structural and functional asymmetries of the cerebellum. In relation to explaining handedness, Michael Peters said that "... the cerebellum does not normally enter the discussion. However, there are good reasons to focus some attention on this structure" [48]. Most researchers tend to assume that functional asymmetries have to be cortical in origin; there is a growing awareness that the cerebellum may also be asymmetric in its functioning [49], and may be related to handedness and other functions, including perhaps handedness [50,51]. It is also possible that symmetries relate from turning tendencies originating in the brain stem [52–54].

The cerebellum has long been known to be involved in motor control, and its relationship to handedness has therefore been of interest. The early morphological study by Snyder et al. in 1995 [55] suggested, in 23 participants, that cerebellar torque was related to handedness, whereas cortical torque was not. Of some interest is a study of the dentate nuclei in the cerebellum, where nine right-handers had a larger left dentate nucleus, but the sole left-hander had a larger right dentate [56]. However, a recent study of 2226 participants found no correlation of cerebellar anatomical asymmetry and handedness [57]. Functional analyses have suggested that handedness is related to contralateral cortical activity and ipsilateral cerebellar activity, with a strong cortico-cerebellar network [58]. The detailed functional study of handedness by Tzourio-Mazoyer and colleagues emphasised, though, "that handedness neural support is complex and not simply based on a mirrored organization of hand motor areas", but with two different mechanisms in right- and left-handers [59]. The fMRI study of Häberling and Corballis [60] suggested two separate patterns of cerebellar activities, with cerebellar asymmetry related to a fronto-temporal cortical language network, and handedness to an action-based parieto-cortical network. Overall, there are undoubtedly functional asymmetries in cerebellar activity, but it is unclear whether these are sometimes independent of cortical asymmetries. In particular, to what extent do individuals with atypically lateralised frontal and temporal dominance also show the same pattern at the cerebellar level; or is it perhaps possible that, on occasion, cortical and cerebellar asymmetries become separated? At present, there are probably insufficient data from individuals to be clear whether cortical and cerebellar asymmetries can become separate, or whether mostly they march in lock-step.

5. The Approach of the Present Paper

Explaining cerebral polymorphisms will not be straightforward, and in part that reflects a problem that besets much of current psychology, that data collection is currently privileged over theory, with many theories being relatively weak, and primarily verbal in structure, making prediction and testing difficulty. This paper will therefore begin by briefly considering the nature and paucity of theory in psychology in general, as theory has been ignored during the important concerns of the replication crisis over the past decade or two, and that is to a large extent also true in neuroscience and neuropsychology.

A particular branch of theory that is likely to be important is genetics, which has had robust theoretical approaches for the past century, typically numerical, from the work of Galton, Pearson, Fisher and Sewall Wright, and in recent years has also had the support of molecular genetics. Theorising in genetics is robust, quantitative and extensive [61,62]. Handedness, in particular, has been suggested for over a century to have a genetic basis [63,64], and I have also long argued that case [12,22,65]. Understanding the genetics of - handedness probably underpins a more general understanding of cerebral polymorphisms.

Although genetic influences on lateralisation have long been controversial, in part because of the absence of formal confirmation of linkages with particular genes, there is now some form of closure occurring as a result both from the discovery of a large set of polygenic markers of handedness [66], and also because of a growing acceptance that variance not accounted for by genes is also unlikely to reflect environmental factors in the traditional sense, but rather is due to what have been called epigenetic effects [67], "a third source of developmental differences" [68], and more recently 'developmental variance' [65,69].

The primary thrust of the present paper will be on genetic modelling, but as ever it is not possible to make predictions on the basis of *genetics* unless there is also a clear understanding of *phenotypics* Lateralised phenotypes have their own specific measurement problems, and in particular have difficulties in statistical analysis due to the presence of mixture distributions, and these will need to be explored as they are likely to confound any fit between model and data. It is not appropriate to reject genetic models solely because phenotypes are ill-described or inappropriately described. Any genetic model inevitably forces questions about molecular and developmental mechanisms, as well as evolutionary origins, and these will therefore also be considered here.

6. Theorising about Theory

This paper is theoretical. Indeed, this Special Issue of *Symmetry* is about theory. The paper itself is about a theory which, for want of a better name, I have called the DC theory. The theory is far from novel, being first put forward in embryonic, unpublished form in 1977 [70], and in 1979 and 1985 described more formally [20,71]. Its age may seem to make the theory of little interest, but I hope not. The DC theory itself has developed, the problems it was trying to solve still exists, the ability to test the model has also progressed, not least because of growing amounts of fMRI and fTCD data, as well as better computational modelling, and there have been developments in the theory itself. Perhaps most crucially, no other theories have replaced it, although the RS model of Annett does have a little overlap [72–74], even if many of the details differ.

Useful theories do not have a clear shelf-life, after which they must be replaced by newer, more modern ones. The scientific literature itself though is dominated by a 'recency effect', with few researchers citing work that is not from the current millennium, however important or interesting it may be (and in Ocklenburg et al.'s horizon-scanning in 2021 of lateralisation's next decade [75], 65% of the citations came from the previous decade, and only 11% from before 2000 [68]). Having said that, the myths, fictions and backward steps of some of the past half century's lateralisation research are undoubtedly better left in obscurity [12].

Psychology, as well as biology and other sciences, have become "hyperempirical science[s]" [76] in the 21st century, hoovering up data in ever greater amounts, and worrying, with good reason, about failures of replication. In biology, Sir Paul Nurse worried in the journal *Nature* that while research talks in biology "unleash a tsunami of data", "researchers are holding back on ideas" [77]. Where would biology be, he asks, "if Darwin had stopped thinking after he had described the shapes and sizes of finch beaks, and not gone on to describe the idea of evolution by natural selection"?

Theory's central role in psychological science has only recently been resurrected and reemphasised, and in July 2021 an entire Special Issue of *Perspectives on Psychological Science* was devoted to the problems of *Theory in Psychological Science*. Theory in a serious sense has become ever more and more ignored, such that now "a growing chorus of researchers has argued that psychological theory is in a state of crisis" [78]. Where there *is* theory it is narrow in scope, and is only set out verbally, allowing at best weak testing [79], with computational modelling rare, so that generalisability and testability are limited for many theories [80].

Almost all criticisms of the state of theory in psychology can be traced back to Paul Meehl's devastating analyses in his various papers [81–83], and almost all of the seventeen *Perspectives* papers cite him. A central criticism, stated below, emphasises that support for a substantive theory cannot be derived from significant tests against a null hypothesis:

> "the almost universal reliance on merely refuting the null hypothesis as the standard method for corroborating substantive theories ... is a terrible mistake, is basically unsound, poor scientific strategy, and one of the worst things that ever happened in the history of psychology [82] (p. 817)".

Meehl compares 'soft psychology', his term, with the exemplar of physics, with its ever more precise testing of point estimates. Psychology has mostly not been cumulative, so that theories "never die, they just slowly fade away [and] finally people just sort of lose interest in the thing and pursue other endeavors" [82] (p. 807). Instead of growth, there is merely change, with different problems now being researched and old problems abandoned.

What is a good theory? "Good theories are [...] *hard to vary*: they explain what they are supposed to explain, they are consistent with other good theories, and they are not easily adaptable to explain [almost] anything" (emphasis in original) [84]; "theoretical analyses can endow a theory with minimal plausibility even before contact with empirical data" [85]; formal theories have "immense deductive fertility ... support clear and demonstrable explanations, [and] supply precise predictions about the behavior expected from the theory" [78]; in addition, good theory should "be independent of its creator ... [as] ... It is difficult to make much use of a theory otherwise". Avoiding "the trap of vagueness is hard" and "usually requires mathematical formalism" [78]. As an exemplar of good theory, Navarro advocates Shepard's 1987 generalization model, which currently has 2711 citations [86,87].

All such requirements of good theory inevitably sound like impossible paragons of perfection, but Meehl was pragmatic and identified twenty approaches to improving the recurrent problems of psychological theory, some of which are particularly relevant to any theory of cerebral polymorphisms [82].

- "Slic[e] up the raw behavioural flux into meaningful intervals identified by causally relevant intervals ... ". As Plato said, nature comes pre-divided, so that the best theory 'carves nature at the joints' [88]. Classifications of handedness often ignore that precept, and thereby provide so many dependent variables that at least one will be statistically significant [89]. To put it another way, "keep it simple".

- Individual differences are both the central core and the central problem of psychology, for as Meehl says, "what is one psychologist's subject matter ... is another psychologist's error term". The tension between similarities and differences between people was perfectly put by Kluckhohn and Murray in 1949, saying that "Every man is in certain respects (a) like all other men, (b) like some other men, (c) like no other man" [90]. Brains fit that description very well, and that is what a theory of cerebral polymorphisms must explain.
- "Most of the attributes studied by ... psychologists are influenced by polygenic systems" and Meehl anticipated the three laws of behaviour genetics by two decades Turkheimer's [91], as well as the subsequent fourth law [92]. Handedness is now undoubtedly seen to be polygenic [65,93].
- Random factors can often cancel out ('convergent causality'), but "there are other systems in which ... slight perturbations are ... amplified over the long run" ('divergent causality'). Meehl's example describes how "an object in unstable equilibrium can lean slightly towards the right instead of the left", resulting in an avalanche. Symmetry, symmetry-breaking, bifurcations and canalization are all fundamental to understanding the nature of lateralisation, as will be discussed.
- "Luck is one of the most important contributions to individual differences ... an embarrassingly 'obvious' point that social scientists readily forget". Meehl looks particularly at discordant MZ twins where no factors explain the difference, and emphasizes that "something akin to the stochastic process known as a 'random walk'". Randomness is the 'third source' of developmental differences [67].

Meehl is essentially a Popperian, arguing that the main feature of scientific theory is the existence of *conjectures*, which are then capable of *falsification*, and thereby of *refutation* [94–96]. Falsification can though be premature, for as Meehl says, the core of a theory is surrounded by auxiliary theories, understanding of the instruments used, and particular experimental conditions, as well as an assumption of 'all other things being equal', failure of any of which can result in a seeming failure of the theory proper [83]. Meehl's 1990 paper follows the approach of Imre Lakatos, who argued that new, tender and vulnerable theories need to be surrounded by a 'protective belt' to prevent premature refutation [97], as I have discussed elsewhere in the context of theories of lateralisation [98]. Although auxiliary hypotheses can be necessary (astronomers require auxiliary theories of optics to explain their telescopes), it is auxiliary hypotheses which are *ad hoc* which are the real problem, being driven by failed predictions of theory and allowing that anything and everything can be explained [99].

This is not the point at which to ask whether the DC theory of cerebral polymorphisms is a good or successful theory, but it will help the reader in asking whether it, or indeed any other theories of the phenomena, show any signs of being good. If we want to have effective theories in lateralisation, as this Special Issue of *Symmetry* is suggesting, then a crucial first step is knowing what good theories are.

Testing theories is never easy, and Meehl follows Popper, who suggested that conjectures or theories (**T**) cannot be proven, but they can be disproved or falsified by appropriate empirical evidence or results, **R**. A strong version of Popper can be written succinctly as follows:

$$\mathbf{T} \rightarrow \mathbf{R};\ {\sim}\mathbf{R};\ \therefore\ {\sim}\mathbf{T}$$

i.e., a theory, **T**, implies a result **R**; not finding **R** (not-**R**; ~**R**) implies the theory is not true (not-**T**; ~**T**). In its strongest form, not finding the predicted result, **R**, means that **T**, is falsified [94].

That model, though, is too simplistic, there often being a convoluted route between a theoretical prediction and an observed result (think of the relatively brief predictions by

Higgs of the boson named after him [100,101], and the eventual testing and discovery in the vastness and complexities of the Large Hadron Collider at CERN).

Kuhn, Lakatos, Feyerabend and others suggested the real world of empirical science was less straightforward than Popper's equation suggested, with premature falsification being a risk because of weaknesses in the theory or the method and data for testing it [97,102,103]. Meehl's summary replaced $\mathbf{T} \rightarrow \mathbf{R}$ with the more realistic,

$$\mathbf{T} \cdot \mathbf{A_t} \cdot \mathbf{C_p} \cdot \mathbf{A_i} \cdot \mathbf{C_n} \rightarrow \mathbf{R}$$

The theory ($\mathbf{T}$) can only be tested in conjunction with the following: $\mathbf{A_t}$, a set of Auxiliary hypotheses which connect the theory to the observations; $\mathbf{C_p}$, the *ceteris paribus* clause of 'all other things being equal'; $\mathbf{A_i}$, a set of auxiliary hypotheses about the instruments necessary for measuring the predictions; and $\mathbf{C_n}$, a set of particulars about how the experiment was actually realised [83].

With the new equation, if $\mathbf{R}$ is not as predicted, then the more complex implication is $\sim(\mathbf{T} \cdot \mathbf{A_t} \cdot \mathbf{C_p} \cdot \mathbf{A_i} \cdot \mathbf{C_n})$, i.e.,

$$\mathbf{T} \cdot \mathbf{A_t} \cdot \mathbf{C_p} \cdot \mathbf{A_i} \cdot \mathbf{C_n} \rightarrow \mathbf{R}; \sim \mathbf{R}; \therefore \sim(\mathbf{T} \cdot \mathbf{A_t} \cdot \mathbf{C_p} \cdot \mathbf{A_i} \cdot \mathbf{C_n})$$

In other words, $\sim\mathbf{R}$ implies something is clearly wrong and something indeed has been falsified, but it is not necessarily $\mathbf{T}$, but something within the conjunction of $\mathbf{T}$, $\mathbf{A_t}$, $\mathbf{C_p}$, $\mathbf{A_i}$ and $\mathbf{C_n}$, of which $\mathbf{T}$ is but a part. That both provides a lot of 'wriggle-room' for theorists and modellers trying to support their theories, but also puts an onus upon those testing models to ensure that the test of a theory really is a test, with $\sim\mathbf{R}$ genuinely occurring not just because of inadequate auxiliary theories, instrumentation, etc., so that the refutation of the theory is compelling.

Within the context of the DC model, as with most of psychology, it is a long way from model predictions themselves to results from dichotic listening, fMRI or fTCD scanners, or patients with brain damage. That is why this section on theories, and the testing of theories, has been included. The paper also includes sections on the nature of lateralisation and its measurement, as well as the biological background to lateralisation. Without considering such factors, data may seem to refute the model when a proper test has not been provided. Not only the theoretical core of a model needs assessing but also the extent to which it has been or can be properly tested.

It is time to stop thinking about what a theory might look like, and instead to delve into the DC model and see what it might explain. It is easiest to begin with the earliest, single-gene (monogenic) version of the model, wrong that it is now known to be, and then move onto the nature of lateralisation and its biological basis, and eventually to reach the polygenic version of the model, along with discussion of its biological basis.

7. Modelling Cerebral Polymorphisms

The Original, Monogenic Version of the DC Model and the Data It Needs to Explain

This section will describe the original DC model, and in particular will unpack some of the maths underlying it, as without an understanding of how the model is doing its calculations there is little hope of the reader understanding what the model is and is not saying.

The DC model (dextral-chance model) has been named after the two alleles, D for dextral and C for chance, which were originally proposed in the 1970s to explain the genetics of handedness and language dominance [22,68,73]. Since that time, the GWAS that we carried out has made it clear that there is no single gene which determined handedness [104], and in 2013 we suggested that "there are probably at least 30–40 loci involved in handedness" [11]. In 2021, an important study reported that 41 loci were associated with right- and left-handedness [66]. Although the original DC model was a single-gene model, in 2014 it was shown that a variant of it, which we called the pathology model, and took primary ciliary dyskinesia as its biological inspiration, broadly predicted the same patterns of handedness as did the original single-gene model [11]. The polygenic DC model will be discussed later, but for simplicity I will describe initially the monogenic, single-gene version of the DC model, which set out to model three features of handedness [70]. This will not be as irrelevant as it may at first seem, since, as often occurs in modelling, simple models often summarise well the essence of more complex models.

The DC model needed to account for several features of handedness and language dominance [12,22,65,73], of which a broad-brush description is:

- Handedness runs in families, but two right-handed parents sometimes have left-handed children, and only about a third or so of children of two left-handed parents are left-handed;
- Identical (monozygotic; MZ) twins are discordant for handedness in about one in five pairs, although somewhat more fraternal (dizygotic; DZ) twin pairs are discordant;
- Cerebral dominance for language is correlated but only partially with handedness, the majority of left-handers having language in the left hemisphere, just as do right-handers.

In addition, the original DC model, in order to be biologically convincing, wanted to take into account the increased rate of left-handedness in conditions such as psychosis and severe learning difficulties, and to be consistent with asymmetries, normal and pathological, in humans and in animals. In particular, the model was very aware of the two influential papers by Michael Morgan and Michael Corballis, not least since Michael Morgan was my PhD supervisor [33,105].

The monogenic DC model is far from being the first genetic model of handedness [64,106–111], although most had not been able to account for family patterns, twins and language dominance [10]. Annett's Right-Shift (RS) model was being developed at much the same time as the DC model [72–74,112]. A key difference between both the DC and RS model and all previous models of handedness was that they invoked randomness, the concept known to biologists as *fluctuating asymmetry*.

8. Fluctuating Asymmetry

That some biological asymmetries can be random was described by Charles Darwin, in the second volume of his 1854 monograph on barnacles (1854), where he said that, in the genus *Verruca*, "Extraordinarily great is the difference between the right and left sides of the whole shell, yet in all of the species it seems to be *entirely a matter of chance whether it be [the right or left side] which become[s] abnormally developed*" [113] (p. 499) (my emphasis). Darwin also notes Crustacea in which "the unequal development of the thoracic limbs seems *quite capriciously* to affect either the left or right side of the body" [113] (p. 499, my emphasis). Overall, Darwin "anticipated that deviations from the law of symmetry would not have been inherited", although he later makes a clear exception for handedness [114] (vol 2, p. 12).

Randomness in asymmetry, fluctuating asymmetry, has been of interest to biologists for a long while [115–119], not merely because of its asymmetry as such, but also as an indirect index of developmental stability resulting from developmental buffering [120]. Comprehensive reviews are available of the origins and nature of fluctuating asymmetry and its applications [121–123].

The simple idea at the centre of the monogenic DC model (and it is similar to the idea at the core of the RS model) is that one of the genes underlying handedness produces randomness. Key biological underpinnings for that position were provided by the following: Layton's 1976 finding of the recessive *iv* gene, which resulted randomly in 50% of mice having situs inversus, their heart, spleen, etc., being on the left, and their liver, appendix, etc. on the right [124]; Afzelius's 1976 demonstration that ciliary inactivity randomly resulted in situs inversus in immotile cilia syndrome (now called primary ciliary dyskinesia; PCD) [125]; and Collins's 1968 and 1969 selective breeding experiments in mice, which resulted in a 50:50 mixture of right- and left-pawedness, with no selectable variance remaining [126,127]. It was therefore biologically plausible that one genotype could produce randomness, while there could still be strong directional asymmetry in the majority 'wild-type' population. That there are demonstrated mechanisms for the generation of randomness will be considered next, to form a foundation for the description of the DC model.

9. The Biological Nature and Origins of Randomness

Randomness is a "fundamental process[...] rooted in the very basis of life" [128] and is a crucial part of biology, evolution needing random mutations, diploid genetics needing random chromosomal assortment into sperm and eggs, and neural functioning being intrinsically noisy [129]. Joober and Karama [128] suggest, however, that random variation may seem antithetical to many scientists who instinctively assume deterministic mechanisms, with the genome seen as a developmental blueprint which unfolds deterministically. That development is not always deterministic was recently made clear by Linneweber et al. [130], who described large individual differences of visual behaviour in genetically identical, inbred strains of *Drosophilia*, resulting in a "non-heritable, temporally stable trait that is independent of sex, genetic background, and genetic diversity". The behavioural differences arose from *an intrinsically stochastic mechanism of brain wiring*, resulting in variation in the asymmetry of dorsal cluster neurons (DCNs). The authors suggested that multiple neural and behavioural phenotypes from a single genotype, as a result of biological noise, may be beneficial under strong selection pressure, and emphasise that "the role of non-heritable noise during brain development ... is understudied".

Developmental mechanisms generating randomness can be dissected in genetically identical individuals, including human monozygotic twins and, most intriguingly, in the monozygotic quadruplets which, uniquely, are the normal pattern of offspring in the nine-banded armadillo. In 1909, Newman and Patterson [131] described armadillo quadruplets as being "practically identical, ... but a more searching comparison ... revealed, as one might expect, slight departures from complete identity". Those differences have recently been studied in fascinating detail by Jesse Gillis and his team [132,133], using five sets of armadillo quadruplets reared in near-identical environments. DNA and transcriptional RNA were sequenced to compare the siblings and partition developmental noise into separate categories. X-chromosome inactivation (lyonization), a source of random developmental variation in XX females, occurred at about the 25-cell stage in females, producing mosaics potentially with 2^{25} phenotypic combinations. Epigenetic effects can, in some sense, be regarded as a form of autosomal lyonization, and have been proposed in some occasions to explain 'partial penetrance' [134], as seems to be the case for heterozygotes in the DC model. Monoallelic expression of autosomes in the armadillo occurs somewhere at about the 150-cell stage, with allelic imbalance of expression of about 700 of the 20,000 genes. The authors estimated that "developmental stochasticity accounts for 20% as much variability as genetics does ... perhaps 10% of total variance". In addition, there is undoubtedly proper environmental variance. Expressing differences between groups as mean $|log_2FCs|$ (mean absolute $\log_2$fold changes), pairs drawn from armadillo quadruplets had a mean difference of 0.16, compared with 0.30 for pairs of unrelated armadillos (and absolute identity would have scored zero). The armadillo quadruplets had very controlled environments and their score of 0.16 compared with a score of 0.38 for human MZ twins who almost certainly had more variable environments. Human DZ twins who shared half of each other's genes scored 0.44, and pairs of unrelated human individuals scored 0.56, reflecting still greater genetic differences. That picture is very different from classical twin modelling, which predicts correlations of 1, 0.5 and 0 for MZ, DX and unrelated pairs (i.e., identity for MZ twins, if there is no measurement error, which is arguably the case here). Clearly, within even MZ twins there is much variation which is not genetic in the classical sense, the authors concluding that "purely stochastic variation in development has a large and permanent impact on gene expression" [132].

Differences in allelic expression in armadillos, due to epigenetic effects such as methylation, appeared to be at random across the genome, within and between the quadruplets. However, it might be that in some conditions methylation may be preferentially associated with particular loci, for instance, as has been reported in MZ twins discordant for schizophrenia and bipolar disorder, where some particular loci appeared to differentially methylated [135].

The Gillis et al. study of armadillo quadruplets ends with the hope that "in time, ... 'noise' will cease to be a catchall term and, instead, be added to the traditional axes of nature and nurture as a principal and well-defined contributor to phenotypic variance" [132]. Expressed more formally, Gillis, in a tweet [136], has said that a statistical model could be expressed as the following:

$$\text{Phenotype} = \text{G} + \text{E} + \text{G} \times \text{E} + \text{Noise}$$

where Noise can be expressed as

$$\text{Noise} = \text{Developmental legacy} + \text{Other effects}$$

so that,

$$\text{Phenotype} = \text{G} + \text{E} + \text{G} \times \text{E} + \text{Developmental legacy} + \text{Other effects}$$

10. The Basic Monogenic DC Model

This section will in part proceed as a tutorial to help those with little experience of genetic calculations, to understand the computational basis of the model.

The DC model has two alleles (genes), D and C. While the D allele (D; dextral) determines right-handedness in a strong sense, the other, (C; chance), results in a 50:50 mixture of right- and left-handers. Genetic models have the advantage that population genetics allows numerical predictions to be made. Since there are two different *alleles*, D and C, there will be three *genotypes*, DD, DC and CC, each person having two alleles, with one acquired from the mother and the other from the father. Let the probability of a C allele in the gene pool (p(C) = c) be 0.2 (i.e., 20%). The proportion of D alleles, p(D) = d, will be $1 - c = 1 - 0.2 = 0.8$ (i.e., 80%). If two alleles are selected from the gene-pool at random to make a genotype then 4% of people will be CC ($c \times c = c^2 = 0.2 \times 0.2$), and 64% will be DD ($d^2 = [1 - c]^2 = 0.8 \times 0.8$). The rest will be DC (heterozygotes) with one of each allele, of whom there will be $2 \cdot c \cdot (1 - c) = 2 \times 0.2 \times 0.8 = 0.32$, i.e., 32% of the population. The three genotypes are therefore in *binomial proportions*, because of the Hardy–Weinberg equilibrium [137], and are shown in Table 1 below.

Table 1. Genotype probabilities and handedness in the monogenic DC model, p(L) = 0.10, P(C) = 0.2.

(1)	(2)	(3)	(4)
Genotypes	P(Genotype) = p(G)	Probability of Phenotype = p(H ǀ G)	
		Right-Handed	Left-Handed
DD	$1 - c^2 = 0.8 \times 0.8 = 0.64$	p(R ǀ DD) = 1	p(L ǀ DD) = 0
DC	$2 \cdot c \cdot (1 - c) = 0.32$	p(R ǀ DC) = 0.75	p(L ǀ DC) = 0.25
CC	$c^2 = 0.04$	p(R ǀ CC) = 0.5	p(L ǀ CC) = 0.5

In a genetic model it needs to be specified how genotypes are related to the phenotypes, and Table 1 gives the probability of being right-handed (in green) or left-handed (in red) for each particular genotype (pH ǀ G). The randomness in the biological models of Layton, Afzelius and Collins suggests that 50% of the CC individuals should be right-handed and the remainder left-handed, which is symbolized as p(L ǀ CC) = 0.5, where p(L ǀ CC) symbolises the conditional probability of being left-handed given that a person is CC (the vertical line [solidus] being read as 'given that'). If P(L ǀ CC) = 0.5 then it will be the case that p(R ǀ CC) = 1 − 0.5 = 0.5. In contrast, the DD genotype will all be right-handed, just as the mice without the *iv* gene are all typical and have their heart on the left, liver on the right, etc., i.e., p(R | DD) = 1 and hence p(L | DD) = 0. The only thing undefined a priori is what happens with the heterozygotes, but the best fit of the model overall seems to be when DC individuals have a 25% chance of being left-handed and a 75% chance of being right-handed, which is called an *additive model*. Model-fitting with recessive and dominant models showed that they were less good fits than the additive model [20,71].

It is now possible to calculate the expected proportion of left-handers. Notice in Table 1 that no DD individuals can be left-handed, only 25% of DCs will be left-handed, whereas 50% of CCs will be left-handed. Unlike most of the previous genetic models of handedness mentioned previously, none of the genotypes produce only left-handers (although DD does produce only right-handers). The proportion of left-handers overall, p(L), will be (0.64 × 0) + (0.32 × 0.25) + (0.04 × 0.5) = 0 + 0.08 + 0.02 = 0.10, i.e., 10%. The proportion of C alleles, c = 0.2, was chosen for this example so that 10% of the population would be left-handed, which is a good approximation to rates of left-handedness in most populations [138], and 10% also makes for numerically easier values for explanations. P(L) = 10% will therefore be used for the rest of this exposition, although in practice other values may be more precise for particular populations.

11. Handedness in Families

The simple genetic model of Table 1 does not have any very obvious, direct use. It forms, however, the basis for asking how handedness would be expected to run in families, with the simplest situation being to calculate the probability of a child being left-handed when the parents are either both right-handed (R × R), one is left-handed (R × L), or both are left-handed (L × L).

The first step, shown in Table 2, is to calculate the probability that an individual is of each of the three genotypes given their handedness (i.e., p(G | H))—note that this is not the same as p(H | G), shown in Table 1, which is the probability of being a particular handedness given that one is a particular genotype. The details of the calculations can be seen in Table 2. Columns 1 and 2 show all possible combinations of genotype and handedness, and column 3 shows the proportions of the genotypes in the population (pG), which are those of column 3 in Table 1. Column 4 is p(H | G), the probability of being a particular handedness given a particular genotype (and is the same as columns 3 and 4 in Table 1).

Table 2. Probability of genotypes by right and left handedness in the monogenic DC model, p(L) = 0.10, P(C) = 0.2.

(1)	(2)	(3)	(4)	(5)	(6)	(7)	(8)	(9)
Genotype (G)	**Hand (H)**	**P(G)**	**P(H | G)**	**P(G&H)**	**RH**	**P(G | RH)**	**LH**	**P(G | LH)**
DD	R	0.64	1	0.64 × 1 = 0.64	0.64	0.64/0.9 = **0.711**		
DD	L	0.64	0	0.64 × 0 = 0			0	0/0.1 = **0**
DC	R	0.32	0.75	0.32 × 0.75 = 0.24	0.24	0.24/0.9 = **0.267**		
DC	L	0.32	0.25	0.32 × 0.25 = 0.08			0.08	0.08/0.1 = **0.800**
CC	R	0.04	0.5	0.04 × 0.5 = 0.02	0.02	0.02/0.09 = **0.022**		
CC	L	0.04	0.5	0.05 × 0.5 = 0.02			0.02	0.02/0.1 = **0.200**
Total	-	-	-	1	0.9	1	0.1	1

Column 5 is the probability across the population of an individual being a particular genotype (G) *and* a particular handedness (H), p(G&H), which is the product of columns 3 and 4. Column 5 should sum to 1, as it includes all possible combinations. Notice that although CC individuals are more likely than DC individuals to be left-handed, most left-handers are heterozygotes, being of genotype DC.

To calculate p(G | H), the probability of being a particular genotype given that an individual has a particular handedness, one needs *to consider only the individuals of that handedness*. Column 6, for instance, shows only the rows for individuals who are right-handed (shown in green), which sum to 0.9, since 90% of individuals are right-handed. The values in column 6 can then be divided by their total to give the proportion of each genotype, which are shown as p(G | RH) in column 7. Amongst right-handers, 71.1% are DD, 26.7% are DC and 2.2% are CC. Finally, the same thing can be done for the left-handers (shown in red), in columns 8 and 9. Notice that with the left-handers there are no individuals who are DD (since all DD individuals in the model are right-handed, and the calculations automatically take care of that). The majority of left-handers, 80%, are of the DC genotype, although intuitively one may have expected most left-handers to be CC.

Although the calculations may seem long-winded for this simple question, the method has great generality, and can readily be programmed, as all one needs to do is tabulate the probabilities of all possible combinations of genotype and phenotype, and then sum those in whom one is interested. Complicated family or twin models with multiple phenotypes may have tens of thousands of combinations, but a series of loops in a computer program will rapidly go through them all, and that is how most of the calculations in this paper have been carried out. The calculations can also be carried out using Bayesian graphical models, but that will not be considered further here. Likewise the calculations can be extended to family trees of different shape and extent [139].

Calculating the probability of the children of two right-handers being left-handed firstly requires the probability of each of the parental combinations of handedness and phenotype of which there are 36 (the mother can be R or L, and of genotype DD, DC or CC, giving 6 combinations, and the father also has 6 combinations, making 36 combinations in total in the parents). One then needs calculates the probability that a child will be of, say, genotype DD, given their parents are particular genotypes. When both parents are DD it is inevitable, given Mendel's laws, that the child must also be DD. However, other combinations give other outcomes, so that if, the parents are DD and DC then half of the offspring will be DD and other half DC, as a child takes one allele by chance from each parent. Additionally, when both parents are DC then a quarter of their children are DD, a quarter are CC, and a half are DC. Finally, each of the children will be one of 6 combinations of genotype and phenotype, meaning that overall there are 216 parents × child combinations to be worked through. To make that clearer, there are 2 maternal phenotypes × 3 maternal genotypes × 2 paternal phenotypes × 3 paternal genotypes × 2 child phenotypes × 3 child phenotypes, i.e., $(2 \times 3)^3 = 6^3 = 216$ combinations. The calculations are straightforward on a computer, and give the results shown in Table 3, with right-handers shown in green and left-handers in red.

Table 3. Proportion of right- and left-handed offspring in relation to parental handedness, for the monogenic DC model, p(L) = 0.10, P(C) = 0.2.

		Offspring	
		% RH	% LH
Parental Handedness	R × R	92.22%	7.78%
	R × L	81.10%	18.90%
	L × L	70.00%	30.00%

The immediate thing to notice about this model from a theoretical perspective is that *qualitatively* it produces the sort of result that one would have expected from looking at patterns of handedness in families, with actual data from families summarised elsewhere [10,20,71]. The key results are that two right-handed parents can sometimes have left-handed children, the proportion of left-handed offspring increases with the number of left-handed parents, and perhaps of particular interest, the majority of the children of two left-handed parents are *right-handed*, left-handers occurring in only 30% of cases.

The model is simple in its conception, having only four parameters, one of which is a guesstimate of the rate of left-handedness (10%), two of which are derived from biology (pure directional asymmetry for DD and pure fluctuating asymmetry for CC), and the fourth comes from the model being additive so that DC is midway between DD and CC in its proportion of left handers (and that parameter was actually derived from model-fitting, with dominant and recessive models not working as well with actual data). A limited number of parameters to be specified is the sort of property that one wants to see in computational models [78].

12. Handedness in Twins

Twins have often been misunderstood by researchers on handedness, the assumption being that identical twins should be identical if a trait is genetic. That is wrong. What genetics says is that if a trait has genetic influences, then MZ twins should be *more similar* than DZ twins, and that is clearly the case as larger and larger meta-analyses of twins have shown [140–144].

Modelling handedness in monozygotic twins is straightforward using the methods described earlier, as one merely has to take a table such as Table 1 and realise that, at the level of the pair, MZ twins have one of three phenotypes—both RH (R-R), both LH (L-L) or discordant with one RH and the other LH (R-L). Table 4 in columns 1 and 2 shows the genotypes and their probabilities, and columns 3 and 4 the probability of right and left-handedness for singletons, as in Table 4, with green for right-handers and red for left-handers. The probability of being right- or left-handed for each twin is the same *within a genotype*, but is independent for each twin in a pair, so that for the CC genotype each twin has a 50% chance of being right-handed and 50% of being left-handed. As a result, for CC twins, 25% of pairs are R-R (0.5 × 0.5), 25% are L-L (0.5 × 0.5), and 50% are R-L (2 × 0.5 × 0.5)—see columns 5, 6 and 7 in Table 4.

Table 4. Handedness of monozygotic twins in the monogenic DC model, p(L) = 0.10, P(C) = 0.2.

(1)	(2)	(3)	(4)	(5)	(6)	(7)
		Probability of Phenotypes for Singletons		**Probability of Phenotypes for MZ Twins**		
Genotypes	P(Genotype)	Right-handed	Left-handed	R-R	R-L	L-L
DD	$1 - c^2 = 0.8 \times 0.8 = 0.64$	p(R∣DD) = 1	p(L∣DD) = 0	100%	0%	0%
DC	$2 \cdot c \cdot (1 - c) = 0.32$	p(R∣DC) = 0.75	p(L∣DC) = 0.25	56.25%	37.50%	6.25%
CC	$c^2 = 0.04$	p(R∣CC) = 0.5	p(L∣CC) = 0.5	25%	50%	25%

Most twin pairs in Table 4 are DD (column 2), and the least frequent genotype is CC. The overall proportion of R-R pairs in the population can be found by multiplying column 2 by column 5 (i.e., 0.64 × 1 + 0.32 × 0.5625 + 0.04 × 0.25 = 0.83 = 83%). Similarly, 14.0% of MZ twin pairs are R-L and 3% are L-L, those figures being shown in Table 5. The proportion of discordant twin pairs (14%), shown in bold in Table 5, is a little below the frequently cited value of about one in five discordant pairs, but it is broadly in the right ballpark given the minimal number of parameters used in the model.

Table 5. Handedness discordance in MZ and DZ twins with the monogenic DC model, p(L) = 0.10, P(C) = 0.2.

(1)	(2)	(3)	(4)
	R-R	**R-L**	**L-L**
MZ twins	83.00%	**14.00%**	3%
DZ twins	82.00%	**16.00%**	2%

12.1. Mirror-Image Twins

Discordance in MZ twin pairs has long been recognised; for instance, Danforth commented in 1919 that "a surprising number of twin pairs seem to be composed of one right and one left handed individual" [145]. Danforth talks of one twin being the "mirror image" of the other, and the idea remained prevalent in the scientific literature [146–149] and still exists in popular culture [150]. There is, though, little biological support for mirror-image twinning, except in conjoined twins [151]. Instead, the DC model very naturally explains apparent mirroring for handedness, for if chance is involved in the handedness of DC and CC twins, then it is inevitable that some MZ twins will sometimes be discordant for handedness.

12.2. DZ Twins

Concordance in dizygotic twins is more complicated to calculate than for monozygotic twins. The calculation is similar to that for Table 4, except that there are two offspring who are not of course genetically identical, although they are more similar than randomly chosen individuals as they share genes with their parents. The calculation involves looking at many more combinations, which will not be given here, but it follows the basic approach used earlier, generating all possible combinations of parental and offspring genotypes and phenotypes. Specifically, there are six hand × genotype combinations for the mother, six for the father, six for the first child and six for a second child, the sibling of the first (and DZ twins can be treated as siblings), so that there are $6^4 = 1296$ combinations. Table 5 summarises the results for MZ and DZ twins, with discordant pairs shown in bold in column 3. The results look superficially similar in MZ and DZ twins, but the important feature of the model is that MZ twins are somewhat more similar (a little less discordant) than DZ twins. That describes well the broad pattern seen in meta-analyses of twin handedness, where the difference is small but statistically robust [138–142]. Once again, the monogenic DC model with its few parameters provides a qualitative similarity to the data.

A surprising feature of almost all twin studies of handedness, and the meta-analyses show there are many such studies, is that, despite the major interest in the genetic basis of handedness, almost all are single-generation studies. Typical family studies (see Table 3) are of course two-generational, and assessing twin handedness in relation to parental handedness would seem an obvious thing to do. The DC model predictions for twin handedness in relation to parental handedness are shown in Table 6, with twin discordance rates shown in column 5 in bold for emphasis. The clearest result is that discordance rates increase dramatically as the number of left-handed parents increases, with the difference between MZ and DZ twins also becoming a little larger in absolute terms. I know of only one large twin study where parental handedness was also measured, the Netherlands Twin Registry, but sadly the key tabulation similar to that of Table 6 is not reported [152]. A later re-analysis of the same data looked only at discordant twin pairs and examined environmental factors, but it did not include familial handedness, which, as Table 6 shows, is a major predictor of twin handedness discordance [153]. The Netherlands Twin Registry now has very many more twins in it, and in conjunction with the Registry I hope soon to be able to report data equivalent to those in Table 6.

Table 6. Discordance and concordance of MZ and DZ twins in relation to parental handedness in the monogenic DC model, p(L) = 0.10, P(C) = 0.2.

(1)	(2)	(3)	(4)	(5)
	Parents	R-R	**R-L**	L-L
MZ twins	All	83.00%	**14.00%**	3.00%
	R × R	86.70%	**11.10%**	2.20%
	R × L	68.10%	**26.00%**	5.90%
	L × L	52.00%	**36.00%**	12.00%
DZ twins	All	82.00%	**16.00%**	2.00%
	R × R	85.90%	**12.70%**	1.40%
	R × L	66.40%	**29.30%**	4.20%
	L × L	49.50%	**41.00%**	9.50%

12.3. Handedness in Twins and in Singletons

A difficult question in studying twins, which needs considering, is whether twins have a higher prevalence of left-handedness than do singletons. In 1973, Nagylaki and Levy asserted very strongly that "it is impossible to assess the heritability of a trait by using twin data if the frequency of the trait among twins differs from that along non-twins" [140,154]. Nagylaki and Levy's simple analysis of studies of twins and singletons suggested twins did indeed have a higher rate of left-handedness. The problem, though, as I put it in a 1980 review, is that in few studies was handedness in twins and singletons "assessed by the same criteria, in the same study, by the same investigators" [138]. The only exceptions then, the studies of Wilson and Jones in 1932 [155] and Zazzo in 1960 [148], reported no differences between singletons and twins. A recent large meta-analysis by Pfeifer et al. [144] addresses the issue once more, but found secular trends in the twin-singleton ratio in handedness, suggesting that there may be ascertainment or other biases. Twins are also more likely to be born prematurely or show birth complications, which need considering as covariates, although the role of birth complications and prematurity in causing left-handedess is not entirely clear [156–160]. The only large population study taking birth weight, Apgar score and gestational age into account, that of Heikkilä et al. [161], reported no twin-singleton difference. As Pfeifer et al. say in their meta-analysis [144], the secular shift and the possible influence of covariates probably make it unsafe to conclude that twinning has a genuine relationship to handedness. Clearly, more research is needed, presumably in some of the large birth cohort studies that now exist.

13. Language Dominance and Handedness

The last of the three desiderata for any broadly acceptable genetic model of handedness is that it accounts for the association of handedness and language dominance. Language dominance, which is typically in the left hemisphere, can be assessed in multiple ways, from unilateral brain damage (as Dax and Broca discovered), though other methods such as dichotic listening, unilateral ECT, and various forms of functional brain scanning including fMRI and fTCD. This is not the place to review them, but they do give somewhat different rates of left dominance. A good working approximation is that about one in ten individuals has language in the right rather than the left hemisphere, and while right-language dominance does occur in a small percentage of right-handers, it occurs much more frequently in left-handers. However, there is still the puzzling and difficult finding that *a clear majority of left-handers are similar to right-handers in having language in the left hemisphere* (i.e., they are not the mirror-image of right-handers). Explaining the numerical relationships of this simplest of cerebral polymorphisms is inevitably a challenge for genetic theories, although as will be seen the DC model can cope with it.

The key theoretical assumption for modelling two separate modules in the DC module is *statistical independence of lateralised modules*. Statistical independence has already been seen in looking at handedness in MZ twins, but there it occurs in two separate individuals, albeit genetically identical. Statistical independence of lateralised modules, however, means *independence within the same individual*. In particular, it means that if, for an individual, the probability of module A being atypical is p, then the same probability, p, applies to module B, with two separate metaphorical coins being tossed to decide on the overall outcome. The result, which comes from the binomial distribution, is that $(1-p)^2$ individuals will be typical (right-handers with left-language dominance), p^2 will have both modules lateralised atypically (left-handers with right-language dominance), and $2 \cdot p \cdot (1-p)$ individuals will have one atypically lateralised module. It should be emphasised that a key feature of the model is that the independence of modules is *not* within the population overall (and there is a clear correlation of handedness and language dominance in the population, so any such model would fail), but within individuals with the same probability of having modules organised in the typical way to the right or left side, i.e., within the three genotypes.

The monogenic DC model for two modules, handedness and language dominance, is summarised in Table 7. Notice that although the model is a genetic model, with different genotypes behaving differently, in Table 7 there is only one-generational data. Later it will be shown what the results look like for family data where there are two generations. Column 1 shows the three genotypes with their population proportions in column 2. Columns 3 and 4 show the two separate handedness phenotypes, H, in column 3 (with right-handers in green and left-handers in red), and language dominance, Lg, in column 4, with right-language dominance shown as shaded rows. Column 5 shows the probability of the handedness phenotypes according to the genotype, and column 6 shows equivalent values for the probability of having language in the right or left hemisphere. Column 7 shows the probability of the various combinations of handedness and language lateralisation for each genotype. Notice that DD individuals are all right-handed and also all are left-language dominant, and therefore three of the four combinations have zero probability. The other two genotypes, DC and CC, show various probabilities, with the CC genotype showing all four combinations in equal proportions. The values in column 7, p(H&Lg | G), can be multiplied by the genotype probabilities in column 2, p(G), to give the overall probability for each of the 12 combinations of being the genotype and the handedness and language phenotypes (p(G&H&Lg) in column 8); in addition, these of course sum to one.

Table 7. Details of calculations for language dominance in relation to handedness for the monogenic DC model, p(L) = 0.10, P(C) = 0.2. Right-handers are shown in green and left-handers in red. Rows with pale green shading indicate right-side language dominance.

(1)	(2)	(3)	(4)	(5)	(6)	(7)	(8)	(9)	(10)	(11)	(12)
Genotype (G)	**P(G)**	**Hand (H)**	**Lang (Lg)**	**P(H | G)**	**P(Lg | G)**	**P(H&Lg | G)**	**P(G&H&Lg)**	**RH**	**P(G&Lg &RH)**	**LH**	**P(G&Lg | LH)**
DD	0.64	R	L	1	1	1	0.64 × 1 × 1 = 0.64	0.64	0.64/0.9 = 0.711		
DD	0.64	R	R	1	0	0	0.64 × 0 × 0 = 0	0	0/0.9 = 0		
DD	0.64	L	L	0	1	0	0.64 × 0 × 0 = 0			0	0/0.1 = 0.0
DD	0.64	L	R	0	0	0	0.64 × 0 × 0 = 0			0	0/0.1 = 0.0
DC	0.32	R	L	0.75	0.075	0.75 × 0.75 = 0.5625	0.5625 × 0.32 = 0.18	0.18	0.18/0.9 = 0.200		
DC	0.32	R	R	0.75	0.25	0.75 × 0.25 = 0.1875	0.1875 × 0.32 = 0.06	0.0	0.06/0.9 = 0.067		
DC	0.32	L	L	0.25	0.75	0.25 × 0.75 = 0.1875	0.1875 × 0.32 = 0.06			0.06	0.06/0.1 = 0.6
DC	0.32	L	R	0.25	0.25	0.25 × 0.25 = 0.0625	0.0625 × 0.32 = 0.02			0.0	0.02/0.1 = 0.2
CC	0.04	R	L	0.5	0.5	0.5 × 0.5 = 0.25	0.25 × 0.04 = 0.01	0.01	0.01/0.09 = 0.011		
CC	0.04	R	R	0.5	0.5	0.5 × 0.5 = 0.25	0.25 × 0.04 = 0.01	0.0	0.01/0.09 = 0.011		
CC	0.04	L	L	0.5	0.5	0.5 × 0.5 = 0.25	0.25 × 0.04 = 0.01			0.01	0.01/0.1 = 0.1
CC	0.04	L	R	0.5	0.5	0.5 × 0.5 = 0.25	0.25 × 0.04 = 0.01			0.0	0.01/0.1 = 0.1
Total	-	-	-	-	-	-	1	0.9	1	0.1	1

The remainder of Table 7 considers just right and just left-handers. Columns 9 and 10 are for right-handers, shown in green. Column 9 sums the appropriate values in column 8, which come to 0.9, the proportion of right-handers. Column 10 gives p(G&Lg | RH), the probability of each combination of genotype and language phenotype for right-handers.

Right-language dominance is shown by rows with page green shading, which represent 0 + 0.067 + 0.011 = 0.078 of the right-handers, so 7.8% of right-handers are right-language dominant. The equivalent values for left-handers, shown in red, are 0 + 0.2 + 0.1 = 0.30, so that 30.0% of left-handers are right-language dominant. These are the values which it was hoped that the model would explain.

Despite the large number of calculations, the model has successfully found that a small proportion of right-handers are right-language dominant, and many more left-handers are right-language dominant; nevertheless, a clear majority of left-handers are left-language dominant. Language dominance was not originally built into the genetic model of handedness (and in that way it differs fundamentally from Annett's RS model [72–74]); instead, the explanation of language dominance emerges merely from the general assumption that language dominance could be explained by the DC model in the same way as is handedness, by random allocation of modules to the hemispheres for the DC and CC genotypes. That simple idea can be extended to greater numbers of modules, which will be done later.

The model in Table 7 is not strictly genetic, being only one-generational. The DC model can readily be extended to predict the proportions of handedness and language dominance in relation to parental handedness, as is shown in Table 8. As the number of left-handed parents increases, so the proportion of right-language dominance increases, although the effect is more dramatic in right-handers (from 6.0% to 25.7%) compared with left-handers (from 28.9% to 40.0%). Most right-handers (shown in green) are not right-language dominant, but having a sinistral parent affects that a lot, whereas many left-handers (shown in red) are right-language dominant, and the marginal effect of having a left-handed parent is then relatively small.

Table 8. Right-hemisphere language dominance in relation to handedness and parental handedness in the monogenic DC model, p(L) = 0.10, P(C) = 0.2.

(1)	(2)	(3)
Parental Handedness	Percent Right-Language Dominance	
	Right-Handers	Left-Handers
All	7.80%	30.00%
R × R	6.00%	28.90%
R × L	16.00%	31.20%
L × L	25.70%	40.00%

Language dominance can also be looked at in twins, where it becomes more complicated, particularly if parental handedness is also included. For completeness, Table 9 summarises the results of the modelling.

Notes: Concordance of language dominance in MZ and DZ twin pairs, where both are typical (L-L), both are atypical (R-R), or there is discordance in language dominance (L-R), as a function of parental handedness, and twin handedness (R-R, R-L(discordant) and L-L). Discordant pairs for language are shown in bold, and discordant pairs for handedness in italics.

A major interest is in whether twin pairs show concordance or discordance for language dominance. The model suggests that if twins are discordant for handedness then they are also likely to be discordant for language dominance in about 40% of cases. That, however, is not mirror-imaging but is a result of random processes in the genetic model. Parental handedness has little effect except when both twins are right-handed, in which case the proportion of language discordant pairs increases with the number of left-handed parents. Discordance for language dominance, overall, as with handedness, is more likely in DZ twins, although the effects are relatively small.

Table 9. Concordance of language dominance in MZ and DZ twin pairs, in relation to twin handedness and parental handedness, p(L) = 0.10, P(C) = 0.2.

(1)	(2)	(3)	(4)	(5)	(6)	(7)	(8)
Handedness		**Language Dominance in MZ Twin Pairs**			**Language Dominance in DZ Twin Pairs**		
Parents	Twins	L-L	**L-R**	R-R	L-L	**L-R**	R-R
All	All	83.00%	**14.00%**	3.00%	82.00%	**16.00%**	2.00%
All	R-R	89.61%	**8.74%**	1.66%	87.34%	**11.53%**	1.13%
	R-L	*51.79%*	***39.29%***	*8.93%*	*59.09%*	***35.56%***	*5.34%*
	L-L	45.83%	**41.67%**	12.50%	46.13%	**42.75%**	11.13%
R × R	R-R	92.02%	**6.74%**	1.24%	90.17%	**9.04%**	0.79%
	R-L	*52.83%*	***38.87%***	*8.30%*	*60.95%*	***34.40%***	*4.64%*
	L-L	47.83%	**40.87%**	11.30%	47.95%	**41.87%**	10.18%
R × L	R-R	76.88%	**19.33%**	3.79%	72.16%	**25.00%**	2.84%
	R-L	*50.64%*	***39.74%***	*9.62%*	*56.63%*	***37.12%***	*6.25%*
	L-L	43.87%	**42.45%**	13.67%	44.73%	**43.42%**	11.84%
L × L	R-R	64.30%	**28.13%**	7.57%	56.75%	**36.49%**	6.76%
	R-L	*40.62%*	***43.75%***	*15.62%*	*44.06%*	***44.81%***	*11.12%*
	L-L	32.81%	**46.88%**	20.31%	35.20%	**40.03%**	16.78%

More complex models could be formulated in which individual modules have different probabilities of being atypical, or in which there is some non-zero association within genotypes between A and B being on the left. The problem of such theoretical gambits is that more and more free parameters are introduced, so that almost any data can eventually be fitted, and the model loses its theoretical elegance. The assumption of Occam's Razor, of keeping models as simple as possible, would be lost. Here, the model has been kept simple for explanatory purposes, but later in the paper the need for more complicated models will be considered.

14. Sex Differences in Handedness and Functional Lateralisation

The original DC model did not model sex differences in handedness. Although many twentieth century studies measured handedness in males and females, results were variable, with no systematic analyses, and the influential 1980 narrative review of McGlone [162,163] merely noted studies both with and without sex differences in handedness prevalence. A meta-analysis of 100 populations from 88 studies was carried out in 1991 with Beatrice Seddon, but only published decades later [164], although the key result was published in 1991 [165]. Overall, there were about five left-handed males for every four left-handed females, meaning that samples of 5000+ were needed for adequately powered comparisons, accounting for some of the confusion in the literature. A larger meta-analysis in 2008 confirmed the existence of sex differences in handedness [166], with an effect size similar to that reported in 1991 [164]. Less successful has been an attempt to suggest that sex differences arise solely because of an X-linked recessive gene [167], which would predict a difference far too large to be compatible with the data [168].

The original versions of neither the Annett nor the McManus genetic models considered sex differences [10], but in 1985, Annett [73] did propose that the right-shift in her model was greater in males than in females (and also in singletons than twins). The DC model had not included sex differences (although it was said that, in principle, some parameters could differ between the sexes, perhaps in heterozygotes [20]). A 1992 review with Phil Bryden [10] looked at data from 64,582 offspring in 25 datasets where both parental and offspring sex were known. As well as a clear excess of male left-handers, there was also a maternal effect, R × L families (right-handed father × left-handed mother) having more left-handed offspring than L × R families (left-handed father × right-handed mother), with a highly significant odds ratio (OR) of 1.387 (SE 0.057). On that basis we speculated that there may be a sex-linked recessive modifier gene, *m*, on the X chromosome, which resulted in a maternal effect of about the correct size [10,165]. Although interesting as a model, further investigation then went into abeyance because of a key criticism from a colleague that the maternal effect could merely be the result of some non-paternity in L × R families. Only in a recent reanalysis by Schmitz et al., of data from the Avon Longitudinal Study of Parents and Children (ALSPAC), have full parental genetic data allowed confirmation of paternity and maternity [169]. The full study, without confirmation of paternity, had 5028 offspring and showed a maternal effect (OR 1.292, SE 0.171), which, although not significant, was compatible with the OR in the McManus and Bryden study. Full parental genetic data were only available for 1161 offspring, and those data also showed a maternal effect (OR 1.208, SE 0.369) which, although not significantly different from one, is also not significantly different from the OR of 1.387 in the McManus and Bryden result. Although larger and more powerful studies with confirmed paternity are required, taken overall, the results suggests that the maternal effect is probably real and not due to non-paternity, and that exploration and modelling of the maternal effect should recommence.

Sex differences in functional cerebral lateralisations are far less clear, despite the much-cited but very misleading 1995 claim by Shaywitz et al. of large sex differences in cerebral lateralisation [170]. A 2009 meta-analysis of language lateralisation, assessed with dichotic listening or with functional imaging, found no evidence for sex differences [171]. If indeed there are sex differences in handedness but not in functional lateralisation, then that raises many difficult theoretical questions, since most approaches, including the DC model, implicitly presume that the underlying genetics of handedness and cerebral lateralisation will be similar in their architecture [166]. However, if effect sizes are similar to those for handedness, then current sample sizes for language lateralization may be underpowered for detecting effects. The idea that cognitive sex differences in general relate to hemispheric asymmetry originated with the work of Jerre Levy in the 1970s [172,173]. A review of research in the four decades since then concluded, from converging evidence, that "the stronger [functional] hemispheric asymmetry in males is *very small* but robust" [my emphasis] [174], with effect sizes of the order of d = 0.01, which requires large sample sizes to be reliably detected. That will be problematic at the present in relation to understanding cerebral polymorphisms.

15. Qualitative Fits, and Levels of Analysis of Lateralisation

The argument used here, that the monogenic DC model is adequate in *qualitative* terms, needs unpacking a little. A starting point is the visionary work by David Marr, rightly called *Vision*, which thought deeply about how to theorise how brains might work, in his case for vision science, but more generally for all aspects of biological science [175]. *Vision* created the area now known as computational neuroscience [176]. Although major advances in vision research occurred during the 1960s and 1970s, particularly in single-cell recording, as a result of the work of Hubel and Wiesel and others, Marr realised that something deep was missing, and that merely knowing firing rates of neurones in the occipital lobe when presented with visual stimuli would not result in an understanding of how vision worked *as a process*. Marr therefore distinguished three very separate levels of analysis, which in the context of vision he called the *computational*, the *representational* and the *hardware implementation*. Those terms are not necessarily appropriate in other areas of biology, as he recognised when he talks about the problems of understanding the flight of birds:

> "trying to understand perception by studying only neurons is like trying to understand bird flight by studying only feathers. It just cannot be done. In order to understand bird flight, we have to understand aerodynamics; only

then do the structure of feathers and the different shapes of birds' wings make sense [175] (p. 27)".

To bring the quote forward to the 21st century, understanding the process of flight also probably cannot be done by looking for genes for feathers, although the evolution of feathers is undoubtedly fascinating [177].

For lateralisation, Marr's levels can be regarded as:

- **Why?** *The biology and evolutionary benefits of asymmetry* (the problems being solved by having lateralised, asymmetric bodies and brains);
- **How?** *The implementation of biological asymmetries* (what are the rules underlying organisms becoming asymmetric, and the variations in that asymmetry?);
- **What?** *The hardware for creating biological asymmetries* (cilia, sub-cellular asymmetries, etc., and their genetics).

The present review will consider all three of these levels, but the DC model at the heart of the paper is mainly at the implementational level—it does not and perhaps cannot yet be related to actual genes, but instead it provides a general approach to thinking about how the biology of asymmetry may be implemented in brains and bodies. It does not need to be exactly correct, therefore, but it needs to fit with what Marr would have called the computational or the algorithmic task—implementing asymmetry in a workable form in the brain. It does not therefore matter for immediate purposes if the DC model is correct in its details, but the DC model does need to be correct in its broad approach and conceptualisation. That biology can almost inevitably provide hardware solutions to problems is effectively a given nowadays. However, what the evolutionary and functional problems are that the hardware is solving is a bigger set of questions to be answered.

16. The Nature of Lateralisation

The previous section has shown that the monogenic DC model is at least plausible as a model, with the DD genotype producing directional lateralisation and the CC genotype producing fluctuating asymmetry. A minimum of free parameters and assumptions produces what are reasonable ballpark estimates of rates of left-handedness and right-language lateralisation in relation to familial handedness and twinning.

At this point it therefore makes sense to explore the deeper nature of fluctuating and directional asymmetry, and how they are described and measured. The biological background is also important for understanding lateralisations, with the genetics of primary ciliary dyskinesia (PCD) providing a well-understood model of the biology of lateralisation in the viscera, which may also provide a model for cerebral lateralisation. PCD will provide a basis for modelling lateralisation using a polygenic rather than a monogenic model, and then for modelling more complex cerebral polymorphisms.

17. Theorising about Lateralisation, Symmetry-Breaking and Phase Transitions

Lateralisation is different from many other behavioural measures as it is a phase transition. Most measures in psychology and cognitive neuroscience, such as extraversion, intelligence, brain volumes, or fMRI blood flow, are continuous, with more simply meaning more. Lateralisation, however, shows a *phase transition*, with different properties arising around a key point on the scale, *zero*, where right and left balance exactly, with zero being a boundary between a phase in which right is greater than left and one in which left is greater than right. Consequently, one needs to talk not simply of 'amount of lateralisation' but of *direction of lateralisation* and *degree of lateralisation*, which can change how one describes and theorises about asymmetries.

Phase transitions are well-shown in terms of the everyday properties of water, H_2O, where there are three phases, ice, liquid water, and steam. At 0 °C, ice melts, and at 100 °C, water boils and becomes steam. The 0 and the 100 are arbitrary, the Celsius temperature scale having been defined that way (and on the Fahrenheit (Kelvin) scales the melting and boiling points are 32 °F (273.15 °K), and 212 °F (373.15 °K), respectively). At 0 °C, water changes its *state* from a solid to a liquid, a *phase transition*. Changing from ice to liquid water is *symmetry-breaking*, the symmetry of the frozen ice crystals being lost, which requires energy. A further symmetry-breaking occurs at 100 °C, when liquid water vaporises to become steam, which requires additional energy. Although the difference between water at −1 °C and +1 °C is only two degrees on the Celsius scale, that change of 2 °C is different qualitatively to, say, the very same temperature difference of 2 °C between water at −12 °C and −10 °C, both of which are water that can be skated on, and water at 10 °C and 12 °C, both of which are water that can be swum in. A major state transition occurs across the phase boundary at 0 °C (32 °F; 273.15 °K), and merely knowing that water has changed its temperature by 2 °C has a very different meaning according to where it occurs on the temperature scale. The same applies to changes in laterality coefficients.

Behavioural and functional asymmetries behave similarly, with major shifts at the phase boundary between them. Unlike with water, where the phase transition and change of symmetry is defined by the physics, the symmetry and the phase change between left and right is intrinsically defined as a transition between symmetry and asymmetry. The transition of asymmetry, *left* > *right*, *left* = *right*, *left* < *right*, moves from a lack of symmetry to an intrinsically defined symmetry, and back to asymmetry once more, with *left* = *right* often being intrinsically unstable. Moving along the scale therefore involves gains and losses of symmetry.

18. How Left and Right Become Differentiated

In a left–right system, phase transitions occur as a system swings from left to right. To see how that happens, consider why traffic on roads drives on one side rather than the other.

For very low traffic density, it matters little on which side of the road a car drives, but as the number of cars increases it becomes both more efficient and safer for cars to drive on one particular side of the road. Although most countries drive either on the right or the left, there are actually three *stable equilibria* for driving, everyone on the right, everyone on the left, or drive on either side at random [178]. All are stable in that it is difficult for an individual driver to alter the overall pattern, change mostly having to be introduced by governmental fiat. Driving on either side at random may be stable, but it is also slower and more dangerous, and therefore legislated against. Shifting between any of the three equilibria requires energy of some form in order to make the process occur.

Most countries enforce by law a 'rule of the road', which for continental Europe is the right, but has not always been so [179,180]. The rule of the road also determines other structural features, roundabouts being anti-clockwise, over-taking occurring on the left, traffic lights and road signs being beside right-hand lanes, etc. Altering the rule of the road is a non-trivial structural transition as much infra-structure has to be changed, and in Europe last occurred in Sweden, in 1967, where, after months of planning, driving was switched from the left to the right. A reason for Sweden changing its rule of the road was that neighbouring countries drove on the right, making interactions at borders complicated, although island populations tend to be isolated from such problems[179]. Individual asymmetries can sometimes therefore interact with other asymmetries.

Switching from right to left is more than merely flipping a single switch to alter just one parameter from positive to negative but involves changes in a set of correlated systems. Biologically, that can be seen in situs inversus totalis, where the most obvious differences is that the heart is on the right-hand side of the chest, rather than the usual left-hand side [9], but there are also multiple reversals in almost all organ systems of the body. When visceral situs is not complete (situs ambiguous) then a host of cardiovascular, respiratory and gastroenterological problems can result.

19. Symmetry, Symmetry-Breaking, Bifurcations and Canalisation

Symmetry, and hence also asymmetry, is a fundamental concept across mathematics and the sciences [181]. Symmetric systems can become *'broken'* when disturbances of some sort impose an asymmetry. Stand a narrow wooden plank on edge on a table and from above it shows *bilateral symmetry*. However, the smallest breath of air or tiniest vibration of the table can make the plank fall to one side or the other, and the bilateral symmetry is lost. If the disturbance is random, then the plank is equally likely to fall to the right or left and therefore *a set of such events* will itself show a symmetry, half the time falling to the right and half to the left, resulting in *fluctuating asymmetry* (and it is an asymmetry as the individual cases are themselves asymmetric despite the overall symmetry of the set). If the current of air comes from a particular direction, then the plank is more likely to fall, say, to the right, and the set of events shows more right falls than left falls, resulting in a *directional asymmetry*. The initial symmetry has been broken, and the individual outcomes are then asymmetric (and individually can only go to left or right), but sets of events can sometimes retain some form of symmetry, so that for fluctuating asymmetry the plank falls equally to right or left.

A well-known representation of symmetry breaking in biology is Waddington's model of the *epigenetic landscape* (Figure 1a) [67,182], symmetry being retained in the system as the ball rolls down the landscape until, at the first bifurcation point, the slightest random deviation will make it go either to the right or the left, with symmetry broken at the *bifurcation point*, and so the system becomes *canalized*. This mechanism has, for many years, been seen as relevant to lateralisation [71]. Particularly relevant to lateralisation is Ferrell's analysis of canalization occurring due to *lateral inhibition* which occurs at a *pitchfork bifurcation* (strictly a supercritical pitchfork bifurcation) [183,184]. A physical example of a pitchfork bifurcation is a slender wooden ruler, fixed at its lower end and a downward pressure exerted at the top end. As the pressure increases the ruler starts to bow either to the right or the left, the greater compression on one side and the stretching of the other side reinforcing the deviation from vertical. The situation is equivalent to two groups of cells, which, for present purposes, can be to the right or the left of the organism, with each cell group inhibiting the other, the end result being that one group of cells dominates and the other disappears (Figure 1b). Even if the system starts out as perfectly symmetric, the tiniest of random fluctuations eventually means that either the right or the left side will dominate, resulting in fluctuating asymmetry, half of cases having only the right side and half having only the left side. If the inhibition from one side, say the right, is systematically greater than that from the left, the right side will always predominate and the left disappear (Figure 1c). The world contains symmetries of many sorts which can be broken in many ways, resulting in phase transitions in complex systems [185], often through processes that are catastrophic in the mathematical sense [186]. For theorising about lateralisation in bilaterally symmetric systems, the pitchfork bifurcation is, though, probably sufficient.

Canalisation therefore results either in fluctuating asymmetry (with a 50:50 mixture of right and left), complete directional asymmetry (with all cases in one direction, be it right or left), or in some cases, partial directional asymmetry (with a P%: (100−P%) mixture).

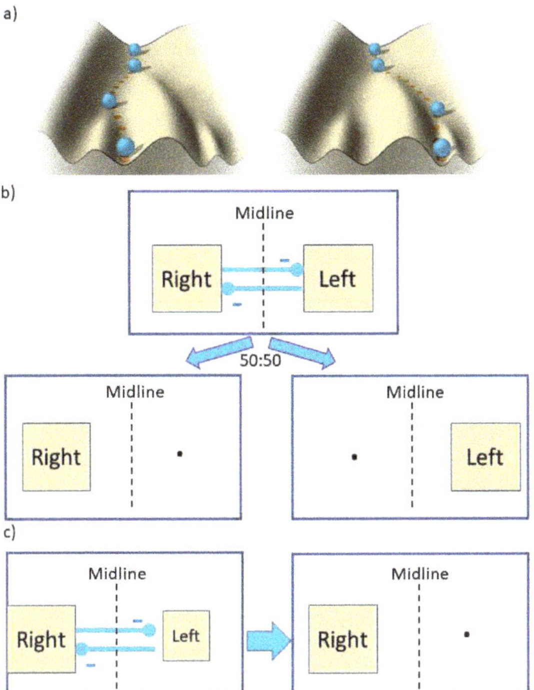

Figure 1. (**a**) Waddington's 1957 epigenetic landscape [67]; (**b**) Canalisation, whereby right and left inhibit each other equally so that there is a 50:50 chance of one side entirely dominating the other; (**c**) Directional asymmetry, where if one side is larger than the other then it always dominates. Figure 1a is slightly redrawn from doi:10.1371/journal.pbio.0050113.g001 (accessed 15 February 2022) to which the Creative Commons Attribution (CC BY) license applies.

20. Describing and Analysing Behavioural Lateralisation

The DC model has looked mainly at handedness, but also at language lateralisation, and the measurement of each requires further analysis.

20.1. Describing Handedness

Handedness typically involves a 90:10 mixture of right and left handers. Handedness can be regarded either as difference in *hand preference* (one hand is preferred to the other for whatever reason), or differences in *hand skill*. Generally, people prefer to use the hand with which they are most skilled, and so the measures are concordant, although that is not always the case [187,188]. Hand skill and hand preference can be seen well in a measure of motor fluency, the tapping task of Tapley and Bryden (T&B), in which participants 'tap' with a pen or pencil on a pre-printed sheet with rows of circles, and on each trial using the right or the left hand to dot as many circles as possible in 20 s—see Figure 2a [189]. The raw data from the T&B study have recently been re-analysed [190].

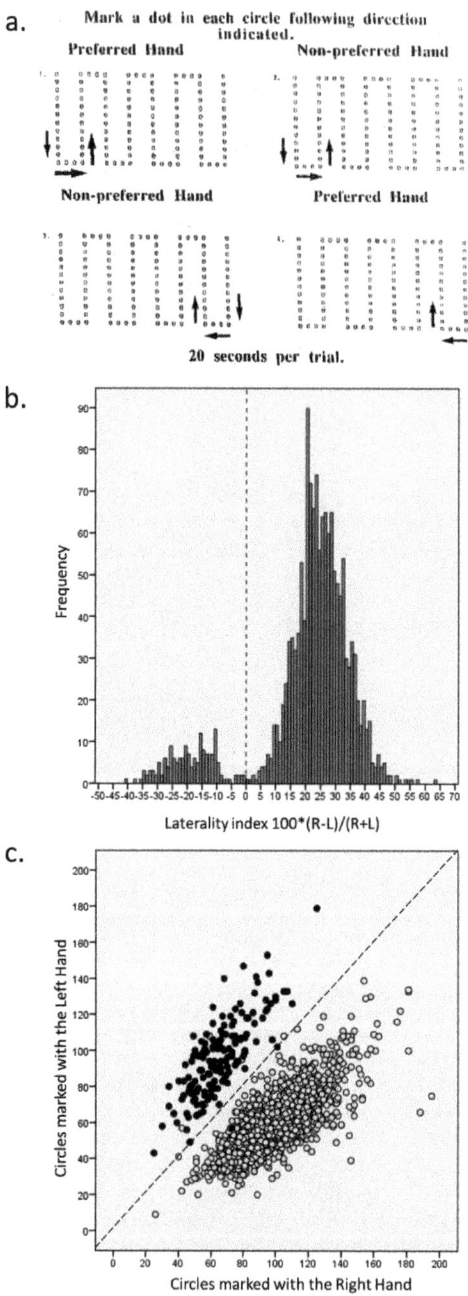

Figure 2. (**a**) Tapley and Bryden's 1985 tapping task; (**b**) recalculated laterality index for the Tapley and Bryden task; (**c**) performance on the Tapley and Bryden task of the right hand (horizontal) and left hand (vertical), for self-declared right-handers (open points) and left-handers (solid points). Note that the quality of (**a**) is low as it is scanned from an old hard copy of low resolution.

Lateralisation data are often summarised using *laterality indices* (LIs) which provide a standardised score, such as $100 \times (R - L)/(R + L)$, where R and L for the tapping task are the number of circles dotted with the right and the left hand. Since for most people the right hand is more proficient, the majority of people have a positive score, with left-handers having a negative score. Figure 2b shows the laterality indices for 1556 participants taking the Tapley and Bryden task, 10.3% of whom self-reported as left-handed. There is a clearly bimodal distribution with the right-handers as the larger distribution and the left-handers the minor distribution, representing about 10% of the participants. Note that the two distributions are separated by the dashed line at exactly zero, which is symmetry. Although it might be tempting to describe these individuals as 'ambidextrous', that can be problematic, not least as with further testing left–right differences almost always emerge.

The metrics for laterality indices can be either *speed (velocity) measures*, i.e., time per item, or can be *quantity (distance) measures*, i.e., items per time interval. In the T&B data, a typical right-hander taps a total of about 100 circles in the two 20 s tests (total time = 40 secs), which is about 2.5 circles per second or 400 mS per circle (see Figure 2c). One metric is the reciprocal of the other, and they are interpreted in opposite directions, better performance corresponding to *more items* carried out but to *less time* to process a single item. That difference can be important in modelling scores.

Laterality indices for speeded measures such as the Annett pegboard task [191] are often calculated as $100 \times (R - L)/(R + L)$, known as *PegQ* [192], the reversal of L and R meaning that right-handedness is still associated with positive scores. When different laterality measures are being compared then all scores should be in the same direction, as in a study using tapping speeds (circles/sec), writing speed (letters/sec), and writing quality (quality units/sec) where positive scores all indicated better performance [193].

In measurement terms, laterality indices such as $100 \times (R - L)/(R + L)$ are dimensionless, being on a pure numerical scale. Sometimes other measures are used, such as $100 \cdot (R/L)$, which is also dimensionless, but it is also possible to use a simple difference score, $100 \times (R - L)$, which is not dimensionless but has units of speed or quantity.

The justification for using indices such as $100 \times (R - L)/R + L)$ is, as Tapley and Bryden put it, "because of a feeling that the difference between 100 and 98 circles filled represented performance similar to that seen in a performance of 200 and 196, rather than 200 and 198." [189] (p. 216). That feeling does indeed seem intuitively sensible, but if one wants an index which is independent of total score, $R + L$, then the much simpler $R - L$ may actually be better [190], the reason being that $R + L$ and $(R - L)/(R + L)$ are not independent measures but are necessarily correlated. Laterality indices can sometimes confuse rather than clarify.

Laterality indices have an inherent problem in that they lose information about overall performance, and that becomes apparent when considering bilateral functions. Figure 2c shows the same data as Figure 2b but with the separate raw scores plotted for the right and left hand of each participant, and self-reported handedness is also indicated (○: right-handers; •: left-handers). Self-reported left- and right-handers show a very clear separation, with two almost entirely separate distributions, and it is also now apparent that participants differ in their overall level of performance, some being faster with each hand than other participants.

20.2. Reliability

Bimodal scores such as laterality indices have *two different types of reliability*, which essentially are within mode and between mode, and can be assessed for the T&B task because participants carried out the task twice with each hand. In simple terms, the *reliability of direction of lateralisation* asks whether an individual has the same direction of lateralisation on two separate occasions (i.e., if *sign[LI]* is positive or negative), and is hence referenced against the absolute score of zero. For the T&B task, the reliability of direction of lateralisation is 0.965, almost no-one being right-handed on one occasion and left-handed on another. The *reliability of degree of lateralisation* looks at the similarity of the degree of lateralisation, measured as the distance of a participant from zero (i.e., *abs[LI]*), irrespective of direction of lateralisation. For the T&B task, this reliability of 0.711 is much lower than for direction of lateralisation, suggesting that degree of lateralisation is a less robust measure than is direction of lateralisation. Studies often calculate reliability based on the correlation of the raw laterality index (i.e., LI) across occasions. For the T&B task, this gives a reliability of 0.908, which is still impressive, but inevitably is mainly dominated by direction rather than degree of lateralisation, as most of the variance is between modes rather than within. It can be very misleading particularly if differences in degree of lateralisation are of interest, as the raw reliability of LI can still be high, due to a high reliability of *sign(LI)*, even if *abs(LI)* has a reliability of zero.

20.3. Measurement Error

Measurement error must be considered when interpreting measures of lateralisation. Although measures such as the Annett pegboard typically result in a bell-shaped distribution [192], much of that distribution reflects a substantial amount of measurement error, rather than any normally distributed latent underlying distribution, as predicted by Annett's RS theory [73].

The simulation in Figure 3 shows what happens as measurement error increases. Figure 3a starts with two distinct categories of individuals, R and L, with 80% R and 20% L (to make the effects easier to see). As each category is measured with greater error (Figure 3b,c), so the two categories spread, although continuing to be a bimodal mixture distribution. As the variance increases, so eventually each distribution substantially crosses the zero line (Figure 3d), and there is only a single mode visible, the presence of two groups only being indicated by the skewness.

Although distributions for pegboard and other motor tasks may look bell-shaped, in practice the two underlying distributions can be separated and they can prove to be mixtures [194]. Laterality indices for tasks such as pegboards are typically unimodal because of having a relatively small number of trials, only ten or so pegs being moved, each having quite a lot of measurement error, particularly due to occasional 'stumbles', and the laterality index therefore depending on the difference between two unreliable measures. In contrast, the Tapley and Bryden tasks involves a hundred or more taps, reducing the proportion of measurement error and separating the distributions.

Figure 3 shows that it can become increasingly difficult to classify individuals categorically as R or L, some actual Rs having negative scores and some actual Ls having positive scores. Within the population there will be more true R cases manifesting as L than true Ls manifesting as R, so that a simple dividing line at zero can give the misleading appearance of more left-handers than is truly the case (and that effect might explain studies using bone asymmetry data to suggest that medieval populations had higher rates of left-handedness than modern populations [195]). The problem is formally similar to Satz's description, in 1972, when modelling pathological left-handedness, which is more frequent than pathological right-handedness [196].

20.4. The Statistics of Lateralisation

Statistical analysis of data such as those shown in Figure 3 is difficult. For Figure 3a, one would normally use a measure such as the chi-square statistic or logistic regression, which compares proportions of the two categories in various groups. One could do the same for Figure 3b, dividing the distribution at zero, but one could then also separately analyse just the right-handers (blue) or the left-handers (red), comparing means across groups using ANOVA or other statistics. The overlapping distributions in Figure 3c,d are, however, more complicated and cannot be analysed properly using conventional statistics, and special methods are needed for mixture distributions [197–200].

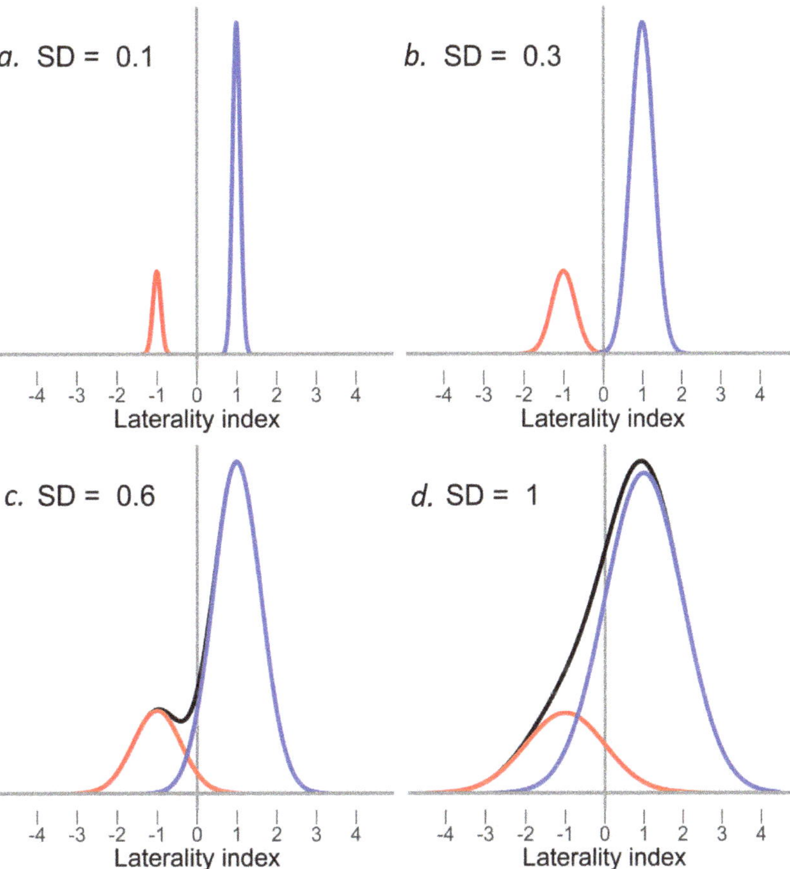

Figure 3. Example of how measurement error affects two discrete distributions, one positive, major distribution (blue) and the other negative, minor distribution (red), equivalent to right and left handers. Means of the distributions are +1 and −1, and the thin vertical line is at zero. The standard deviations increase, due to increasing amounts of measurement error being added, so that what starts in (**a**) as two almost categorical variables (SDs = 0.1), becomes in (**b**) two entirely separate normal distributions (SD = 0.3), in (**c**) a distribution that is still bimodal but with substantial overlap between the distributions (SD = 0.6), and finally, in (**d**) a single, unimodal distribution where the minor distribution is entirely absorbed into the tail of the major distribution (SD = 1).

Figure 4a shows a simulated set of typical data from a small lateralisation study in which 20 right-handers (•) and 20 left-handers (○) are tested on some measure for which a laterality index is calculated [197]. The vertical line indicates zero on the score, and it seems that there are more left-handers than right-handers with negative scores. A conventional statistical analysis would calculate the mean and standard deviation within each of the two groups and use a t-test to compare the means (Figure 3b), but the difference in means is not quite significant ($p = 0.063$), although there is a significant difference in variances ($p = 0.027$). The usual interpretation would then be that left-handers are more variable than right-handers. In fact, the simulated data were drawn from two mixture distributions shown in Figure 4c, the means for each mode being symmetric around zero, and standard deviations for each mode being identical in right- and left-handers. The only difference by handedness is that the left-handers show more cases in the minor distribution than do the right-handers. Without going into the details, the symmetric bimodal model with different proportions (Figure 4c) fits the data very much better than the ANOVA model shown in Figure 4b) ($p < 0.001$), with the proportion in the minor distribution being higher for left-handers than right-handers ($p = 0.048$) [195]. The full analysis shows that the right- and left-handers therefore differ only in direction of lateralisation and not in degree of lateralisation [197]. Similar calculations can be done using modern software. However, one has to be careful with typical packages in *R* such as *flexmix*, etc., as although they fit multiple normal distributions it is not usually possible to constrain the two distributions in data, as in Figure 4c, so that the parameters are mirror-images. That can, however, be done in *R* using *OpenMx* [201,202].

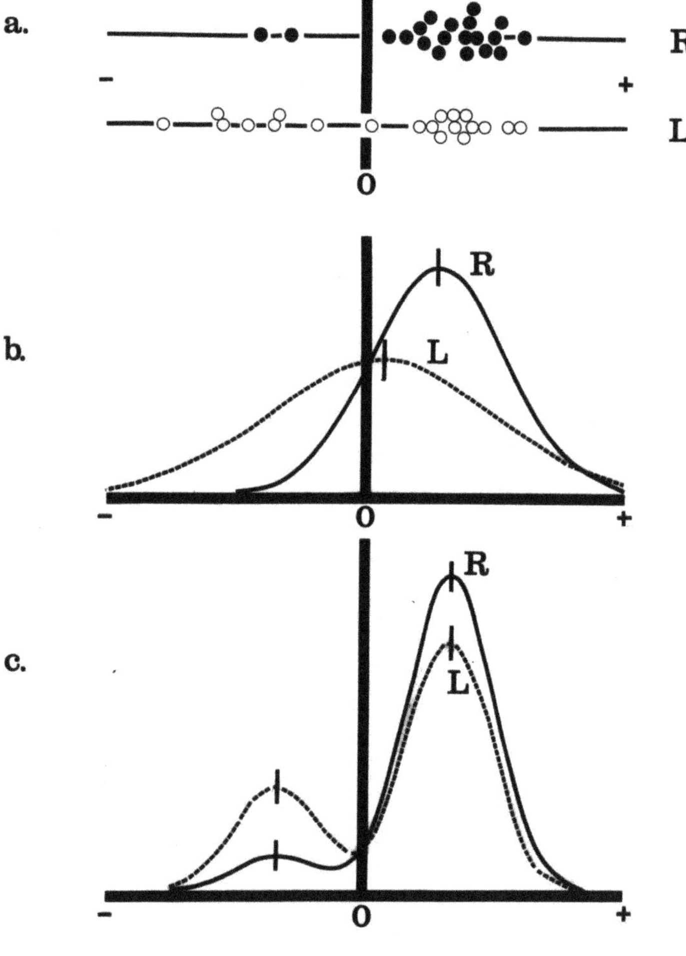

Figure 4. Example of problems of interpreting laterality scores [197]: (**a**) scores of 20 right-handers (solid points) and 20 left-handers (open points); (**b**) fitted normal distributions for right- and left-handers showing different means (vertical lines) and greater variance in left-handers; (**c**) the symmetric bimodal distributions from which the data were actually taken, with bimodal distributions, identical means and SDs for the major and minor distributions in right and left handers, and differing only in the proportions of left-handers in the minor distribution.

The approach shown in Figure 4 can be extended using maximum-likelihood methods to a range of other designs, both of within- and between-group differences in means and variances, and the assessment of within- and between-group reliability [197]. Power calculations are also available [203].

A more complicated example is shown in Figure 5a with a bivariate mixture distribution for two measures, A and B, which could be different tasks. The top right quadrant shows individuals who have positive scores on A and B, the lower left quadrant shows individuals with negative scores on A and B, and the other two quadrants shows individuals who are negative for one score and positive for the other. The quadrants correspond to directions of lateralisation, with the proportions in the quadrants given in terms of π_A, π_B, π_{AB} and $1 - \pi_A - \pi_B - \pi_{AB}$. Within each quadrant is a bivariate normal distribution with a degree of lateralisation described in terms of mean scores for A and B (μ_A and μ_B), standard deviations (σ_A and σ_B), and the correlation of A and B (ρ). The challenge in modelling terms is to estimate all the various parameters, which is possible [197], while simplifying where necessary by fixing parameters to be equal where there is no evidence to the contrary. In the study itself, Figure 5b, A and B were data from a 1980 study by Gordon [204], who administered a dichotic listening test in English and in Hebrew. The statistical analysis found individuals who had opposite lateralisation in the two languages, shown in the upper left and lower right quadrants. Whether such a pattern occurs elsewhere might perhaps become apparent in the meta-analysis of neural activity in L1 and L2 in bilinguals [205].

20.5. Tetrachoric Correlations

Often, it is the case that all one knows about the relationship between two lateralities is the proportions to right and left of the zero line which distinguishes right and left. An example might be fMRI measures of functional lateralisations, as shown in the examples of cerebral polymorphisms in the introduction to this paper. Although conventional, Pearson, correlations can be used, but they are not good with binary variables (and hence the emergence of spurious 'difficulty factors' in factor analyses of item-correct scores in educational testing [206]), so that tetrachoric correlations are undoubtedly preferred. Essentially these consider what the correlation of an underlying bivariate normal distribution would have to be if the distribution were sliced horizontally and vertically to give four categories, categories in the case of laterality, which are sliced at zero, the phase boundary. The calculations are readily available in *R* packages such as *psych* or *polycor* [207]. For the data in Figure 5b, the estimated proportions in the four quadrants are 0.6818, 0.0328, 0.0389 and 0.2465. The tetrachoric correlation is 0.965, compared with a Pearson correlation for the same values of 0.822. For the raw data, which include the within-group randomness visible in Figure 6b, r = 0.78. The tetrachoric correlation is probably the best descriptor here, with an estimate of about 7% of participants having English and Hebrew in different hemispheres, at least in terms of the dichotic listening results.

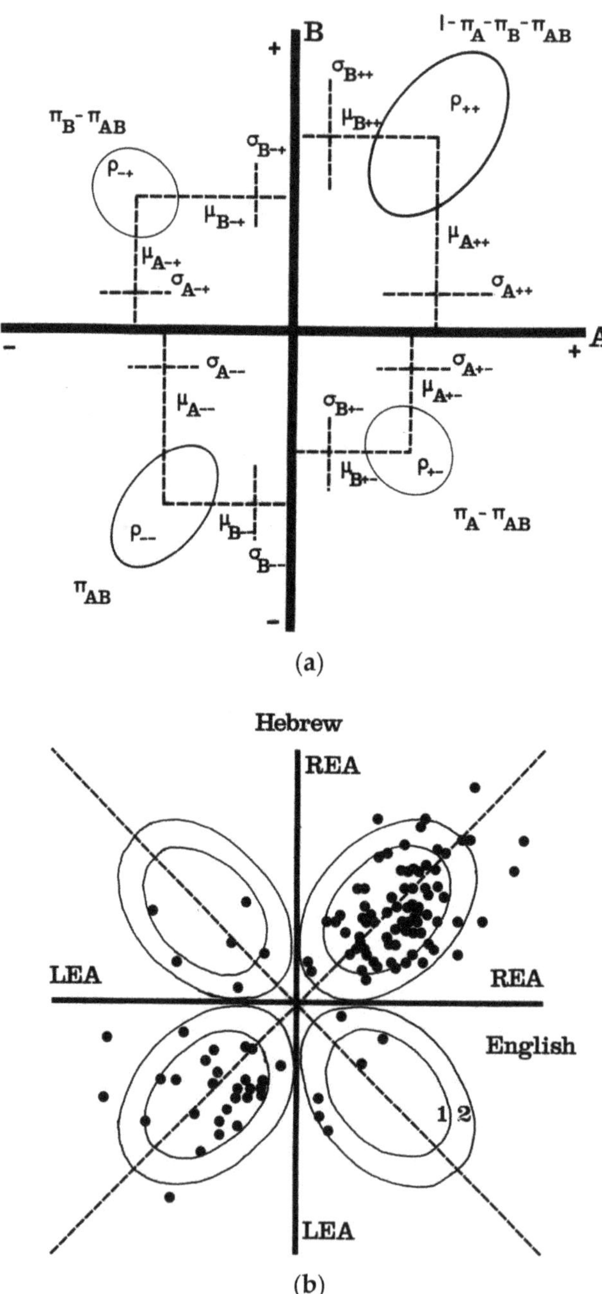

Figure 5. (**a**) (Upper) symmetric bimodal model for two separate tasks, A and B, with bimodal distributions for A and for B, with separate correlations of A and B within each quadrant, and separate proportions in the four quadrants; (**b**) (lower) model fitted to 1980 data of Gordon for dichotic listening tests carried out in Hebrew and English.

a)

	(1)	(2)	(3)	(4)	(5)	(6)
(1) Atypical handedness	1.	**0.662**	0.210	**0.466**	**0.376**	**0.467**
(2) Atypical verbal fluency	**0.662**	1.	**0.673**	**0.556**	0.278	**0.513**
(3) Atypical face perception	0.210	**0.673**	1.	**0.716**	0.167	**0.322**
(4) Atypical body perception	**0.466**	**0.556**	**0.716**	1.	**0.449**	0.068
(5) Atypical emotional prosody	**0.376**	0.278	0.167	**0.449**	1.	0.124
(6) Atypical emotional vocalisations	**0.467**	**0.513**	**0.322**	0.068	0.124	1.

b)

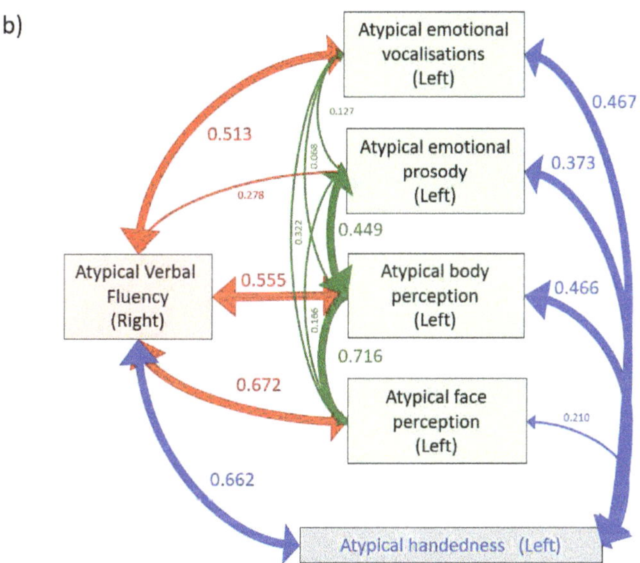

Figure 6. Associations of atypical lateralisation for the six modules described by Karlsson in 2020: (**a**) tetrachoric correlations with significant correlations $p < 0.05$ shown in bold; (**b**) the inter-relations shown diagrammatically, with red for the typically left-hemisphere function, green for the typically right-hemisphere functions, and blue for handedness.

20.6. Laterality Coefficients for Assessment of Language Lateralisation

Assessing language lateralisation in modern studies is mostly carried out using fMRI, fTCD, or, in patient populations, the Wada technique. Lateralisation can also be inferred from rates of aphasia after cerebral lesions [6]. fMRI laterality indices have used various approaches, reflecting differences in threshold and the number and particular voxels included [208], with *LI-tool* being a popular method [209]. In one study using whole hemispheres, the majority of participants showed left-hemisphere dominance (67/82; 81.7%), with a higher rate of left-hemisphere dominance amongst right-handers than left-handers [210]. fTCD methods have mostly used the approach of Deppe and colleagues [211], but recent work has raised the question of whether searching for the peak blood flow difference artificially creates distributions that are more bimodal, resulting in a 'notch' at zero, so that it is better to use mean blood flow difference [212] (see https://osf.io/tfyk3/ accessed on 3 November 2021). Wada results have been interpreted in terms of the WLI,

Wada Laterality Index, although there are several versions of the index [213]. A potential problem of all methods of calculating laterality indices, be it from fMRI, fTCD or Wada tests, is that they inevitably lose information on activity in the two separate hemispheres, so that while they provide information equivalent to those shown for motor skill in the present Figure 2b, almost no studies show plots equivalent to Figure 2c. An important exception is the study of Wegrzyn et al. [214], who plot L − R against L + R, their Figures 8 and 9 being equivalent to Figure 2c rotated 45 degrees. Unlike Figure 2c there is though no clear divide between the various groups. The authors conclude that "none of the approaches presented here showed a satisfying sensitivity regarding the detection of bilateral cases ... [which] might reflect that many instances of bilaterality cannot be well-expressed with a simple LI", and they particularly mention "crossed lateralisations with left-sided activity in Broca's area and right-sided activity in Wernicke's area [current reference [215]] might by definition be unsuitable to be represented by a simple score ... " (p. 14).

21. The Biological Background: Primary Ciliary Dyskinesia (PCD) as a Model Condition

Right- and left-handedness may be the most immediately visible signs of lateralisation in humans, but within biology there are numerous other examples [179,216,217] with one of the most encyclopaedic collections still being the 1932 book by Ludwig [119]. Many biological asymmetries are often curiosities, intriguing in many ways, but not opening up a clear route to a deeper biological understanding. The major exception to that is the asymmetry of the heart, which in vertebrates is to the left of the body, and is associated with other asymmetries such as of the lungs, stomach, spleen, liver, appendix and colon, as well as a host of associated minor asymmetries of the skeleton, etc. In about 1 in 10,000 people, though, and also in many animals, the situation is reversed in the condition of *situs inversus totalis*, in which the heart is on the right, and most other organs are reversed, making a near mirror-image of the typical condition.

Understanding situs inversus has been a major success for biomedical research in the last twenty-five years, and the condition of primary ciliary dyskinesia provides a clear conceptual and biological model for the genetics of handedness, and therefore will be described in detail here.

Individual cases of situs inversus have been known since antiquity, and considered scientifically since the early nineteenth century [218]. Sporadic cases, with no other symptoms, mostly seem to have a normal life-span and no obvious pathologies. The prevalence was estimated in the 1950s using images taken during mass population chest X-ray screening for tuberculosis, the occasional reversal being readily spotted in the millions of diagnostic images [219].

In 1904, A.K. Siewert, a physician in Kyiv, now in Ukraine, described an unusual syndrome in a patient with bronchiectasis and chronic sinusitis who showed situs inversus [220]. In 1933, the condition was described again, became known as Kartagener's syndrome [221], and in addition partial cases without situs inversus were also described.

Two important breakthroughs occurred in 1976. Almost exactly 300 years after van Leeuwenhoek first saw cilia beating under his primitive microscopes, Afzelius showed that patients with Kartagener's syndrome had defective cilia [125], their cilia missing the internal dynein arms so that they could not beat, and therefore patients were unable to remove detritus from their lungs and sinuses using the muco-ciliary escalator [222]. The background to the discovery has been described elsewhere [223,224]. Subsequently renamed 'immotile cilia syndrome', the syndrome is now known as primary ciliary dyskinesia, although the name is slightly confusing, with 'primary' referring to the disease not being secondary to smoking or other conditions, rather than to the 'primary cilia' in the nervous system which mainly have a sensory role [225]. The other major breakthrough in 1976 was Layton's exploration of a recessive gene, previously hinted at in two earlier studies and named the *v* and then the *iv* gene [226,227], in which 50% of mice showed situs inversus [124]. The title of Layton's paper emphasised that the 50% rate of situs inversus in *iv* mice was not due to partial penetrance, as originally suggested, but was the result of "random determination of a developmental process". A key experiment in Layton's study used an inbred strain of mice, all of which were *iv* homozygotes, and Layton crossed all possible combinations of situs inversus and situs *solitus*. Each combination had 50% of offspring with situs inversus, implying that there was no selectable variance remaining [124]. As Afzelius later put it, this was "the inheritance of randomness" [228]. It is exactly analogous in the DC model to the mating of two homozygous CC parents.

22. The Inheritance of Randomness

Unpicking the mechanism of inherited randomness took a long while. Much hard searching of the *iv* mouse genome eventually found that the mutation was, as Afzelius had suspected, in a dynein, a molecular motor, although how that then determined situs was still unclear [229]. On another front, developmental biologists had been looking hard at the very early events in embryogenesis that make the heart become asymmetric, with it being found that in the chick the side of the chick heart could be experimentally altered by applying the signalling molecules *activin* and *sonic hedgehog* to the very early, and visually entirely symmetric, embryo [230]. A key part of the jigsaw fell into place in 1998 with the remarkably surprising discovery that cilia in the nodal region of the early mouse embryo were motile, rotating anti-clockwise, the motion creating a fluid flow which pushed signalling molecules such as activin to be concentrated on the left side of the embryo [231]. To clinch the mechanism, the process could be reversed by experimentally pushing fluid in the other direction, resulting in embryos with situs inversus [232]. Rotation of the cilia was the symmetry-breaking event which usually put the heart on the left side (directional asymmetry), with immotile cilia resulting in random symmetry-breaking (fluctuating asymmetry) [233]. As to why the cilia always rotate anti-clockwise, that is almost certainly due to the fact that they, like almost all proteins, are built from L-amino acids (rather than the biologically rare D-amino acids), and, hence, the cilia rotate in one particular direction [179]. The implication is that if cilia could be built from D-amino acids and L-sugars, then they would rotate in the opposite direction.

Enticing though cilia are as the fons et origo of asymmetries of situs, there remain difficult things still to be explained, not least that left–right body patterning in a range of asymmetries does not seem to be related to cilia, and cilia do not seem to be important in some species such as chicks. Since cilial proteins are very widespread intracellularly, it is possible that sub-cellular polarity can result in asymmetry and can also co-associate with cilial action [234]. Clearly, there are still many questions to be answered on the origins of situs and asymmetries, and the field remains buoyant and exciting [235].

22.1. The Inheritance of PCD

The molecular genetic basis of PCD had been sought since Afzelius's discovery of immotile cilia, with a locus eventually being identified in 2000 on chromosome 19q, although apparently with different genes in five separate families [236]. Since then, there has been much progress in understanding the biology of the cilium, an essential and omnipresent piece of molecular machinery throughout the biological world, using the cilium of the protozoon *Paramecium* as a model organism [237]. The complex molecular machinery of the cilium has at least 600 proteins [238], which undoubtedly are generated and influenced by many more genes, with cilia liable to dysfunction as a result of a host of possible mutations, many of which could also cause the growing number of diseases and conditions now recognised as ciliopathies [239,240]. Nearly 50 separate genes have now been found that cause PCD [241], acting mostly but not entirely on the dynein molecules at the centre of the 9 + 2 structure, and, as a result, diagnosis is being revolutionised [242].

PCD is a fascinating condition in many ways, and is a tribute to modern molecular medicine, and one in which I have had an interest. The relatively large numbers of patients with PCD make it possible to study the relationship of situs inversus and handedness [243], but also to study other aspects of the psychology of being a patient with the condition [244–246]. However, that is not why this paper has such a long account of PCD, but rather it is because *PCD represents a very close biological parallel to the DC model of lateralisation.*

22.2. The Genetic Architecture of PCD

Although PCD is polygenic, its architecture is subtly different to many other polygenic conditions, and in some ways the architecture is more similar to the monogenic inborn errors of metabolism. The error is not in a single enzyme but rather in a single organelle, the cilium. There is a partial parallel to the myriad defects resulting from errors within another cellular organelle, the mitochondrion (although mitochondrial DNA is, of course, inherited entirely maternally) [247]. To work properly, a cilium needs all of its proteins to be assembled correctly, and failure of any one can results in partial or total inactivity of the cilium. The proteins can be considered conceptually as 'wired in series', any problem anywhere preventing the entire chain from functioning.

The classical genetics of PCD in many ways looks like a monogenic disorder. Within families, the condition acts as a monogenic autosomal recessive, in most cases neither parent being affected but one in four of the offspring being affected. Cases are more common in inbred families, and offspring of homozygous individuals who are affected can have normal cilia if the second parent is unaffected and is not carrying the same gene. That is very similar to the classical inheritance of deafness, which also acts a recessive, and one in four siblings in a family is affected. There is, however, strong assortative mating for deafness, the deaf often marrying the deaf. As a result, the offspring of two deaf parents are very often not deaf because the homozygous parents are homozygous for different genes, so that the children are then heterozygous for each gene, and they have normal hearing. That is the basis for the calculation reported in 1971 of at least 15–25 recessive genes for deafness [137] (p. 375). Now, autosomal recessive deafness is thought to be due at least 700 mutations in 42 different genes, with some more common than others [248]. Additionally, there are many other types of deafness which are also under various forms of genetic control [249].

A key feature of the genetic architecture of PCD, as well as that of deafness, is that *within families* the trait runs in a classical Mendelian recessive fashion, but *in different families* PCD and deafness *run in families for different reasons because they have different genes*. That is the reason that the children of deaf parents (and presumably of PCD parents) can often have unaffected children (and that forms the basis for genetic counselling). *The principle that a trait can run in different families for different reasons underpins the polygenic DC model.*

22.3. Phenotypes and Endophenotypes

A reviewer asked about the role of endophenotypes in the DC model. Given that definitions of even standard genetic terms such as phenotype, genotype and genome, as well as phenome, are confusing and confused [250], the term 'endophenotype' (EP) is also less than clear and used inconsistently. EP was first used in insect genetics in 1966 [251], referring to a phenotype which was not visible from the outside (unlike a typical exophenotype) but appeared only, say, by examining chromosomes under a microscope. In that sense, trisomy 21 is clearly an endophenotype for Down's syndrome. Gottesman and Shields briefly discussed EPs in psychiatric genetics in 1973 [252], describing them as "only knowable after aid to the naked eye, e.g., a biochemical test or a microscopic examination of chromosome morphology". The much cited 2003 review of Gottesman and Gould [253] still referred to "components unseen by the unaided eye", a category that now in the vast number of publications on EPs [254] seems to include almost anything from fMRI scans to DNA sequences, and questionnaires, and behavioural/psychophysiological measures. Gottesman and Gould also provided six formal criteria for an EP [253], although there are other related definitions [255]. Walters and Owen suggested that a "critical assumption is that the genetic architecture of the endophenotype is simpler than that of the disease phenotype" [256], so that it is closer to the level of gene action than is a diagnosis. EPs have shown much conceptual slippage in the many papers using the term, and Lenzenweger [257] has carefully distinguished EPS from intermediate phenotypes and biomarkers. Kendler and Neale [255] have used the language of structural equation modelling to say that EPs are properly mediating variables, rather than being mere correlates or epiphenomema [256]). EPs as mediating variables can themselves also have EPs and be EPs for other EPs, as shown in the complex causal chains for ApoE in relation to a range of neurological conditions [258]. For PCD, it is clear that ciliary motility is a major EP, being genetically determined and responsible for the exophenotype, the symptoms, of the disease itself, and PCD is also a good example where the EP has a simpler inheritance than the more complex phenomenology of a disease entity [259].

Atypical handedness has been invoked on several occasions as an EP for psychiatric and developmental conditions, most notably for schizophrenia, with a higher rate in cases than relatives [259] (and a meta-analysis confirms higher rates of atypical handedness in schizophrenia [260]). Altered brain asymmetry for pseudoneglect has also been invoked as an EP for schizophrenia [261]. Mixed-handedness has been proposed as an EP for developmental language disturbances [262], although Bishop [263] has argued that the argument for weak laterality being an EP for specific language impairment is "unconvincing", because "there is little support for strong genetic differences on individual differences in cerebral asymmetry". Left-handedness has also been invoked as an EP for anorexia nervosa [264], and atypical lateralization is also reported in autism [265].

The reviewer asked specifically about EPs for handedness itself, and Ocklenburg et al. [266], in response to the study of Guadalupe et al. of differences in cortical anatomy in right- and left-handers [267], hoped that "identifying structural markers for left- or right-handedness may provide endophenotypes that aid the ongoing quest to identify the genetic, epigenetic, and environmental influences that determine handedness". That is possible, but it might also be that such associations are epiphenomena, or correlates, or even consequences of handedness [256]. If there is an endophenotype for handedness then its purest form it is probably fluctuating asymmetry (for left-handedness) or directional asymmetry (for right-handedness), in the same way that the fundamental EP for situs inversus in PCD is fluctuating asymmetry, driven in the case of PCD by ciliary immotility (which may itself be caused by many genetic errors). However, it seems unlikely that the randomness from fluctuating asymmetry resulting in situs inversus in PCD and the randomness from fluctuating asymmetry resulting in left-handedness are the same randomness, since individuals with PCD and situs inversus are no more likely than controls to show left-handedness [243]. To put it simply, two separate coins are being tossed, rather than a single coin toss determining both asymmetries. Where the separate coin is that which determines handedness, the endophenotype for handedness is currently unclear, but it might be hoped that recent discoveries of genes for handedness [66] might eventually point at a biological answer.

23. Molecular Genetics of Handedness and Language Dominance

Darwin, while recognising that many asymmetries are not heritable, also said that "A man who is left-handed, and a shell in which the spire turns in the wrong directions, are departures from the normal asymmetrical condition, and they are *well-known to be inherited*" [114] (vol 2, p. 12, my emphasis). Darwin was correct about snails, the genetic basis of left and right spiralling being known now to depend on formins, molecules found in all eukaryotes which affect actin and cytoskeleton development, and also have been implicated in vertebrate body asymmetry [268,269]. Darwin was also correct about left-handedness being inherited.

Handedness undoubtedly runs in families [10,20,71,165,270,271], and many genetic theories have been proposed [64,106–111,272]. All of these models are classical, in the sense that genes are hypothesised on the basis of phenotypic patterns in populations. With the advent of molecular genetics, there had long been a desire to find an actual gene for handedness (and once, when asked in 1997 "What is your unrealised ambition?", I replied "Finding the gene for left-handedness and cerebral language dominance" [273]). I was not the only one, and between 1998 and 2011 there were at least eight attempts, most of which were small by modern standards and mostly had inadequate power [11]. The only two exceptions were a GWAS based on 4,268 *23andMe* customers, and a meta-analytic combination of 12 GWASs with N = 23,443, published only as an abstract, neither finding any significant associations with handedness [274,275]. Our own GWAS, published in 2013, also found no association with handedness, but an important feature of it was the calculation showing there *was* sufficient power to have found any locus corresponding to the monogenic DC model (or Annett's RS model) [104]. The study also concluded from the negative results of the then largest GWAS [274], that "there are probably at least 30–40 loci involved in handedness" [11].

Sample sizes since then have been growing ever larger, a 2019 study using 331,037 participants from the UK Biobank suggested three associations with handedness [276]. Eventually, in 2021, the prediction of 30–40 loci was shown to be correct (although ever larger studies will undoubtedly find even more loci). The 2021 study of Cuellar-Partida et al. had 1,766,671 participants from 34 studies, the vast majority coming from *23andMe* or UK Biobank, finally provided convincing evidence of genes for handedness [66]. The main analysis compared left-handers with other participants and found 41 loci meeting the standard criterion of $p < 5 \times 10^{-8}$, with loci on 18 of the 22 autosomes. Estimates of SNP-based heritability were low (3.45%, 5.87%) which are much lower than estimates from twin studies, of about 25% [142,143]. However, the UK Biobank data allowed estimates of genes identical by descent, giving an additive variance of 19.7% (95% CI 13.6–25.7%), which is compatible with twin studies. For an introductory account of identity by descent, and differences between twin and SNP methods, see Harden [277].

24. Ambidexterity/Mixed-Handedness

Ambidexterity and mixed-handedness are often treated as being equivalent, particularly when assessed by single item questionnaires which have categories 'right', 'left' and then 'mixed' or 'either'. Ambidexterity, strictly, is performing equally well with either hand, and there is little sign of such individuals in performance data such as Figure 2c. Mixed-handedness is usually considered to be the use of different hands for different tasks [278]. Most single item handedness questionnaires probably confound ambidexterity with mixed-handedness, making almost all large-scale studies ambiguous in meaning. The Cuellar-Partida et al. study reported on 'ambidexterity', the *23andMe* and UK Biobank studies having a category of 'either hand' for writing. The study is therefore in the embarrassing situation of having detailed genotyping but a poorly defined phenotype. There were seven significant loci for 'ambidexterity', with some association with the loci identified for left-handedness. Heritability for ambidexterity was estimated at 15%, but there was only a moderate genetic correlation of 0.24 with the analysis for left-handedness. The result is consistent with a Korean study finding some familiality for ambidexterity/mixed-handedness [278]. It has long been unclear whether ambidexters are genuinely different in their lateralisation, or perhaps those choosing to respond 'either' or 'mixed' have personality or other differences. Alternatively, ambidexterity may reflect unintentional errors in responding. Response errors (or idiosyncrasy) are supported by the finding that of the 2% of UK Biobank participants describing their writing hand as 'either', 41% gave a different response on subsequent testing, compared with <2% of changed responses for those answering right or left hand on the first survey [279]. Overall, ambidexterity is probably separate genetically from right- and left-handedness, but may not be a stable measurement construct in terms of lateralised performance. Ambidexterity will therefore not be discussed further here. Mixed-handedness, though, is of more interest, particularly for writing and throwing with different hands, which occurs in about 30% of left-handers and 3% of right-handers [280–283]. Writing and throwing hand appear to be co-inherited, with proportions compatible with the DC model [283].

The Cuellar-Partida et al. study is likely to transform research into handedness. The 41 loci it has identified will be important in understanding the neural basis of handedness, with links to the central nervous system, microtubules and brain morphology. Combining the 41 loci into a 'polygenic score' [284,285] differentiates right- and left-handers in the UK Biobank [279], and such an approach will allow analysis of potential links to handedness in other genetic studies where handedness itself was not measured. Having said that, though, the Cuellar-Partida et al. study also reports the association of each of the 41 loci with handedness. Combining these gives an overall odds-ratio of 3.16×, which is some distance from the odds in the population of about 9× (90/10%), suggesting there may be other loci or factors still to be identified.

25. Modelling Polygenic Effects on Discrete Traits

Handedness is a discrete, categorical trait, with essentially two forms, right and left. The Cuellar-Partida et al. study suggests that at least 41 genes are involved, while the monogenic DC model, with its single locus determining handedness, is effective at explaining how handedness runs in families, in twins and in relation to language dominance. That sentence seems to show a fundamental incompatibility between its two halves, which clearly needs resolving.

The next sections will firstly consider polygenic influences on phenotypes more generally, many of which are categorical or occur in small numbers of forms, and then will use insights gleaned from PCD, to explore how a polygenic DC model may be compatible with the data and give broadly similar predictions to the monogenic model.

26. Polygenic Models in General

At the symptomatic level, diseases and conditions tend to be described by discrete diagnostic categories, sometimes sub-divided into groups, as with Type 1 and Type 2 diabetes. However, the two types of diabetes are influenced by large numbers of gene loci, which together account for much of the variance in the condition [286]. Digging back into medical history, diabetes was recognised as a condition because of an excess production of urine (from the Greek for 'to go through'). At some point, it was recognised that some diabetics had urine which tasted sweet (*diabetes mellitus*) whereas others did not (*diabetes insipidus*), making the latter a separate condition. For the everyday sense of 'diabetes' (i.e., *mellitus*), the key feature is glucose in the urine (and later, it was found, high levels of blood glucose), and glucosuria/hyperglycaemia is in effect the phenotype being analysed in most genetic studies. The metabolism of glucose and the physiology of glucose regulation are complex [287], and since the entire ensemble is needed to work properly it is hardly surprising that many genetic or other factors can alter glucose metabolism and result in diabetes, as every step in the process is potentially vulnerable to a greater or lesser degree, sometimes as the result of a defect in a single gene, as in monogenic diabetes [288]. Whether diseases are indeed 'natural kinds' is a matter of some controversy, and Mitchell has argued that ultimately the concept of disease may be replaced in clinics by "a genetic diagnosis, indexing the primary cause of the disease [a rare, recent mutation] and not merely the surface symptoms" [289].

Occasionally, there is one specific step where vulnerabilities can occur, as with the monogenic condition, phenylketonuria, PKU, which nowadays typifies what Garrod called 'in-born errors of metabolism' [290,291]. In PKU, there is a defect in the enzyme phenylalanine hydroxylase, PAH, which means that phenylalanine can then not be metabolised properly, phenylalanine builds up in the blood, damaging the developing brain, and a metabolic side-product, phenylketones, also appears in the urine. Although only a single gene is involved, there are nearly a thousand variants of the defective PAH gene, some resulting in severe dysfunction and others in mild or minimal dysfunction [292].

Diseases or syndromes or traits that are more broadly defined, and often are continuous in their description, are more likely to be influenced by many genes as there are more causal pathways to the eventual outcome. Height, weight, intelligence, education, and neuroticism are typical of such traits, since they combine very many separate but related processes and, hence, many factors can influence them, making them polygenic [277,284]. Being polygenic does not mean though that all genes have the same effect. More than 250 loci affect body weight and obesity. Rarer genes though tend to have larger effects on body mass index, with the scarce MC4R locus resulting in the largest influence on obesity of a seven kilogram increase in weight per allele, an effect that in longitudinal studies is visible during development [293].

Height is one of the most investigated continuous traits, being approximately normally distributed, which is affected by at least 697 common genetic variants (minor allele frequency [MAF] > 5%) at 423 loci reaching genome-wide significance, accounting for about 20% of phenotypic variance [294,295]. Height is therefore clearly polygenic, with the average effect of genes being about 0.14 mm/allele [296,297]. Further analysis, however, suggests that larger studies would find many more genes, with it being probable that height is not merely polygenic, but *omnigenic*, such that "a substantial fraction of all genes contribute to variation in [phenotype]" [297]. Importantly, most genes involved in complex phenotypes are broadly expressed, rather than being tissue or function specific [297]. Boyle et al. suggest that the architecture consists of a small number of core genes, with the expression of those core genes influenced by other genes, which themselves are influenced by further genes, in a small-world network that rapidly incorporates indirect effects from most of the genome. Not all influences on height are small, though; exome-sequencing looking for rare variants (MAF < 1%) or low-frequency variants (1% < MAF < 5%) found a further 83 influences on height, many of which had large effect sizes (1–2 cms/allele) and influenced insulin-like growth factors. Although polygenes mostly have small effects, that is not true of all of them; the implication is that some genes influence core underlying biological processes. Not all polygenes are therefore equivalent.

This brief review hopefully makes clear that merely knowing that a trait or character is influenced by many genes tells one little about either the phenotypic architecture or the genetic architecture. For handedness, the 41 loci vary in their frequencies across the whole range from 0.04 to 0.91, with most having small effects on handedness, the odds of left-handedness being in the range 1.02 to 1.06 for increased rates of left-handedness and 0.94 to 0.98 for lower rates of left-handedness, all of which are small effects [66]. Unlike obesity or height, there does not at present seem to be a small number of rare loci which have the largest effects on the inheritance of handedness. However, the most recent release of exomes from UK Biobank [298] has not yet been analysed for relationships to handedness, and may provide a different picture.

27. The Polygenic Version of the DC Model of Lateralisation

The monogenic DC model is successful at giving a broad-brush description of handedness in families, twins and in relation to language dominance. It is, however, undoubtedly wrong, since molecular genetics shows there is no single autosomal gene underlying handedness. The polygenic model adopts many of the principles used by the monogenic model, combining them with the sort of inheritance found for the multiple genes identified in PCD.

Despite having theorised about a monogenic DC model since 1977, that the monogenic model could not be correct hardly came as a surprise, as since the millennium there had been many studies where researchers had hoped to find 'the gene' for some condition, but no single gene emerged. Even with conditions such as cystic fibrosis, the classic recessive disease described in many basic science textbooks, it has become ever clearer that there are large numbers of mutations at the single locus which causes the condition [299]. Eye colour, another simple condition which is also a staple of introductory textbooks, with its recessive gene for blue eyes against brown eyes, is in reality far more complex, with 124 genetic associations from 61 discrete regions of the genome [300]. For complex traits, such as height, weight, intelligence, schizophrenia and autism, there are associations with dozens or hundreds of genes, numbers increasing as sample size increases [301]. For some conditions such as diabetes, there are mostly very large numbers of genes of relatively small effect, whereas lower-level, more specific, biomedical traits, such as levels of Vitamin D or LDL cholesterol, also have many genes, but have a few genes of larger effect.

28. The Polygenic DC Model

The monogenic DC model has a single locus with two alleles, D and C, with three genotypes, DD, DC and CC, and the frequency of the C allele being p(C) = c. The probability of being left-handed is 0, 0.25 and 0.5 for DD, DC and CC, respectively.

The polygenic DC model extends the monogenic model by saying that there are n loci, with the ith locus (i = 1:n) having alleles D_i and C_i, with c_i being the frequency of the C_i allele. The phenotypes are related to left-handedness in an analogous way to that of the multiple genetic variants of PCD being related to failure of the muco-ciliary escalator [222], any single defect in the chain resulting in a problem [11]. The model formally says the following:

- *The equivalent of the monogenic DD genotype.* If *all* of the n loci have only D_i alleles, i.e., for all loci the ith genotype is D_iD_i, then there is a 0% probability of being left-handed. In this case, the chain is intact.
- *The equivalent of the monogenic CC genotype.* If for *any* of the n loci the genotype is C_iC_i then the chain is broken and the probability of being left-handed is 50%. The analogy is with the way that PCD occurs if there is homozygosity at any of the many genes producing the cilium, and result in defective ciliary motility.
- *The equivalent of the DC genotype.* For heterozygotes, if for *any* of the n loci the genotype is D_iC_i, as well as *none* of the n loci having a genotype of C_iC_i then the probability of being left-handed is 0.25. As with the monogenic model, the heterozygote is additive in its relation to the equivalents of the homozygotes.

Note that if n = 1 then the model is identical to the monogenic *DC* model. The frequencies of the C_i alleles at each of the loci, c_i, can be set to any value, but the distribution in practice makes relatively little difference, and for simplicity one can set $c_1 = c_i = c_n$. Details of the calculations are described elsewhere [11].

To visualise how the polygenic model works, consider the simplest case with just two loci. Table 10 summarises the calculations. For simplicity, c_1 and c_2 are set equal, and for an overall value of 10% of left-handers that means $c_1 = c_2$ = 0.111036. For the first locus, the three genotypes D_1D_1, D_1C_1 and C_1C_1, are shown in column 2, with their proportions in column 1 (calculated in the usual way as $(1 - c)^2$, $2 \cdot c \cdot (1 - c)$ and c^2. Similarly, the three genotypes for locus 2, D_2D_2, D_2C_2 and C_2C_2, are shown in row 2. The 3 × 3 matrix in rows 3 to 5 and columns 3 to 5 shows the proportions of the various combinations of the genotypes of the two loci, the proportions being estimated by multiplying the row and column proportions in column 2 and row 1.

Table 10. Calculations for the proportion of left-handers for a two-locus model.

	(1)	(2)	(3)	(4)	(5)
(1)		*p(C_2):*	*0.79026*	*0.19741*	*0.01233*
(2)	*p(C_1)*		D_2D_2	D_2C_2	C_2C_2
(3)	*0.79026*	D_1D_1	**0.624506**	0.156008	0.009743
(4)	*0.19741*	D_1C_1	0.156008	0.038972	0.002434
(5)	*0.01233*	C_1C_1	0.009743	0.002434	0.000152

Row 3 column 3, shaded in green, shows individuals who are DD for each of the two loci, giving a proportion of 0.790257 × 0.790257 = 0.624506. These individuals given the model will all be right-handed. The five cells shaded in blue, in row 5 and column 5 all contain at least one CC genotype, and therefore will have a 50% probability of being left-handed; together they total 0.024506 of the population. The remaining three cells, shaded in yellow, contain at least one DC but no CCs, so they have a 25% probability of being left-handed, and comprise 0.350988 of the population. Overall, the proportion of left-handers is 0.624506 × 0 + 0.350988 × 0.25 + 0.024506 × 0.5 = 0.1, which is the required 10% of left-handers. Similar principles apply when there are three or more loci, although there are many more combinations.

Calculations can be carried out for increasing numbers of loci, for individuals in families and in twin pair, with the predictions shown in Table 11, which is taken from a previous paper [11].

Table 11. Handedness in families and MZ twins for the polygenic DC model with varying numbers of loci. Predicted handedness in families and twins for the polygenic DC model, with N loci = 1 (the monogenic DC model) to 1000. All estimates use a Monte Carlo method (see text) except for the second row for N loci=1, in brackets, which are analytic solutions.

N Loci	c_i	Per Cent Left-Handedness by Parental Phenotype			Percent Concordance and Discordance in Monozygotic Twins		
		$R \times R$	$R \times L$	$L \times L$	*R-R*	R-L	L-L
1	*0.2*	7.82% (7.78%)	18.90% (18.89%)	30.63% (30.00%)	83.00% (83.00%)	**14.00%** **(14.00%)**	3.00% (3.00%)
2	*0.1111*	8.15%	17.74%	25.56%	82.80%	**14.40%**	2.83%
3	*0.07715*	8.19%	17.24%	24.17%	82.74%	**14.55%**	2.71%
4	*0.05916*	8.29%	17.01%	22.88%	82.70%	**14.64%**	2.66%
5	*0.0478*	8.35%	16.79%	22.60%	82.65%	**14.69%**	2.66%
10	*0.02473*	8.38%	16.45%	21.86%	82.60%	**14.85%**	2.55%
20	*0.01256*	8.46%	16.50%	21.53%	82.50%	**14.95%**	2.55%
50	*0.00507*	8.52%	16.09%	20.02%	82.51%	**14.97%**	2.53%
100	*0.00254*	8.48%	16.10%	20.28%	82.52%	**14.98%**	2.49%
200	*0.00127*	8.55%	16.30%	20.83%	82.39%	**15.09%**	2.51%
500	*0.00051*	8.56%	16.12%	21.06%	82.54%	**14.91%**	2.55%
1000	*0.00026*	8.52%	16.26%	20.29%	82.46%	**15.03%**	2.51%
Approximate CI		0.05%	0.07%	0.08%	0.08%	**0.07%**	0.03%

Analytic calculations for multiple loci rapidly become very complicated as the numbers of combinations increase, and, therefore, Table 11 is based on a Monte Carlo analysis with a million replications for each number of loci. The accuracy is sufficient to show the effects with confidence intervals estimated at the bottom of the columns. Several things can be noticed from Table 11, as follows:

- Overall, the broad pattern is very similar, however many loci there are. The number of left-handed offspring (in red) increases as the number of left-handed parents increases, and monozygotic twin pairs can be discordant. However, there are some subtle differences.
- Handedness runs slightly less in families as the number of loci increases. The effect is particularly noticeable from one locus to five loci. With one locus, two left-handed parents must each being carrying a C allele. If they both have the CC genotype, then each has a 50% chance of left-handedness, and their children must all be CC, with a 50% chance of left-handedness. With two loci, though, one parent may be, say, C_1C_1:D_2D_2 and the other D_1D_1:C_2C_2, and each, as in the one locus case, has a 50% chance of being left-handed. However, the offspring must all be heterozygotes at the two loci, D_1C_1:D_2C_2, giving them only a 25% chance of being left-handed. As the number of loci increases, so there is a greater possibility that parents are not carrying the same genes for left-handedness, and, therefore, left-handedness is somewhat less likely as children may be heterozygotes or DD homozygotes.
- Discordance in twins (shown in bold) increases very slightly in rate as the number of loci increases, although the effect is so small as to be barely visible even in the largest sample sizes that might occur. DZ twins are not shown in Table 11, but they will

show a slightly greater discordance than MZ twins as they may be carrying different combinations of C alleles at various loci.

Although Table 11 is based on the simplifying assumption that all C alleles have the same frequency, $c_1 = c_i = c_n$, relaxing that assumption has little effect, the assumption not being particularly restricting. Elsewhere, we recalculated the model for 100 loci but with c_i in a triangular distribution from 0.0000508 (1/50 of the 0.00254 for equal c_i) through, in equal steps, to 0.00508 (twice the equal value c_i of 0.00254). For 1,000,000 replications, the proportions of left-handers in $R \times R$, $R \times L$ and $L \times L$ families are 8.47% (8.48%), 16.26% (16.10%), 20.73% (20.28%), respectively, the values in parentheses being for equal c_i from Table 2 in [11]. Therefore, the distribution of values of c_i seems to have little impact on the outcomes of the polygenic DC model.

Taken overall, the broad conclusion is that the overall pattern of results for the polygenic DC model is similar to that for the monogenic DC model; indeed, the similarities are far more apparent than the differences. In practical terms, and certainly at a qualitative level, the monogenic DC model can be used for calculating the likely effects of random variation in genetic models of handedness. As more precise molecular data become available, better models may become available.

For the present, it makes sense to continue with the monogenic DC model, for its computational or algorithmic simplicity, knowing that its broad predictions are likely to be similar to those of the polygenic DC model. It is time, therefore, to return to the cerebral polymorphisms with which this paper began, but with a better sense of how lateralisation can be handled in theoretical terms, what a genetic model may look like, and reassured that the monogenic DC model will be adequate in the first place for understanding variability in brains, and allowing a broader analysis than the merely descriptive.

29. Cerebral Polymorphisms in More Detail

The studies summarised in the introduction to this paper have explored various numbers of modules. Karlsson assessed six modules, one of which was handedness, and the theoretical modelling here will consider five modules plus handedness [25].

The modelling extends the relationship given earlier between handedness and language dominance (Table 3), with additional phenotypes added. For simplicity, given that 10% of individuals have atypical handedness, that 10% of individuals are also atypical for each of the other five modules. The model, as previously, assumes that lateralisation of the modules is independent within the genotypes, DD, DC and CC, with a probability of being atypical of 0%, 25% and 50%, respectively. As emphasised earlier, the modelling is primarily qualitative in its approach, with a minimum of free parameters. All assumptions can be relaxed, if necessary, for more precise model-fitting, but the conceptual force of the model primarily comes from the relative simplicity of the core idea (in just the same way as Newton's laws of motion can be tweaked and altered to take account of, say, air resistance, but the central, conceptual ideas remain simple).

There are five non-handedness modules in Karlsson's data, the lateralisation of each of which can be typical (T) or atypical (A), the 'typical/atypical' nomenclature avoiding the confusion of considering modules some of which are typically on the right and others on the left. Table 12 summarises the modelling. With five modules, each individual can have between 0 and 5 atypically lateralised modules, shown in column (1). The different types are shown in column (2), so that, for instance, there are five organisations with only one atypical module. Column (3) shows the numbers of combinations that can be found with 0, 1, 2, 3, 4 or 5 atypical modules, which are 1, 5, 10, 10, 5 and 1, respectively. Mathematically, these are $n!/(k!(n-k)!)$, the number of combinations of k, the number of atypical modules, from n, and the total number of modules.

Table 12. Numbers of atypical modules in the population and for the data of Karlsson [25].

(1)	(2)	(3)	(4)	(5)	(6)	(7)	(8)	(9)	(10)	(11)	(12)	(13)
		Model Predictions: p(L) = 10%							Karlsson N (%) p. 114			
			Population		Right-Handers		Left-Handers		Right-Handers		Left-Handers	
N (Atypical modules)	Types (T = Typical; A = Atypical)	N (types)	P (type)	P (total types)	P (type)	**P (total types)**	P (type)	**P (total types)**	N (types)	**P (total types)**	N	**P (total types)**
0	**TTTTT**	1	71.72%	71.72%	77.51%	**77.51%**	19.69%	**19.69%**	13	**54%**	4	**9.3%**
1	**ATTTT; TATTT; TTATT; TTTAT; TTTTA**	5	2.66%	13.30%	2.18%	**10.89%**	6.96%	**37.46%**	3	**13%**	12	**28%**
2	**AATTT; ATATT; ATTAT; ATTTA; TAATT; TATAT; TATTA; TTAAT; TTATA; TTTAA**	10	0.97%	9.69%	0.772%	**7.72%**	2.73%	**27.34%**	6	**25%**	8	**19%**
3	**AAATT; AATAT; AATTA; ATAAT; ATATA; ATTAA; TAAAT; TAATA; TATAA; TTAAA;**	10	0.41%	4.06%	0.304%	**3.04%**	1.33%	**13.28%**	1	**4%**	10	**23%**
4	**TAAAA; ATAAA; AATAA; AAATA; AAAAT**	5	0.22%	1.09%	0.148%	**0.738%**	0.859%	**4.30%**	0	**0%**	6	**14%**
5	**AAAAA**	1	0.16%	0.16%	0.095%	**0.095%**	0.703%	**0.703%**	1	**4%**	3	**7%**
Sum			-	100%	-	**100%**	-	**100%**	24	**100%**	43	**100%**

Considering any individual type of atypical organisation of modules, column (4) shows the probability in the population as a whole. There is a 2.66% probability of, say, type ATTTT, which has one atypical module. Overall, there are five different ways of having one atypical module, and therefore the proportion of the population who have one atypical module is 2.66% × 5 = 13.30% (column 5). The majority of the population have no atypical modules (71.72%), with decreasing proportions with 1, 2, 3 or 4 atypical modules (13.30%, 9.69%, 4.06% and 1.09%, respectively), with only 0.16%, about 1 in 600 individuals, having all five modules organised atypically (so-called mirror-organisation).

The sixth module is handedness and it can be modelled along with the other five modules. Columns (6) and (7) show in green the predictions for right-handers. A higher proportion, 77.5%, have no atypical modules, but it is still the case that the remaining 22.5% have at least one atypical module, although having all five typical is expected to be very rare (0.095%). Left-handers, in red in columns (8) and (9), show a very different pattern. The majority have at least one of the five modules (excluding handedness itself) lateralised atypically, and indeed only 19.69% are organised in a typical way (TTTTT). Many left-handers have several atypical modules, and 18.3%, nearly one in five, have three or more atypically lateralised modules.

There are few population datasets with which to compare the modelling of Table 12, and none of the studies mentioned at the beginning of the paper is entirely satisfactory. Here, I will particularly consider the data of Karlsson, who studied 67 participants, assessing *handedness* and five other modules (*Verbal fluency*, *Face recognition*, *Perception of neutral bodies*, *Emotional prosody*, and *Emotional vocalisation*). The study intentionally over-sampled individuals who were likely to have atypical lateralisation, with the following three groups of participants: language typical right-handers (N = 23), language typical non-right-handers (N = 22), and language atypical individuals (N = 22) [25].

A detailed comparison of the model with Karlsson's data is not straightforward because of the intentional oversampling. Columns (10) to (13) in Table 12 show the number of individuals with varying numbers of atypical lateralisations, separately for the right-handers (columns 10 and 11), and left-handers (columns 12 and 13). The pattern in the right-handers is broadly as for the predictions, but with a somewhat lower proportion having no atypical lateralisations (54% vs. 77%), and relatively few with 3 or more atypical lateralities, although there is one case with all five atypical lateralities. The left-handers in columns (12) and (13) show a pattern broadly similar to that of the modelling in columns (8) and (9). Individuals with no atypical lateralities (except left-handedness) are relatively rare (9%), whereas 42% of left-handers have 3 or more atypical lateralities, and three individuals had all six lateralities reversed (including handedness). More precise modelling could perhaps be carried out in relation to the method of over-sampling, but the important message is that the model broadly predicts the sort of rich variation in the number of atypical lateralities which are found not only in left-handers but also in right-handers. Cerebral polymorphisms are clearly common in the population.

Other predictions could be made, but the dataset is not large enough to be able to test the model properly, but there is a broad comparability. Column (2) of Table 12 suggests that within rows with a particular number of atypical modules that the numbers of the variants should be the same (e.g., for those with one atypical module, there should be equal population proportions of ATTTT, TATTT, TTATT, TTTAT, and TTTTA, both overall and also within right-handers and left-handers). Such predictions are, in principle, testable.

Twin and Family Twin Data

The model of Table 12 can readily be extended to more complicated situations, although there is rapidly a combinatorial explosion in the numbers of genotypic and phenotypic combinations to be considered. The calculations and results will not be provided here, but their pattern can be gleaned from Table 9, provided earlier, in which discordance for twin language dominance is provided in relation to twin handedness (which may be discordant), parental handedness, and zygosity. Discordance of language dominance in MZ twins is a marker for the number of atypical lateralities to be found. In the model, atypical lateralities are more frequent in twins discordant for handedness, in twins with a left-handed parent, and also in DZ twins (since they can have different genotypes).

Studies of cerebral lateralisation in twins are unusual, the few exceptions emphasising differences between MZ twins who are discordant for handedness (MZHd) [50,302–307]. Studies often only look at MZHd pairs, and not MZ twins who are both right- or both left-handed, and theorising, often based on small numbers of cases or indeed single cases [308], often invokes mirror-imaging, perinatal brain damage or birth order effects, mostly in the absence of any direct evidence [309]. Needless to say, with enough such factors, any small dataset can be explained. Discordance is then seen as evidence against genetic models, without the predictions of discordance of handedness and language in genetic models, as in Table 9, being taken into account, despite genetic models for twins being in the literature for a long time [20].

Potentially, there is much to be gained from the study of discordances in MZ twins, using co-twins as controls in fMRI (or fTCD), not least as the twins are necessarily matched genetically. However, as Ooki says in a review, "sample sizes are still not very large and hence the statistical power is insufficient" [310] (p. 4). The heritability of language dominance in twins has been assessed using fTCD, but the confidence interval is very wide, and is compatible with that of handedness in general, but the authors also emphasise its relatively low effect size with a confidence interval ranging upwards from zero [212].

The very sparse literature on handedness and cerebral polymorphisms in twins means that it is not easy to test the DC genetic model. This is surely an area where, in the future, collaboration with large twin banks, coupled with relatively cheap brain scanning, could provide much understanding of cerebral polymorphisms. In particular, a cost-effective strategy could be to use fTCD in the first instance to find atypical cases in the population, with fMRI then used for more detailed investigation. Such an approach would particularly allow population base-line estimates so that proper genetic modelling could be carried out. The 2020s is surely an appropriate time to think in such terms, with relatively large, well-characterised twin populations being available for study. One other possibility is to consider UK Biobank data where, although there are only 179 MZ twin pairs, there are 6276 parent-offspring pairs, 22,666 full siblings, 11,113 2nd degree relatives and 66,928 relatives [311], which could presumably be modelled for the many neural phenotypes which have been assessed, including handedness.

30. How Many Independent Lateralities Are There?

The DC model assumes that there are multiple lateralised modular traits, but with many of them influenced by the single random process that is built into each module. A key question for theoretical modelling is the number of statistically independent lateralities. Much depends on how one defines 'independent'. True population-level independence perhaps exists for behavioural lateralities such as handedness, hand-clasping and arm-folding, which have only minimal correlations in populations [312]. To my knowledge, there is no understanding of the neural basis of either hand-clasping or arm-folding. Other lateralities, such as handedness and language dominance, often show correlations with each other at the population level, but genetic modelling suggests that within genotypes there is probably statistical independence (and the occurrence of the population-level correlation is a variant of Simpson's paradox in statistics, where combining several contingency tables can results in reversed or absent correlations) [313,314].

Identifying the lateral anatomical architecture of the cortex is not straightforward. A study by Liu et al. measured intrinsic activity at rest in two samples totalling 300 adults, and calculated laterality indices for 84 cortical regions, which were factor-analysed and four principle components identified therein with eigenvalues of greater than one [315]. The four factors identified relatively small areas in the visual system, namely, the default network, angular gyrus/isthmus, and fronto-temporal area. However, further analysis suggested that 20 factors might be needed to account for 70% or more of the variance. The factors showed some correlations, but the picture overall was unclear. Analysis of intrinsic activity may allow assessment of the dimensionality of lateralisation, but the conclusions are far from clear at present.

A different approach is the large-scale meta-analyses by Vigneau et al. of 129 studies reporting fMRI scans for the left hemisphere [316] and for the right hemisphere [317], considering only individuals who are right-handed. Multiple clusters of left-hemisphere language-related activity were found, phonology having five frontal and six temporal clusters, four frontal and seven temporal clusters for semantic clusters, and three frontal and five temporal clusters for sentence processing. A total of 59 studies reported right-hemisphere involvement, with mostly homotopic areas from left-hemisphere areas. Many activations were bilateral, but unilateral right-hemispheric activations were also found. Taken overall, there seem to be at least a minimum of three language-related clusters (phonology, lexico-semantic, and sentence/text), but each has multiple sub-clusters, perhaps 30 in total, with many having evidence of bilateral activity, or, in some cases, just right-hemisphere activity. It should be remembered that the study concentrated entirely on normal right-handers, and although relatively little right-hemisphere activation might have been expected when only looking at right-handers, it was found in 33% of studies. The analysis of the DC model suggests that multiple atypical asymmetries can be found in individuals, with Karlsson's data suggesting there might be at least six independent asymmetries, all of which can be considered as being independently determined by the DC genotypes. The following three questions arise: what are the associations between the various lateralities? How many separate lateralities might there be controlled by the DC genes? Are there are other cerebral lateralities which are independent of the DC system?

Associations between Functional Lateralities

Tetrachoric correlations are the best way to assess associations between asymmetries (see earlier). Palmer has argued that right–left asymmetries, because they are primarily concerned with binary categories, allow meaningful absolute comparisons in many situations, including across different species in his particular case [318], but in our case across different neural modules. The logic though requires that all measures are referenced relative to symmetry itself, to the midline, with deviations therefore always to left or right. Figure 6 shows the tetrachoric correlations between the six modules in the Karlsson data, as well as a diagrammatic summary.

Figure 6a shows for the Karlsson data that the atypicality of any one module is positively correlated with atypicality in all of the other modules, suggesting an underlying single factor (presumably the DC genes). The overall N is only 67, and so standard errors are large, and some correlations are not significantly different from zero. The correlations are shown diagrammatically in Figure 6b, with the typical left-hemisphere functions in red, the typical right-hemisphere functions in green, and handedness in blue. Significant correlations ($p < 0.05$) are shown as thick lines, and non-significant correlations are thin lines. Overall, the pattern is consistent with the six modular functions between correlated but also being independent to a large extent. In particular, it might be thought that the four right-hemisphere tasks could all be a single function, but if that were the case then the correlations between them should be *high*, but several of them are *low* (0.127, 0.68, 0.322, 0.166). Clearly, more data are needed, but the implication is that there are probably at least six statistically independent modules.

A different and equally important dataset is that of Woodhead et al., who used fTCD to study 43 right-handers and 31 left-handers on six putatively left-hemisphere language functions, studied in an elegant design in which retesting allowed reliabilities of measures to be calculated [27]. The study suggested there were at least two dimensions, but it is possible, once retest data are properly taken into account, that all six measures could be statistically independent. Of the many strengths of this study, one is that the raw data are all available in open-source files, which makes more detailed analysis possible.

The number of lateralities in the DC system is therefore unclear, but there seems to be no reason for it not being at least a dozen, with Karlsson finding four separate typically right-hemisphere tasks, and the Woodhead et al. study suggesting perhaps six language-associated tasks, so that there could easily be as many as twelve or fifteen overall. A key feature for all modules would be their association with each other and with handedness.

A modular system *without* an association with handedness was found in the important and large study of Badzekova-Trajkov et al., which looked at four modular functions [24]. *Word generation*, a typically left-hemisphere task, and a *faces task*, a typically right-hemisphere task, were correlated with each other (r = −0.339), and 0.357 and −0.236, respectively, and with *handedness*, which is compatible with being part of the DC system. The fourth task, though, was the *landmark task of spatial attention*, involving bisection of a horizontal line, which typically involves fronto-parietal cortex in the right hemisphere. The landmark task was not correlated at all with handedness (r = 0.001), and it had a correlation of −0.176 with the word task, which was just significant ($p = 0.029$), and a non-significant correlation of 0.164 with the faces task. The landmark test therefore appears at population level to be independent of the other three tasks, with about 21% of the participants atypical for the landmark task, showing left-hemisphere dominance. The independence of handedness and the landmark (line bisection) tasks in experimental studies (non-fMRI) is also confirmed in meta-analyses in children and in adults [319,320].

An extension of the previous study included *gesture*, and using factor analysis found three independent factors, one language-related, one linked to handedness, and one that was handedness independent [321], although a separate analysis of the same data found only two factors, but also had the interesting feature that it found cerebellar asymmetries which showed correlations with cortical and other asymmetries [60] (see earlier for the possible role of the cerebellum).

The independence of the landmark task suggests that there could be many more phenotypic variants of cerebral polymorphism. If the Karlsson study, with its six modules and a probable $2^6 = 64$ combinations of its five tasks plus handedness, had also included the landmark task and gesture tasks then it seems reasonable to assume there could have been 256 combinations. If, as speculated earlier, the DC system could have 12 to 15 modules, then that would make 14 to 17 modules, giving up to perhaps $2^{17} = 131{,}072$ possible variations, albeit that some would only occur rarely, and appear only in very large samples (or as isolated case studies in the literature).

31. Bilateral Lateralisation, Particularly of Language

Functional asymmetries are often described as 'bilateral', but it is often unclear what is meant by that, particularly if only laterality indices are available. As Carey and Johnstone have said for fMRI, "many neuroimagers are struck by bilateral activations in any language-related task", but on the other hand "several methodological issues make simple left, right, bilateral classifications more contentious [requiring] hard decisions about regions of interest and thresholding, … equating regions from each hemisphere which are not structurally identical … and the nature of baseline conditions" [6] (p. 14). The problem is seen in Figure 2b, where it is difficult to know which participants are 'bilateral' for the tapping task, although Figure 2c resolves the problem, participants all being bilateral in that they are capable of tapping with just their right hand or just their left hand (even if mostly they are better with one hand than the other). The Wada test (see below) can be conceptualised in a similar way, with language facility after right hemisphere injection on one axis and after left injection on the other axis in a bivariate analysis. The study of Wegrzyn et al. [212], mentioned earlier, with their bivariate plots, suggest there is no clear divide between bilateral and other groups. However, they do refer to cases in which Broca's and Wernicke's area are in different hemispheres, and presumably, therefore, some components of language can be truly regarded as bilateral [213]. The study of Bernal and Ardila specifically rejects the idea "that language lateralisation is a matter of all or nothing", and provides a detailed analysis of five unusual cases analysed with fMRI, suggesting that language representation should be considered in terms of receptive vs. expressive and phonology vs. semantics, each potentially being unilateral or bilateral, with many possible combinations [322]. How common such 'unusual cases' are is a key question that needs answering.

31.1. Modelling Bilateral Language

The 1985 monograph on the DC model suggested a simple model for bilateral language in which there are two separate but otherwise equivalent language centres, language A (LA) and language B (LB), which are determined independently; LA and LB are in the left hemisphere for all DD individuals, whereas DC and DD have LA and LB randomly and independently randomised to the right, with the usual 25% or 50% probability [20]. The predictions are shown at the bottom of Table 13 below.

Table 13. Summary of left-, bilateral- and right-hemispheric language dominance in right and left handers, assessed using intra-carotid sodium amytal.

	All Participants				Right-Handers				Left- or Mixed-Handers			
Language	**Left**	**Bilateral**	**Right**	**N**	**Left**	**Bilateral**	**Right**	**N**	**Left**	**Bilateral**	**Right**	**N**
Rasmussen and Milner (1977) [323]	220 (84%)	18 (7%)	24 (9%)	262	134 (96%)	0 (0%)	6 (4%)	140	86 (70%)	18 (15%)	18 (15%)	122
Kurthen et al. (1994) [215]	116 (67%)	38 (27%)	19 (11%)	173	109 (77%)	27 (19%)	6 (4%)	142	7 (23%)	11 (35%)	13 (42%)	31
Loring et al. (1990) [324]	442 (80%)	61 (11%)	48 (9%)	551	403 (86%)	42 (9%)	24 (5%)	469	39 (48%)	19 (23%)	24 (5%)	82

Table 13. *Cont.*

	All Participants				Right-Handers				Left- or Mixed-Handers			
Risse et al. (1997) [325]	304 (83%)	40 (11%)	24 (7%)	368	265 (87%)	27 (9%)	12 (4%)	304	39 (62%)	13 (20%)	12 (19%)	64
Möddel et al. (2009) [326]	356 (80%)	71 (16%)	28 (6%)	445	320 (82%)	55 (14%)	16 (4%)	391	26 (48%)	16 (30%)	12 (22%)	54
Bauer et al. (2013) [327]	382 (76%)	44 (9%)	78 (15%)	504	na	na	na	na	na	na	na	na
Janecek et al. (2013) [328]	184 (80%)	30 (13%)	15 (7%)	229	na	na	na	na	na	na	na	na
Total	**1994 (78.80%**	**302 (11.90%)**	**236 (9.30%)**	**2532**	**1231 (85.10%**	**151 (10.40%)**	**64 (4.40%)**	**1446**	**197 (55.80**	**77 (21.80%)**	**79 (22.40%**	**353**
Model	**83.00%**	**14.00%**	**3.00%**		87.00%	11.00%	2.00%		50.00%	40.00%	10.00%	

31.2. The Wada Test

Perhaps the clearest example of bilateral language/speech representation is with the Wada test, in which intra-carotid sodium amytal is used to anaesthetise just one hemisphere [329–331]. In 1964, Branch et al. described a series of 114 cases, with 77 (67.5%) having left-hemisphere language, 27 (23.7%) right-hemisphere language, and 10 (8.8%) bilateral language, with no data given on the precise proportion of bilateral language in right- and left-handers, although 43 (89.5%) of 48 right-handers and 28 (63.6%) of 44 left-handers had language in the left hemisphere [332]. The more recent data of Rasmussen and Milner, which may include the data from the earlier Branch et al. study, sub-divided patients into those who had or did not have early evidence of left-hemisphere brain damage [323]. The results for early left-hemisphere damage are not reported here.

More recent amytal studies are also summarised in Table 13 [215,323–328], with some values taken from the summary table of Bernal and Ardila. Meta-analyses have compared the Wada test with fMRI, although one did not attempt to distinguish bilateral language from right-hemisphere language [333]. The other meta-analysis stated that when defining bilateral language using fMRI, "arbitrary decisions have to be made" [327]. The meta-analysis of Wada testing by Carey and Johnstone did not report bilateral language separately from right language, but they do mention the criterion problems of distinguishing 'good bilateral' and 'bad bilateral' [6]. It was clear, though, that atypical dominance was more common in left-handers than right-handers.

Although not a formal meta-analysis, Table 13 summarises seven reasonably large Wada studies which have reported bilateral language, five of which also classify results by handedness. The influence of early damage is mostly not reported consistently, and the cases of Rasmussen et al. [323] with early left-hemisphere damage have been omitted. In 2532 cases, language was left-hemispheric (L) in 78.8% of cases, bilateral (B) in 11.9% of cases and right-hemispheric (R) in 9.3% of cases, which are broadly compatible with the DC model predictions of 83.0%, 14.0% and 3.0%, respectively. In the 1446 right-handers there were 85.1%, 10.4% and 4.4% L, B and R cases, respectively, which is similar to the predictions of 87.0, 11.0% and 2.0%, respectively, whereas in the 353 left-handers there were 55.8%, 21.8% and 22.4% in the L, B and R groups, respectively, compared with 50.0%, 40.0% and 10.0% predicted proportions, respectively.

The DC predictions fit the data for right-handers reasonably well, with bilateral language (10.4%) being more prevalent than right-hemisphere language (4.4%), which fits reasonably with the model predictions (11.0% and 2.0%, respectively). The data for left-handers fit the model less well, with bilateral language (21.8%) being about the same rate as right-hemisphere language (22.4%), while the model predicts 40.0% of bilateral cases and 10.0% of right-hemisphere cases. However, the studies are quite variable and not always well defined. The most successful feature of the model is that bilateral language is undoubtedly more prevalent in left-handers than right-handers. There may be problems of criterion in all the studies. Alternatively, the model may be mis-specified, either LA and LB not being completely independent, or there being other language modules (LC, LD, etc). Bilateral language remains a problem at many levels, including measurement and modelling, which has not yet been properly explored in large numbers of individuals with typical functioning.

32. Recovery from Aphasia

A simple account of aphasia would be to say that language is in one hemisphere, and if large-scale damage occurs to that hemisphere, as with a middle cerebral artery occlusion or haemorrhage, then aphasia results, along with a hemiplegia or hemiparesis. That model works for many cases, but an immediate problem is that patients with aphasia sometimes recover, which needs explanation, particularly with lesions that are unlikely to be reversible neurologically. There is also the problem that although the simple model of Table 8 implies that about 7.8% of right-handers should have right-hemisphere language dominance, cases of so-called 'crossed aphasia' are rare, perhaps 1% of right-handed cases, although a prevalence is hard to estimate accurately [334].

Handedness shows the following three features in relation to aphasia, which the 1985 monograph wanted to explain: (i) acute aphasia is more likely to occur in left-handers, but, (ii) left-handers are also more likely to recover from aphasia [335–340], and (iii) a family history of sinistrality is associated with recovery from aphasia [340,341]. Data were available for modelling from 13 large series published between 1949 and 1981.

The 1985 monograph presented a basic model that could account for the observations, proposing that there are two language modules, LA and LB, which could be in one or both hemispheres, the modal form being to have both LA and LB in the left hemisphere (Figure 7). Figure 7a shows the four types of individuals, with the modal type being more common in right-handers than left-handers. Rows b and c show the effects of a large left or large right hemisphere lesion, indicated by the big red X. The difference between acute and chronic aphasia, along with an explanation for recovery, is provided by von Monakow's concept of diaschisis [342–344], whereby brain damage in one part of the brain can cause remote effects in other parts of the brain, particularly across the corpus callosum. Although out of fashion for a long while, diaschisis has been having a renaissance in recent years, not least as scanning has allowed possibilities for observing it more directly [345,346].

The role of diaschisis in the model is shown by considering the second column of Figure 7b. The large red cross shows damage to the left hemisphere and, hence, the module LA is damaged. LB remains intact, but diaschisis from the damaged LA inhibits the action of LB in the short-term and there is therefore an acute aphasia. However, as the diaschisis wears off so recovery occurs. That is contrasted with the situation in the first column of Figure 7b, where LA and LB are both damaged, so that not only is there an acute aphasia, but the aphasia is permanent, there being no functioning language module remaining.

Without going into the numerical details, which can be found in the 1985 monograph [20], if the model does work in this way, then left-handers will be more likely to suffer an acute aphasia, their greater likelihood of having bilateral language meaning that it is more likely that damage will affect one of them. However, recovery is also more likely as the diaschisis dissipates, and the one remaining language module can then take over language. An interesting prediction of such a model is that it explains occasional cases in which there is an acute aphasia after a stroke (say due to LA being damaged), recovery occurs with LB functioning, but then a lesion in the contralateral hemisphere results in a permanent aphasia as LB is damaged. Such cases were reported by Gowers in 1887 [4,347], with a range of cases reported since, sometimes with Wada and other data showing that recovered language after a left-hemisphere stroke originated in the undamaged right hemisphere [347,348]. The model also predicts that familial sinistrality would result in a greater likelihood of recovery, as it would be associated with a higher likelihood of bilateral language modules.

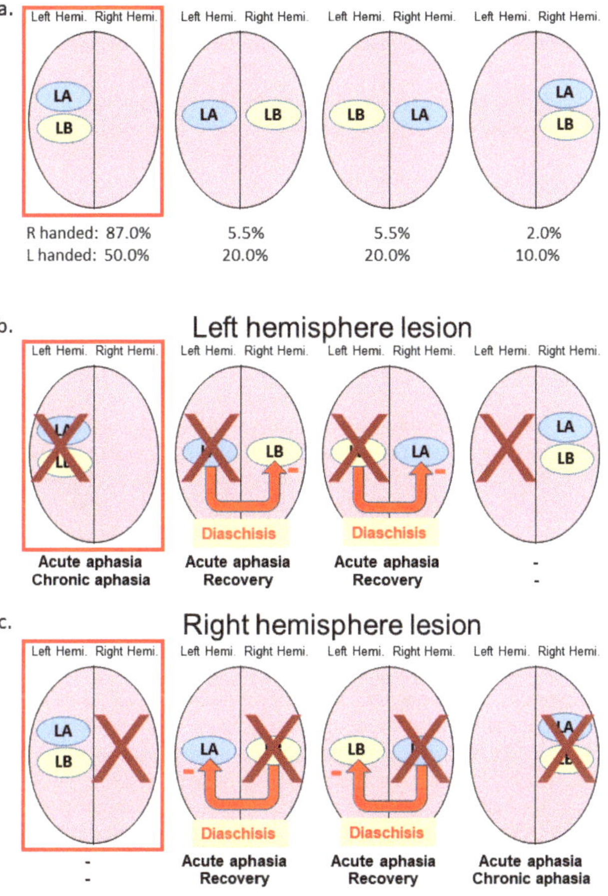

Figure 7. Schematic model for two equivalent language modules, LA and LB, which can be in the same or different hemispheres, and can account for acute and chronic aphasia differing in rates and rate of recovery between right- and left-handers (**a**) shows the four different modular combinations in the population, (**b**) shows the effect of a large left hemisphere lesion (X) and (**c**) shows the effect of a large right hemisphere lesion (X). See text for further details..

The analysis of aphasia shows the potential for the DC model for explaining cerebral polymorphisms and neurological damage, but there is a clear shortage of high-quality modern data to allow proper testing. Detailed series of patients with aphasia do not seem to have been published in recent decades, despite there being at least 28 stroke registries in 26 countries [349], although "few variables were measured consistently among the registries". Handedness seems rarely to be reported, with left-handedness seeming to be rare (e.g., in the Lausanne Stroke Registry, where of 1541 stroke cases with aphasia, only 2% were left-handed, but handedness was not known in 16% of cases, making interpretation difficult [350]). A similar pattern of a very low rate of left-handedness was reported in a series of 1000 stroke patients in South Africa with cognitive deficits, of whom only 18 (1.8%) were left-handed (N = 13) or ambidextrous (N = 5) [351]. Such low rates of left-handedness suggest difficulties in ascertainment of handedness in stroke patients.

33. The Functional Consequences of Cerebral Polymorphisms

A polymorphism composed of multiple modules organised across the hemispheres potentially has functional consequences for the integration of modules for particular tasks. A common assumption might be that reorganising the layout of one or several functional modules between the cerebral hemispheres might be assumed to have consequences, and these are generally assumed to be negative, although that is not necessarily the case. The commonest explanation is in terms of *crowding*.

33.1. *Crowding*

The first formal use of the term crowding was by Teuber, who suggested that after early left-hemisphere damage, which caused language to move to the right hemisphere, there might be problems with non-speech functions as a result of "competition in the developing brain for terminal space, with consequent crowding when one hemisphere tries to do more than it had originally been meant to do" [39,352]. The hypothesis was investigated in 27 adults who had early refectory epileptic seizures from an early age, whose language lateralisation was assessed using intra-carotid amytal testing [352]. Patients with bilateral, or particularly right-hemisphere language, showed poorer non-verbal performance using the Weschler Intelligence Scales. Whether the effect is due to interference of speech functions on non-verbal processes was unclear, and "limited capacity or incompatibility ... may be equally likely explanations" [352] (p. 1226). A recent study of children with intractable epilepsy found similar results to the earlier study [353].

Crowding is more usually invoked in a situation in which there is no early damage and consequent rearrangement, but in cases where it happens to be found that language or some other function is not located in its typical location (e.g., language on the right rather than the left-hand side). Groen et al., when reviewing the literature, found only one relatively small study on the 'parallel processing hypothesis', but they did find some processing deficiency in participants with atypical lateralisation [354]. The study of Groen et al. used fTCD to assess the lateralisation of language and visuo-spatial memory in 55 typically developing children, and while most children were left-lateralised for language, and right-lateralised for visuo-spatial memory, many showed reverse lateralisation, with 3 having right-language and left visuo-spatial memory. Cognitive ability, assessed on tests for non-verbal ability, vocabulary, reading, and phonological short-term ability, showed no difference in ability between children with language and visuo-spatial memory in the same hemisphere (who should have been subject to crowding) and those with the functions in different hemispheres. The study therefore provided "no evidence for this 'functional crowding' hypothesis" [40] (p. 256).

Reviewing the literature, Bishop concluded that, while "It would seem that atypical lateralisation is compatible with normal or even above-average cognitive function", it was also the case that in "studies that oversample those with developmental difficulties, an association with [atypical lateralisation of] language and literacy skills becomes apparent" [259].

Taken overall, there is probably only weak evidence that atypical patterns of cerebral lateralisation are associated with cognitive deficits, and that crowding, although plausible as a mechanism, seems not to be supported by robust evidence.

33.2. *Intra- and Inter-Hemispheric Connectivity*

Although it may not matter for isolated language functions as such whether they are located in the left or right hemisphere, cognition alone does not depend on language, and integration with other functions is necessary for some skilled activities. The connectivity of functional modules needs therefore to be considered. It may well be that the reason that language-related modules are mostly in the left hemisphere is because they need to have rapid access to one another, as may also the modules for visuo-spatial functions which are mostly in the right hemisphere. This has been referred to as 'pseudo-crowding' [22,47].

Estimating long-range connectivity in the human brain is not easy, but the inter-hemispheric connections of the corpus callosum are typically said to comprise about 2×10^8

(200 million) fibres [355,356]. Intra-hemispheric connections are harder to assess, but in the two hemispheres together there are of the order of 6×10^8 (600 million) long-range 'Compartment C' fibres, fibres which do not follow the folding of the cortical gyri [357] (pp. 381–382). While callosal fibres necessarily cross between the hemispheres, albeit mostly between homotopic areas, intra-hemispheric fibres are of variable lengths connecting near and distant areas of the cortex. Medium-range B ('U'-shaped fibres) and shorter A fibres (about 9×10^{10} and 8×10^{11} fibres, respectively) also contribute to connections within hemispheres, particularly for more adjacent areas, making it plausible that intra-hemispheric connectivity, with its network structure, is more efficient than the mainly homotopic (point-to-point) inter-hemispheric connections.

34. Disconnection, Hyperconnection and Hypoconnection

Norman Geschwind, in two famous and lengthy papers in 1965 [358,359], in what has been described as "the most influential work ever published in the discipline that became known as behavioural neurology" [360], showed how "disconnexion syndromes" can explain a wide range of symptoms occurring after brain damage. In 2005, Catani and ffytche [361] developed the concept further, with the concepts of hyperconnectivity, and regions becoming hyperfunctional or hypofunctional. This paper will consider the potential consequences of functional modules becoming better connected, perhaps by being located within the same rather than different hemispheres, or less well connected, as a result of being in different hemispheres. It should be emphasised that these postulated differences are not the consequence of lesions or brain damage, but are broadly within the typical spectrum of neural functioning, but might explain individual differences in talents and deficits.

35. Talents and Deficits

The potential of module connectivity, the connectome, for understanding the functional consequences of cerebral polymorphisms, can be shown with a simple 'toy' model. Figure 8 shows a very a ic brain containing just three functional modules, H (hand control), L (language), and Vs (visuo-spatial ability). The modal combination in the population is shown at the top left, with H and L in the left hemisphere and Vs in the right hemisphere, which can be called LLR. The DC model suggests that about 78% of the population will show this modal pattern assuming a 10% rate of left-handedness; these individuals show the pattern of cerebral lateralisation described in introductory textbooks of neuropsychology. There are, however, seven other ways in which H, L and Vs can be distributed across the hemispheres, representing 22% of the population, and they are shown in the rest of Figure 8. Modules which are atypically lateralised are shown with thick red borders. Some variants are scarcer than others. There are three ways in which one module can be atypical in its lateralisation, RLR, LRR and LLL (where the underlining indicates the atypical module compared with the typical pattern of LLR), with 5% in each combination, three ways in which two modules can be atypically located, LRL, RLL, RRR, with 2% in each combination, and just one way for all three modules to be atypical, RRL, which is 1% of the population. The handedness of the individuals is shown in colour, with green for the right-handers (H in the left hemisphere), and red for the left-handers (H in the right hemisphere). Overall, 5% + 2% + 2% + 1% = 10% are left-handed.

The interesting thing about Figure 8 is in the potential problems and benefits of variation in interconnectivity. Purely as a sketch of the possibilities, consider a skill such as writing, which may require connections between hand control and language. The curved arrows in Figure 9 connect language and hand control. For the modal combination, H and L are in the same hemisphere, which is also the case for three other combinations, and their intra-hemispheric connections are shown as black arrows. However, the other four combinations have H and L in opposite hemispheres, with the inter-hemispheric connections shown in red. Inter-hemispheric connections will require both homotopic callosal connections as well as ordinary intra-hemispheric fibres, and overall may be less efficient resulting in hypofunction.

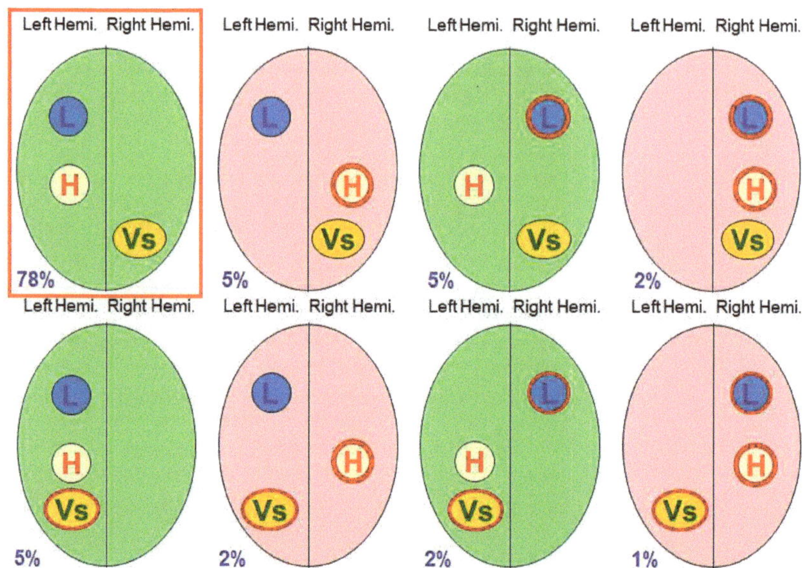

Figure 8. Schematic 'toy model' of three lateralised modules for handedness (H), Language (L) and Visuo-spatial ability (Vs) placed in the Left or the Right Hemisphere. The top left diagram shows the modal form, with H and L in the left hemisphere and Vs in the right hemisphere. Green and red shading indicates right- and left-handers. Red borders on modules indicate atypical locations. The population proportions of the eight types are shown in blue as percentages.

Taken overall, 14% of the individuals in Figure 9 have language and hand control in opposite hemispheres, but amongst the right-handers (green), 7.8% have inter-hemispheric connections between L and H, whereas amongst the left-handers (red), 70% have inter-hemispheric connections. This simple, indeed simplistic, model suggests that the left-handers may be much more likely to have problems. It should not be assumed that this is a model of dyslexia—which is almost certainly more complex than the connections of just two modules—but it is of note that meta-analyses have shown that dyslexia is more prevalent in left-handers [15], albeit with an odds ratio based on 45 studies of 1.57 [15,362]. The qualitative picture is, though, plausible and intriguing.

What about the other connections between possible pairs of the three modules? Figure 10 provides a simple interpretation of what may be helpful for someone who is skilled at what loosely can be conceptualised as 'manual craftsmanship'—perhaps wood carving, or throwing or hitting a ball accurately, or maybe skills such as computer gaming. For such skills, it may be beneficial for hand control, H, to be in the same hemisphere as visuo-spatial skill, Vs, and 14% of the population show that pattern, which is not the modal form, where H and Vs are in opposite hemispheres. Figure 10 also suggests that H and Vs may be in the same hemisphere in 70% of left-handers, compared with 7.8% of right-handers. Atypical forms are again more frequent in left-handers.

Consider next Figure 11, where connections between language, L, and visuo-spatial ability, V, are shown. Skills such as storytelling or poetry may be enhanced with such connectivity, and perhaps also mathematical or other systems involving symbolic and spatial constructions. Once again, 14% of individuals overall have unusual patterns of intra-cortical connections, but that now occurs in 40% of left-handers and 11.1% of right-handers.

Note that these proportions are different from those in Figures 9 and 10 since handedness there forms part of the pair being connected, whereas that is not the case in Figure 11.

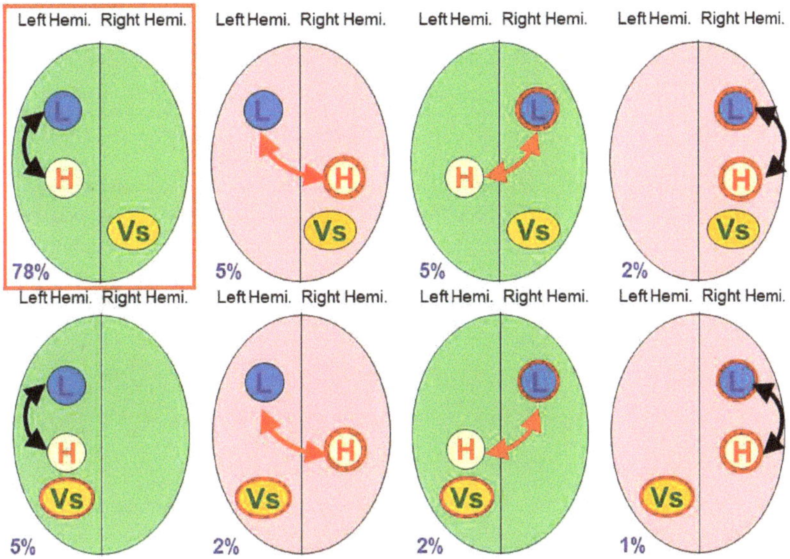

Figure 9. An extension of Figure 8 to show connections between the handedness (H) and language (L) modules, which conceptually may be needed for producing written language. The connections, shown as double-headed arrows, are in black for intra-hemispheric connections, which is the modal form. Connections in red are inter-hemispheric and involve in part the corpus callosum.

Thus far, the module connections have been between pairs of modules, H and L, H and Vs, and L and Vs. It may be, though, that some skills are benefitted by having all three modules, H, L and Vs, in the same hemisphere. Figure 12 shows a diagram in which, for simplicity, the two types are highlighted in which H, L and Vs are in the same hemisphere. Only 7% of the population show that pattern, and once more they are more frequent in left-handers (20%) than in right-handers (5.6%).

The model, as presented, is clearly over-simple—just three modules in the two hemispheres, whereas in reality there are probably many—but some patterns of connection are perhaps more likely to result in *deficits*, including perhaps dyslexia, but those same combinations may also result in *talents*. The literature is replete with claims of left-handers having special skills, with repeated, although not particularly robust, claims that left-handers are over-represented amongst musicians [363] and mathematicians [364,365]. There are claims that some talented groups are more likely to have some cognitive deficits, as in the suggestion that people with dyslexia are more likely to have superior visuo-spatial skills and be skilled architects [366,367]. Rarely are such claims properly investigated, and the association of visuo-spatial ability with dyslexia was complex in our meta-analysis [368]. Whether architects are more likely to be left-handed is also unclear [369,370]. What is clear is that there are many individuals with talents, such as architecture, and there is little solid underlying theory or evidence as to why that may be the case. Systematic fMRI or fTCD scanning may be of help in exploring such talents and deficits.

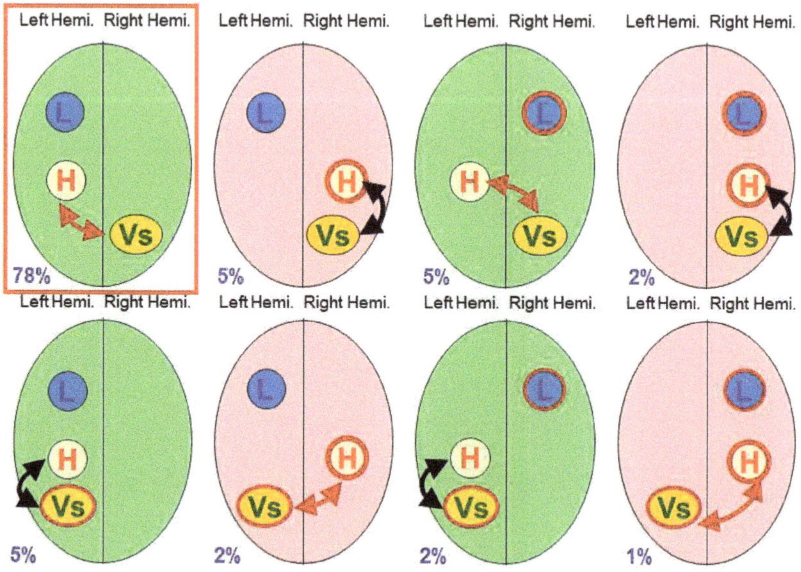

Figure 10. An extension of Figure 8 to show connections between the handedness (H) and visuo-spatial (Vs) modules, which conceptually may be needed for manual craftsmanship in its various forms. The connections, shown as double-headed arrows, are in red for inter-hemispheric connections, the modal form for H and Vs, which involve the corpus callosum. Connections in black are intra-hemispheric.

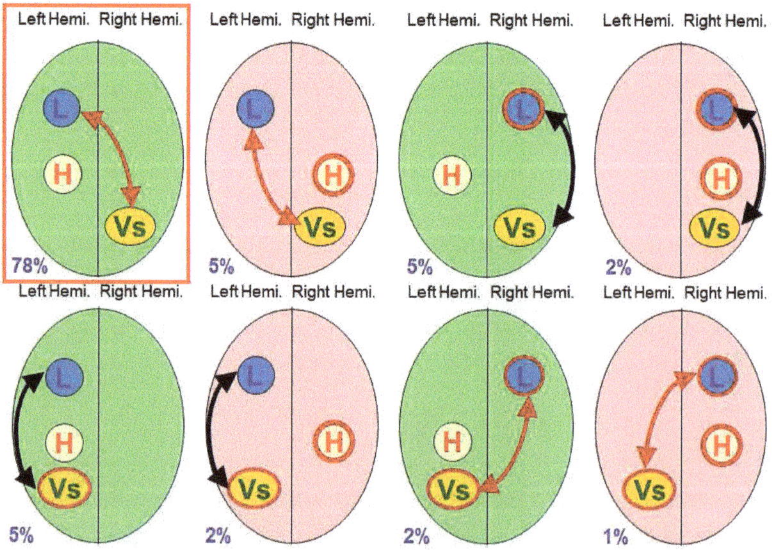

Figure 11. An extension of Figure 8 to show connections between the language (L) and visuo-spatial (Vs) modules, which conceptually may be needed for poetry, mathematics or other uses of language integrating symbolic systems and geometry. The connections, shown as double-headed arrows, are in red for inter-hemispheric connections, the modal form for H and Vs, which involve the corpus callosum. Connections in black are intra-hemispheric.

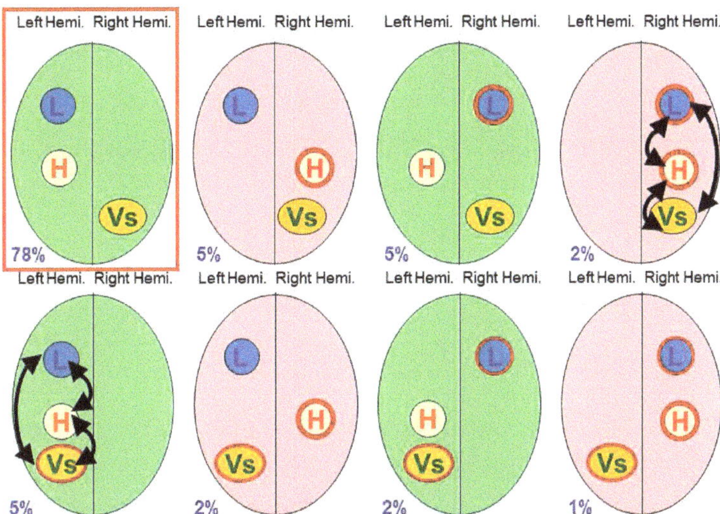

Figure 12. An extension of Figure 8 to show three-way interconnections between the handedness (H), language (L) and visuo-spatial (Vs) modules, which conceptually may be needed for, say, music performance or similar skills. The only connections shown for visual simplicity, in black, are where all connections between the three modules are intra-hemispheric connections.

The models of Figures 8–12 also make other predictions which, in principle, are testable. Consider the two groups of individuals in Figure 9 who are right-handed but have language in the right-hemisphere, and who have an inter-hemispheric connection between handedness and language. Such individuals presumably put more emphasis upon cross-callosal connections than do those in the modal group, and it is therefore of interest that the corpus callosum is larger in individuals identified by intra-carotid amytal with right-hemisphere language [371]. Associations between handedness and corpus callosum size have found a rather mixed picture [372], but a meta-analysis found left-handers had a larger corpus callosum [373], with a suggestion that degree of handedness may also relate to corpus callosum size [374]. The study of handedness discordant MZ twins has also found evidence of a larger callosum in left-handers [305,307].

An unusual study looking at cortical tracts is that of Häberling et al., which assessed the arcuate fasciculus in MZ twins who were discordant for handedness and found some differences [302], although the theoretical conclusion was a little weak, concluding that there was "a strong non-genetic component" as twins were identical, but that omitted to take into account that the DC gene system itself would have produced randomness which could differ between the twins (and presumably comes under the heading of developmental variance [65]). Interestingly, the latter interpretation is used for a related study of cerebral asymmetries in MZ twins [303]. It should also be noted that there appears to be clear inter-individual variation within the arcuate fasciculus [375], and the possibility has been raised of assessing individual variation in arcuate fasciculus asymmetry [376].

The DC model makes some straightforward predictions about variation in connectivity and its relation to cognitive talents and deficits, with a little supporting evidence from corpus callosum thickness in particular. However, the key thing is that little proper evidence has been collected which might be able to test it. What is probably required are studies of populations in which cerebral polymorphisms have been identified (such as those in the opening paragraphs of this paper, for which the study by Karlsson provides a model example [25]), but, in addition, anatomical and functional measures need to be collected, particularly of the corpus callosum, but also assessing tractography in general. Sample sizes may need to be larger than those currently used, but the principle of upweighting individuals likely to have atypical cerebral polymorphisms should provide statistical power.

Mapping the human connectome is possible but complex [377], and the Human Connectome Project has made good progress [378]. A large study of the heritability of asymmetry in MZ and DZ twins DTI identified large numbers of asymmetries, but sadly the analysis was restricted to right-handers [379], reflecting many neuroscience studies which have intentionally omitted the study of left-handers [380,381]. Although it might be expected that handedness could relate to cortico-spinal fibre density in the internal capsule, no such difference has been found using tractography [382]. A preliminary but important study did assess both arcuate fasciculus structure using DTI and functional language activation using fMRI, but with only 25 participants it is difficult to come to any major conclusions, but the approach is undoubtedly important [383]. There seems little doubt that large and systematic connectome mapping will eventually provide insights to connections between functional modules, both within and between hemispheres. Vingerhoets has emphasised "the need for a more thorough investigation of the behavioural relevance of atypical functional segregation in the general population" [17].

36. Selection and Evolution

Handedness, irrespective of whether it is environmental or genetic in origin, is a *stable polymorphism*, the proportions of right- and left-handedness seeming to be stable over long periods of historical time, in the past century or two [384], the last five millennia [385], since the upper palaeolithic [386], or half a million years ago [387,388]. Going back further than that is difficult, but there does seem to be evidence that the majority of humans were right-handed perhaps two to three million years ago [389]. Prior to that, it is likely that the majority of our non-human ancestors had a 50:50 mix of right- and left-handers [390], although the proportion of right-handers in some great apes may be as high as 65–70%, with variation between different species [391].

If a polymorphism is stable for long periods of time, then something must maintain that stability, or random drift, be it either genetic or environmental, will remove one or other morph. Since handedness does seem to be under genetic control, genetic mechanisms for maintaining the polymorphism seem the most likely explanation. A classical monogenic explanation for a *balanced polymorphism* is *heterozygote advantage* with the best known example being sickle cell anaemia, where affected homozygotes have a high mortality, but heterozygotes, carrying just one copy of the gene, are protected against malaria, which means the gene continues to survive in the population [137]. An alternative mechanism is *frequency-dependent selection*, where rarer phenotypes have an advantage precisely because of their scarcity, a mechanism found in the lateralisation of scale-eating cichlid fish [392], and often invoked for left-handers being better at fighting or sport [393,394], although there is controversy over the strength of the effects [394–396]. The only other major method of polymorphisms being maintained in the population is by *mutation*, and it is clear that new mutations are responsible for maintaining the relatively rare disease, haemophilia [397]. However, in general, geneticists have argued that mutation at a single locus cannot maintain a polymorphism at more than about 1% of the population [137].

The monogenic DC model would seem to be an obvious candidate for a balanced polymorphism, and for many years that was a view that I espoused [390]. It seemed plausible that the relatively small amount of extra variation in DC genotypes, with occasional modules being moved to the opposite hemisphere from that which is typical ("nudged"), might sometimes confer selective advantages, perhaps because of unusual organisations that created novel cognitive skills (much as manual dexterity or poetry, as suggested in the previous section). In contrast, the CC genotype would more likely result in many rearrangements of modules ("all shook up") with the likelihood that most rearrangements would be disadvantageous, just as it rarely benefits complex systems for all the components to be taken out and replaced in random order. The monogenic DC model with balanced polymorphism was an elegant model, and it had very much to commend it, except that it was almost certainly wrong. Handedness is clearly not under monogenic control, and while, as we have seen, it may act *as if it is monogenic* for the purposes of calculating handedness in families and twins, that is not the same as saying that it *is* monogenic. Therefore, there cannot be a simple balanced polymorphism. The molecular genetics of handedness, which says there are at least 40 genes [65], and possibly many more yet to be found [93], is problematic for stabilising selection, particularly involving a balanced polymorphism.

Although *stabilising selection* is readily maintained with two or more alleles at a single locus, any more than two loci becomes very complicated. As Walsh and Lynch say in their massive and encyclopaedic *Evolution and selection of quantitative traits*, "one of the most perplexing observations in quantitative genetics [is] the maintenance of high levels of genetic variation for most traits under apparent genetic variation at a number of loci" [62] (p. 132). Much later in their book, a variety of complex models is presented, and "to aid the more casual reader, [Walsh and Lynch's Table 28.3] ... summarises the major inconsistencies for each model ... " (p. 1018). Seventy pages later, after presenting that table, the chapter reports that, in 2005, Johnson and Barton [398] ... stated that

> "it is puzzling that levels of heritability are so pervasive, so high and roughly constant', [meaning] that, 'we are in the somewhat embarrassing position of observing some remarkably robust patterns ... and yet seeing no compelling explanation for them [62] (pp. 1069–1070)".

As the chapter finishes, Walsh and Lynch (p. 1078) refer to the insight of Barton (1990) by saying "that much of the variation associated with a trait is likely maintained for reasons independent of that trait's fitness ... " [62]. Barton puts forward a role for *mutation-selection balance* (MSB), in which there are large numbers of mutations which are only mildly deleterious [399]. Over recent years it has become more and more apparent that rare mutations are very frequent, with each of us carrying thousands of them [289], with many rare variants being found on deep sequencing [400], which may contribute to diseases such as diabetes [401].

Mutation–selection balance has been invoked to explain the inheritance and maintenance of traits such as intelligence and education, and to some degree neuroticism and extraversion [402,403]. As Hill et al. explain,

> "mutation-selection balance provides an explanation of how genetic variation can be maintained for quantitative traits that are under directional selective pressure. Mutation-selection balance describes instances where mutations that are deleterious to the phenotype occur within a population at the same rate that they are removed through the effects of selective pressure. Due to the removal of variants with deleterious effects on the phenotype, the existence of common variants with medium to large effects is not expected under mutation-selection balance. This is consistent with the current findings from large genome-wide association studies (GWAS) on cognitive phenotypes, including general intelligence and education, where common SNPs collectively explain a substantial proportion of phenotypic variance, but the individual effect size of each genome-wide significant SNP discovered so far is around 0.02% [402] (p. 2348)".

Essentially there is a definite benefit to having high levels of intelligence, or other cognitive traits, just as the there is a benefit in having a functioning muco-ciliary escalator to remove detritus from the lungs and sinuses. Mutations at many loci can, however, disrupt the process, and that results either in lower intelligence in the one case, or impaired lung function in the other. The deleterious effect of any individual mutation is relatively low, but eventually they are removed from the gene-pool by selection and drift. However, new mutations inevitably replace them, with selection and mutation in long-term balance.

At present, the most plausible explanation for the inheritance of right- and left-handedness seems to be something like mutation–selection balance. Individuals with all D_i alleles in the polygenic DC model have the standard, 'modal', pattern of cerebral organisation, and their neural function is effective (if perhaps unoriginal or uninspired). New

C_i alleles mutate occasionally, so a locus has a D_iC_i genotype, resulting in left-handedness which can then run in families as a result of heterozygosity at a locus. The D_iC_i genotypes result occasionally in traits which are either beneficial (talents) or deleterious (deficits), which may increase or decrease the fitness of the individual carrying them. Eventually, the locus develops a second mutation, giving a C_iC_i genotype at a locus, which then results in the random location of all modules to right or left. Presumably, the latter is deleterious to a greater extent. If the homozygous C_iC_i genotype also happens to be at a locus where there is a pleotropic effect, having effects on other bodily systems, those pleiotropic effects being on systems which may be entirely unrelated to neural functioning, then those C_iC_i genotypes at the locus disappear from the gene-pool. More mutations are, however, on their way, and so the cycle continues, with rates of handedness being maintained at broadly constant levels across time. It should be noted that this model does not explain the original origins of the baseline genotype with D_i alleles at all of the loci.

37. The Evolutionary Origins of the Asymmetric Brain

A genetic model accounting for why some people are left-handed tells us nothing about the much deeper question of why humans have brain asymmetry at all. That is similar to an understanding of the condition phenylketonuria telling us little about how and why the enzyme phenylalanine hydroxylase evolved in the first place, or an understanding of PCD telling us little of when, where and why organisms developed motile cilia, and eventually a muco-ciliary staircase, or knowing about sickle-cell anaemia helps us to understand the evolution of haemoglobin itself, that wonderful molecule which carries oxygen to every cell in our body. "Genes for" some things are not easily studied—genes for a head, genes for a liver, genes for a heart, exist in some sense, but if they go badly wrong then an embryo fails to develop, even if it gets that far. Making genetic sense of the big things in biology is helped only occasionally by subtle pointers and hints from when those big things go wrong, but ultimately an understanding has to come from evolutionary studies, and particularly molecular evolution which can compare important genes across species, looking for homologues and changes. Genes for brain asymmetry are nowhere near being able to be studied in that way, although understanding the origins of the genes involved in handedness may one day point to their underpinnings, and their earliest beginnings.

Understanding bird flight, to return to the comment made earlier by David Marr [175], means understanding much more than feathers. There is a gene in chickens called *scaleless*, *sc/sc* homozygotes having neither scales nor feathers [404]. The original mutation arose spontaneously in 1954 in the University of California chicken flock, probably in a single male [405]. While those *scaleless* chickens undoubtedly could not fly, neither do they really contribute much to an understanding of the evolutionary origins of bird flight, except to suggest that the protein FGF20 is important in creating vertebrate skin appendages, including feathers and scales. Feathers are highly multi-functional, helping with flight, but they are also involved with water-proofing, buoyancy, protection, balance, avoidance of parasites, camouflage, mimicry, eye and orifice protection, sexual display, male–male competition, distraction, odour production, moulting and renewal, prey capture, and migration, which together as a set resemble a pleiotropy [406]. Birds differ from one another with a DNA-based phylogenetic tree available for 198 species [407], most of which can fly, but a number of which have lost that ability. Birds, in some sense, are living dinosaurs, but their evolution from the dinosaurs raises many and difficult palaeontological questions, not least concerning the apparent coupling of the origin of flight with the origin of birds [408]. Birds are fascinating, but clearly are not the main interest of this paper. They do, however, show the orders of magnitude differences that are present between our understanding of cerebral lateralisation, or indeed vertebrate lateralisation in general, and biologists' understanding of the evolution of birds, feathers and flight.

38. The Evolution of Heart Asymmetry and Brain Asymmetry

Heart asymmetry seems ancient, going back to the beginning of the vertebrates, whereas brain asymmetry feels the newcomer on the block, certainly in terms of handedness and language dominance, although behavioural asymmetries are also ancient in phylogenetic terms [190,409], with an early behavioural asymmetry shown in trilobites from the Cambrian of 500 million years ago [410]. Certainly, when speculating on how cerebral lateralisation may have evolved, it had seemed plausible that the gene for handedness and language dominance would have been a mutated copy of the gene for situs [165]. In retrospect, that may have been back-to-front.

Richard Palmer, in a series of important papers, has transformed thinking about asymmetry and its evolution [217,318,411]. Palmer has distinguished two possible ways in which anatomical asymmetry might evolve in genetic terms. The conventional evolutionary route would be that a symmetrical ancestor, S, has a mutation which results directly in directional asymmetry, D, so that symmetry precedes directional asymmetry as follows: $S \rightarrow D$. An alternative route, which Palmer calls Genetic Assimilation, involves a symmetric ancestor having a mutation which produces *anti-symmetry*, in which offspring occur in both of two mirror-image forms (right-handed and left-handed enantiomorphs, $A + A^T$). In response to selection, a second mutation results in one of the two forms of A being selected, leaving D, directional asymmetry, with the anti-symmetric pair being a transitional stage: $S \rightarrow (A + A^T) \rightarrow D$.

Anti-symmetric forms can sometimes be spotted in the fossil record, with Palmer citing a clear example in the evolution in phallostethid fish of the priapium, which is used for clasping during intromission. Of 21 species, 17 species are anti-symmetric, individuals within the species having the priapium either on the left or the right side. Of the remaining four species, two solely have a left priapium and two solely have a right priapium. The phylogenetic relationships are understood, and the four species with directional asymmetry have evolved from antisymmetric species, down three separate branches of the tree [318]. A similar situation can perhaps also be seen in the turning tendencies of twenty fish species during a detour, where anti-symmetry seems to be ancestral to directional asymmetry in one or other direction [412], as is also the case for hand usage in seven species of marsupials [413]; in both cases directional bias seems to be driven either by social or ecological factors. The situation in anthropoid primates shows some similarities, although the pattern is confusing. Meguerditchian et al. [414], in an analysis similar to that for marsupials, found that ecology rather than phylogeny seemed important, arboreal species showing a left-hand preference for bimanual coordinated tasks, whereas terrestrial species tended to be right-handed. The more extensive study of 38 species of anthropoid primates by Caspar et al. [415] found little phylogenetic influence on direction of handedness, suggesting "unique selective pressures gave rise to ... hand preferences", which is similar to the picture in other phylogenies. There was, however, evidence of phylogenetic and ecological influences on degree of hand lateralisation, stronger lateralisation in the New World monkeys, and terrestrial species generally having weaker hand preferences.

Once directional asymmetry has been set up then it can be fixed in place by what Waddington called *canalization* [416], whereby "the form which occurs in Nature ... is much less variable in appearance than the majority of the mutant races", a process whereby buffering maintains developmental stability [417]. In the case of the heart, Palmer presents compelling data showing that the rate of spontaneous situs inversus is about 5% in fish, 1–2% in amphibians and birds, and about 0.1% in mammals, with a rate of 0.01% in humans, suggesting a progressive increase in buffering of asymmetric heart development [318]. Why asymmetric heart development is necessary has been little explored, but there is a suggestion that an asymmetric, spiral flow of blood through the heart may reduce turbulence and hence blood clotting [418], although it is unclear whether there may be more turbulence in the less buffered situation of situs inversus.

Earlier in this paper, it was suggested that cilial rotation was necessary for determining situs. That is probably correct for humans and mice, but many species do not seem to have rotatory cilia, with chicks and frogs being the most notable exception. Many genes or traits are also expressed asymmetrically during early development, and Palmer tabulates 29 different genes or traits involved in the nodal cascade, across six groups of species, including mammals, birds, amphibians, fish, lancelets (amphioxus) and ascidians (sea squirts and tunicates), in relation to three anatomical asymmetries—coiling of the gut, asymmetry of the heart, and brain asymmetry (particularly of the habenular nucleus). The key finding is that while lancelets and ascidians do not have an asymmetric heart, they do have asymmetries of the gut and also of the brain. Therefore, Palmer concludes that "brain asymmetry seems a more likely ancestral target [of asymmetrically expressed genes]", with heart asymmetry only being a secondary or derived character [313].

Asymmetry of the brain may therefore precede heart asymmetry. The anatomical asymmetry of the brain is mostly in the habenular nuclei in the epithalamus, the asymmetry being found across all vertebrates [419] and has been shown to relate to asymmetric behaviour in fish [420]. The habenulae are asymmetric in humans [421], but are extremely difficult to image accurately using fMRI, with assessment of asymmetries and functional activity being problematic [422]. The functions of the habenula are slowly becoming more apparent, with reward processing seeming to be important [423], with suggestions that disruption can be important in major depression [424]. Having said that, it is interesting to return to the two large review articles on laterality by Corballis and Morgan in the 1970s [33,103], where despite discussing the asymmetry of the habenular nuclei, in the discussion they do comment that "we do not know what the habenular ganglion of amphibia does, still less whether its asymmetry has any effect on behaviour" [103] (p. 326). That, to some extent, still summarises the current situation, even if it may be the case that the habenular is the *Ur*-asymmetry from which other major human anatomical and behavioural asymmetries have developed.

Even if habenular asymmetry is where everything comes from, that is still not an explanation for why handedness and language dominance are inherited, and there are many details that are not accounted for. Consider, say, handedness in mice and other non-human mammals, where hand or paw preference seems to be effectively random. If there are genes determining habenular asymmetry, they do not seem to be affecting handedness. The conclusion has to be that, at some point, mechanisms for making the habenular asymmetric have been co-opted to make cortical tissue or other cerebral tissue asymmetric in its functioning, allowing hand, language and other functional asymmetries to develop. When and how is still though very unclear.

39. Conclusions: Broken Symmetry and 'the Method of Brocan Doubt'

The theoretical challenges raised by cerebral polymorphisms have inevitably raised many questions covering a wide range of areas, but that is the nature of both biology and of theory. Single biological organs, biological functions or even biological molecules do not exist independently of other components of complex organisms, but all are necessarily interdependent. That is shown very clearly in the suggestion by Boyle et al. [297] that complex traits are not just polygenic but in some sense *omnigenic*, with almost all genes having some influence on any complex trait, albeit small. Making sense of polygenic associations with phenotypes, inevitably therefore requires a very broadly based approach. Theory has therefore to consider the bigger picture, drilling down to possible underlying mechanisms, and standing back to consider evolutionary origins and functions. Such an approach is relatively rare in the study of lateralisation, and, as a consequence, this paper has become very long.

In one of the most famous philosophical passages ever written, Rene Descartes began to doubt almost everything that he believed, until he had the sudden insight that the one thing he could not doubt was that he was thinking: *Cogito, ergo sum: I think, therefore I am.* The method of Cartesian doubt had eventually found solid bedrock, a conclusion whose existence could not be doubted, and from which theorising could then properly begin.

Sometimes, when thinking about theories of the broken asymmetries of lateralisation, and particularly about some of the wilder and more eccentric theories that exist [12], I return to what I call '*the method of Brocan doubt*'. When faced with inconsistent laterality data, incompatible predictions or incomprehensible theories, I return to the one thing which is indubitable in trying to make sense of the broken symmetry demonstrated by the human brain—as Broca's seminal research showed, *patients who become aphasic after unilateral brain damage, mostly have damage in the left hemisphere*. Additionally, that can be affirmed by anyone who has seen patients with aphasia, most of whom have a right-sided hemiplegia. Any theory of cerebral lateralisation has to accept and attempt to explain both that one indubitable fact shown by Broca along with its occasional but important exceptions. The equivalent key finding for cerebral polymorphisms, is the 1983 study by Phil Bryden, Henri Hécaen, and Maria De Agostini [21], which identified all eight independent combinations of aphasia, agnosia and hand dominance, thereby extending Broca's work to three modules. I often discussed those 1983 results with Phil Bryden, and this paper is dedicated to his memory, with thanks for his continual and continuing inspiration.

Funding: This research received no external funding.

Institutional Review Board Statement: Not applicable.

Informed Consent Statement: Not applicable.

Data Availability Statement: Not applicable.

Acknowledgments: Many people over many years have discussed with me the theoretical ideas in this paper, and have stimulated, interested and sometimes provoked me, and I am grateful to them all. I particularly thank the instigators and organisers since 2010 of the various North Sea Laterality conferences, and their many participants. Individuals who have particularly influenced me are listed below in alphabetical order, although none of course necessarily approves of anything in the paper, they are definitely not responsible for the inevitable errors, and I apologise to those whom I have omitted. In alphabetical order, my particular thanks to Dorothy Bishop, Dorret Boomsma, David Carey, Clyde Francks, Robin Gerrits, Emma Karlsson, Veronika Odintsova, Sylvia Paracchini, and Guy Vingerhoets, as well to many unnamed others. Finally, of course, I thank Onur Güntürkün and Sebastian Ocklenburg for conceiving of and editing this Special Issue.

Conflicts of Interest: The author declares no conflict of interest.

References

1. Dax, M. Lesions de la moitié gauche de l'encéphale coincidant avec l'oubli des signes de la pensee. *Gaz. Hebd. Med. Chir.* **1865**, *2*, 259–262.
2. Broca, P. Remarques sur le siège de la faculté du langage articulé, suivies d'une observation d'aphémie (perte de la parole). *Bull. Société D'anatomique Paris* **1861**, *6*, 330–357.
3. Leblanc, R. *Fearful Asymmetry: Bouillaud, Dax, Broca, and the Localization of Language, Paris, 1825–1879*; McGill-Queen's Press-MQUP: Kingston, ON, Canada, 2017.
4. Gowers, W.R. *Lectures on the Diagnosis of Diseases of the Brain, delivered at University College Hospital*, 2nd ed.; Churchill: London, UK, 1887.
5. McManus, I.C. Pathological left-handedness: Does it exist? *J. Commun. Disord.* **1983**, *16*, 315–344. [CrossRef]
6. Carey, D.P.; Johnstone, L.T. Quantifying cerebral asymmetries for language in dextrals and adextrals with random-effects meta analysis. *Front. Psychol.* **2014**, *5*, 1128. [CrossRef]
7. Fodor, J.A. *The Modularity of Mind*; MIT Press: Cambridge, MA, USA, 1983.
8. McManus, I.C. The history and geography of human handedness. In *Language Lateralisation and Psychosis*; Sommer, I., Khan, R.S., Eds.; Cambridge University Press: Cambridge, UK, 2009; pp. 37–58.
9. Vingerhoets, G.; Gerrits, R.; Verhelst, H. Atypical Brain Asymmetry in Human Situs Inversus: Gut Feeling or Real Evidence? *Symmetry* **2021**, *13*, 695. [CrossRef]
10. McManus, I.C.; Bryden, M.P. The genetics of handedness, cerebral dominance and lateralization. In *Handbook of Neuropsychology, Volume 6, Section 10: Child Neuropsychology (Part 1)*; Rapin, I., Segalowitz, S.J., Eds.; Elsevier: Amsterdam, The Netherlands, 1992; pp. 115–144.
11. McManus, I.C.; Davison, A.; Armour, J.A.L. Multi-locus genetic models of handedness closely resemble single locus models in explaining family data and are compatible with genome-wide association studies. *Ann. N. Y. Acad. Sci.* **2013**, *1288*, 48–58. [CrossRef]

12. McManus, I.C. Half a century of handedness research: Myths, truths; fictions, facts; backwards, but mostly forward. *Brain Neurosci. Adv.* **2019**, *3*, 2398212818820513. [CrossRef]
13. Orton, S.T. 'Word-blindness' in school-children. *Arch. Neurol. Psychiatry* **1925**, *14*, 581–615. [CrossRef]
14. Orton, S.T. *Reading, Writing and Speech Problems in Children*; Chapman and Hall: London, UK, 1937.
15. Eglinton, E.; Annett, M. Handedness and dyslexia: A meta-analysis. *Percept. Motor Skills* **1994**, *79*, 1611–1616. [CrossRef]
16. Gerrits, R.; Verhelst, H.; Vingerhoets, G. Mirrored brain organization: Statistical anomaly or reversal of hemispheric functional segregation bias? *Proc. Natl. Acad. Sci. USA* **2020**, *117*, 14057–14065. [CrossRef]
17. Vingerhoets, G. Phenotypes in hemispheric functional segregation? Perspectives and challenges. *Phys. Life Rev.* **2019**, *30*, 1–18. [CrossRef] [PubMed]
18. McGlone, J.; Davidson, W. The relationship between cerebral speech laterality and spatial ability with special reference to sex and hand preference. *Neuropsychologia* **1973**, *11*, 105–113. [CrossRef]
19. Levy, J. Cerebral lateralization and spatial ability. *Behav. Genet.* **1976**, *6*, 171–188. [CrossRef] [PubMed]
20. McManus, I.C. *Handedness, Language Dominance and Aphasia: A Genetic Model*; Psychological Medicine, Monograph Supplement No.8; Cambridge University Press: Cambridge, UK, 1985.
21. Bryden, M.P.; Hécaen, H.; De Agostini, M. Patterns of cerebral organisation. *Brain Lang.* **1983**, *20*, 249–262. [CrossRef]
22. Kroliczak, G.; Buchwald, M.; Kleka, P.; Klichowski, M.; Potok, W.; Nowik, A.M.; Randerath, J.; Piper, B.J. Manual praxis and language-production networks, and their links to handedness. *Cortex* **2021**, *140*, 110–127. [CrossRef]
23. Beking, T. Two Sides to Every Story: Sex Hormones, Brain Lateralization and Gender Development. Ph.D. Thesis, University of Groningen, Groningen, The Netherlands, 2018.
24. Badzakova-Trajkov, G.; Häberling, I.S.; Roberts, R.P.; Corballis, M.C. Cerebral Asymmetries: Complementary and Independent Processes. *PLoS ONE* **2010**, *5*, e9682. [CrossRef]
25. Karlsson, E. *Cerebral Asymmetries: Handedness and the Right Hemisphere*; Bangor University: Bangor, UK, 2020.
26. Woodhead, Z.; Bradshaw, A.; Wilson, A.; Thompson, P.; Bishop, D. Testing the unitary theory of language lateralization using functional transcranial Doppler sonography in adults. *R. Soc. Open Sci.* **2019**, *6*, 181801. [CrossRef]
27. Woodhead, Z.; Thompson, P.; Karlsson, E.; Bishop, D. An updated investigation of the multidimensional structure of language lateralization in left-and right-handed adults: A test–retest functional transcranial Doppler sonography study with six language tasks. *R. Soc. Open Sci.* **2021**, *8*, 200696. [CrossRef]
28. Weis, S.; Hausmann, M.; Stoffers, B.; Sturm, W. Dynamic changes in functional cerebral connectivity of spatial cognition during the menstrual cycle. *Hum. Brain Mapp.* **2011**, *32*, 1544–1556. [CrossRef]
29. Weis, S.; Hodgetts, S.; Hausmann, M. Sex differences and menstrual cycle effects in cognitive and sensory resting state networks. *Brain Cogn.* **2019**, *131*, 66–73. [CrossRef]
30. Demareva, V.; Mukhina, E.; Bobro, T.; Abitov, I. Does Double Biofeedback Affect Functional Hemispheric Asymmetry and Activity? A Pilot Study. *Symmetry* **2021**, *13*, 937. [CrossRef]
31. Floegel, M.; Kell, C.A. Functional hemispheric asymmetries during the planning and manual control of virtual avatar movements. *PLoS ONE* **2017**, *12*, e0185152. [CrossRef] [PubMed]
32. Bradshaw, J.L.; Nettleton, N.C. The nature of hemispheric specialization in man. *Behav. Brain Sci.* **1981**, *4*, 51–91. [CrossRef]
33. Corballis, M.C.; Morgan, M.J. On the biological basis of human laterality: I. Evidence for a maturational left-right gradient. *Behav. Brain Sci.* **1978**, *2*, 261–269. [CrossRef]
34. Bryden, M.P. Choosing sides: The left and right of the normal brain. *Can. Psychol.* **1990**, *31*, 297–309. [CrossRef]
35. Bramwell, B. A remarkable case of aphasia. *Brain* **1898**, *21*, 343–373. [CrossRef]
36. Bramwell, B. On 'crossed' aphasia. *Lancet* **1899**, *1*, 1473–1479. [CrossRef]
37. Lansdell, H. Verbal and nonverbal factors in right-hemisphere speech: Relation to early neurological history. *J. Comp. Physiol. Psychol.* **1969**, *69*, 734. [CrossRef]
38. Levy, J. Possible basis for the evolution of lateral specialization of the human brain. *Nature* **1969**, *224*, 614–615. [CrossRef]
39. Teuber, H.L. Why two brains? In *The Neurosciences: Third Study Program*; Schmitt, F.O., Worden, F.G., Eds.; MIT Press: Cambridge, MA, USA, 1974.
40. Groen, M.A.; Whitehouse, A.J.; Badcock, N.A.; Bishop, D.V. Does cerebral lateralization develop? A study using functional transcranial Doppler ultrasound assessing lateralization for language production and visuospatial memory. *Brain Behav.* **2012**, *2*, 256–269. [CrossRef]
41. Illingworth, S.; Bishop, D.V. Atypical cerebral lateralisation in adults with compensated developmental dyslexia demonstrated using functional transcranial Doppler ultrasound. *Brain Lang.* **2009**, *111*, 61–65. [CrossRef] [PubMed]
42. Yeo, R.A.; Gangestad, S.W.; Thoma, R.; Shaw, P.; Repa, K. Developmental instability and cerebral lateralization. *Neuropsychology* **1997**, *11*, 552–561. [CrossRef] [PubMed]
43. Rogers, L.J. Evolution of hemispheric specialization: Advantages and disadvantages. *Brain Lang.* **2000**, *73*, 236–253. [CrossRef] [PubMed]
44. Hirnstein, M.; Hausmann, M.; Güntürkün, O. The evolutionary origins of functional cerebral asymmetries in humans: Does lateralization enhance parallel processing? *Behav. Brain Res.* **2008**, *187*, 297–303. [CrossRef] [PubMed]
45. Zickert, N.; Geuze, R.H.; Beking, T.; Groothuis, T.G. Testing the Darwinian function of lateralization. Does separation of workload between brain hemispheres increase cognitive performance? *Neuropsychologia* **2021**, *159*, 107884. [CrossRef] [PubMed]

46. Häberling, I.S.; Badzakova-Trajkov, G.; Corballis, M.C. Callosal tracts and patterns of hemispheric dominance: A combined fMRI and DTI study. *Neuroimage* **2011**, *54*, 779–786. [CrossRef]
47. Króliczak, G.; Piper, B.J.; Potok, W.; Buchwald, M.; Kleka, P.; Przybylski, Ł.; Styrkowiec, P.P. Praxis and language organisation in left-handers. *Acta Neuropsychol.* **2020**, *18*, 15–28. [CrossRef]
48. Peters, M. Contributions of imaging techniques to our understanding of handedness. In *Side Bias: A Neuropsychological Perspective*; Mandal, M.K., Bulman-Fleming, M.B., Tiwari, G., Eds.; Kluwer: Dordrecht, The Netherlands, 2000; pp. 191–222.
49. Paquier, P.; Mariker, M.L. Review. A synthesis of the role of the cerebellum in cognition. *Aphasiology* **2005**, *19*, 3–19.
50. Rosch, R.; Cowell, P.; Gurd, J. Cerebellar asymmetry and cortical connectivity in monozygotic twins with discordant handedness. *Cerebellum* **2018**, *17*, 191–203. [CrossRef]
51. McManus, I.C.; Cornish, K.M. Fractionating handedness in mental retardation: What is the role of the cerebellum? *Laterality* **1997**, *2*, 81–90. [CrossRef]
52. Uguru-Okorie, D.C.; Arbuthnott, G.W. Altered paw preference after unilateral 6-hydroxydopamine injections into lateral hypothalamus. *Neuropsychologia* **1981**, *19*, 463–467. [CrossRef]
53. Previc, F.H.; Saucedo, J.C. The relationship between turning behavior and motoric dominance in humans. *Percept. Motor Skills* **1992**, *75*, 935–944. [CrossRef] [PubMed]
54. Previc, F.H. A general theory concerning the prenatal origins of cerebral lateralization in humans. *Psychol. Rev.* **1991**, *98*, 299–334. [CrossRef] [PubMed]
55. Snyder, P.J.; Bilder, R.M.; Wu, H.; Bogerts, B.; Lieberman, J.A. Cerebellar volume asymmetries are related to handedness: A quantitative MRI study. *Neuropsychologia* **1995**, *33*, 407–419. [CrossRef]
56. Deoni, S.C.; Catani, M. Visualization of the deep cerebellar nuclei using quantitative T1 and ρ magnetic resonance imaging at 3 Tesla. *Neuroimage* **2007**, *37*, 1260–1266. [CrossRef]
57. Kavaklioglu, T.; Guadalupe, T.; Zwiers, M.; Marquand, A.F.; Onnink, M.; Shumskaya, E.; Brunner, H.; Fernandez, G.; Fisher, S.E.; Francks, C. Structural asymmetries of the human cerebellum in relation to cerebral cortical asymmetries and handedness. *Brain Struct. Funct.* **2017**, *222*, 1611–1623. [CrossRef]
58. Wang, D.; Buckner, R.L.; Liu, H. Cerebellar asymmetry and its relation to cerebral asymmetry estimated by intrinsic functional connectivity. *J. Neurophysiol.* **2013**, *109*, 46–57. [CrossRef]
59. Tzourio-Mazoyer, N.; Labache, L.; Zago, L.; Hesling, I.; Mazoyer, B. Neural support of manual preference revealed by BOLD variations during right and left finger-tapping in a sample of 287 healthy adults balanced for handedness. *Laterality* **2021**, *26*, 398–420. [CrossRef]
60. Häberling, I.S.; Corballis, M.C. Cerebellar asymmetry, cortical asymmetry and handedness: Two independent networks. *Laterality Asymmetries Body Brain Cogn.* **2016**, *21*, 397–414. [CrossRef]
61. Lynch, M.; Walsh, B. *Genetics and Analysis of Quantitative Traits*; Sinauer Associates, Inc.: Sunderland, MA, USA, 1998.
62. Walsh, B.; Lynch, M. *Evolution and Selection of Quantitative Traits*; Oxford University Press: Oxford, UK, 2018.
63. Jordan, H.E. The inheritance of left-handedness. *Am. Breeders Mag.* **1911**, *2*, 113–124. [CrossRef]
64. Ramaley, F. Inheritance of left-handedness. *Am. Nat.* **1913**, *47*, 730–739. [CrossRef]
65. McManus, I.C. Is any but a tiny fraction of handedness variance likely to be due to the external environment? *Laterality* **2021**, *26*, 310–314. [CrossRef] [PubMed]
66. Cuellar-Partida, G.; Tung, J.Y.; Eriksson, N.; Albrecht, E.; Medland, S.E. Genome-wide association study identifies 48 common genetic variants associated with handedness. *Nat. Hum. Behav.* **2021**, *5*, 59–70. [CrossRef] [PubMed]
67. Waddington, C.H. *The Strategy of the Genes*; MacMillan: London, UK, 1957.
68. Molenaar, P.C.; Boomsma, D.I.; Dolan, C.V. A third source of developmental differences. *Behav. Genet.* **1993**, *23*, 519–524. [CrossRef] [PubMed]
69. Mitchell, K.J. *Innate: How the Wiring of Our Brains Shapes Who We Are*; Princeton University Press: Princeton, NJ, USA, 2018.
70. McManus, I.C. *A Model of the Genetics and Epigenetics of Handedness and Cerebral Dominance*; The Psychological Laboratory, University of Cambridge: Cambridge, UK, 1977; Unpublished manuscript.
71. McManus, I.C. Determinants of Laterality in Man. Unpublished. Ph.D. Thesis, University of Cambridge, Cambridge, UK, 1979. Available online: http://www.ucl.ac.uk/medical-education/publications/phd (accessed on 15 February 2022).
72. Annett, M. *A Single Gene Explanation of Right and Left Handedness and Brainedness*; Lanchester Polytechnic: Coventry, UK, 1978.
73. Annett, M. *Left, Right, Hand and Brain: The Right Shift Theory*; Lawrence Erlbaum: Mahwah, NJ, USA, 1985.
74. Annett, M. *Handedness and Brain Asymmetry: The Right Shift Theory*; Psychology Press: Hove, UK, 2002.
75. Ocklenburg, S.; Berretz, G.; Packheister, J.; Friedrich, P. Laterality 2020: Entering the next decade. *Laterality Asymmetries Brain Behav. Cogn.* **2021**, *26*, 265–297. [CrossRef]
76. Grahek, I.; Schaller, M.; Tackett, J.L. Anatomy of a psychological theory: Integrating construct-validation and computational-modeling methods to advance theorizing. *Perspect. Psychol. Sci.* **2021**, *16*, 1745691620966794. [CrossRef]
77. Nurse, P. Biology must generate ideas as well as data. *Nature* **2021**, *597*, 305. [CrossRef]
78. Robinaugh, D.J.; Haslbeck, J.M.; Ryan, O.; Fried, E.I.; Waldorp, L.J. Invisible hands and fine calipers: A call to use formal theory as a toolkit for theory construction. *Perspect. Psychol. Sci.* **2020**, *16*, 725–743. [CrossRef]
79. Broers, N.J. When the numbers do not add up: The practical limits of stochastologicals for soft psychology. *Perspect. Psychol. Sci.* **2021**, *16*, 1745691620970557. [CrossRef]

80. Guest, O.; Martin, A.E. How computational modeling can force theory building in psychological science. *Perspect. Psychol. Sci.* **2021**, *16*, 789–802. [CrossRef]
81. Meehl, P.E. Theory-testing in psychology and physics: A methodological paradox. *Philos. Sci.* **1967**, *34*, 103–115. [CrossRef]
82. Meehl, P.E. Theoretical risks and tabular asterisks: Sir Karl, Sir Ronald, and the slow progress of soft psychology. *J. Consult. Clin. Psychol.* **1978**, *46*, 806. [CrossRef]
83. Meehl, P.E. Appraising and amending theories: The strategy of Lakatosian defense and two principles that warrant it. *Psychol. Inq.* **1990**, *1*, 108–141. [CrossRef]
84. Szollosi, A.; Donkin, C. Arrested theory development: The misguided distinction between exploratory and confirmatory research. *Perspect. Psychol. Sci.* **2021**, *16*, 717–724. [CrossRef]
85. Van Rooij, I.; Baggio, G. Theory before the test: How to build high-verisimilitude explanatory theories in psychological science. *Perspect. Psychol. Sci.* **2021**, *16*, 682–697. [CrossRef]
86. Navarro, D.J. If mathematical psychology did not exist we might need to invent it: A comment on theory building in psychology. *Perspect. Psychol. Sci.* **2021**, *16*, 1745691620974769. [CrossRef]
87. Shepard, R.N. Toward a universal law of generalization for psychological science. *Science* **1987**, *237*, 1317–1323. [CrossRef]
88. Slater, M.H.; Borghini, A. *Introduction: Lessons from the Scientific Butchery*; MIT Press: Cambridge, MA, USA, 2013.
89. Bishop, D.V.M. How to increase your chances of obtaining a significant association between handedness and disorder. *J. Clin. Exp. Neuropsychol.* **1990**, *12*, 812–816. [CrossRef]
90. Kluckhohn, C.K.M.; Murrray, H.A. *Personality in Nature, Society, and Culture*; Knopf: New York, NY, USA, 1948.
91. Turkheimer, E. Three laws of behavior genetics and what they mean. *Curr. Dir. Psychol. Sci.* **2000**, *9*, 160–164. [CrossRef]
92. Chabris, C.F.; Lee, J.J.; Cesarini, D.; Benjamin, D.J.; Laibson, D.I. The fourth law of behavior genetics. *Curr. Dir. Psychol. Sci.* **2015**, *24*, 304–312. [CrossRef]
93. Paracchini, S. Recent Advances in Handedness Genetics. *Symmetry* **2021**, *13*, 1792. [CrossRef]
94. Popper, K.R. *Conjectures and Refutations: The Growth of Scientific Knowledge*; Routledge: London, UK, 1963; pp. 104–110.
95. Popper, K.R. *The Logic of Scientific Discovery*; Routledge: London, UK, 1934.
96. Popper, K.R. *Objective Knowledge*; Oxford University Press: Oxford, UK, 1972; Volume 360.
97. Lakatos, I. *The Methodology of Scientific Research Programmes*; Cambridge Univeristy Press: Cambridge, UK, 1978.
98. McManus, I.C. Book review: 'Left, right, hand and brain: The right shift theory' (Annett, M). *Psychol. Med.* **1987**, *17*, 523–525. [CrossRef]
99. Gershman, S.J. How to never be wrong. *Psychonom. Bull. Rev.* **2019**, *26*, 13–28. [CrossRef]
100. Higgs, P.W. Broken asymmetries, massless particles, and gauge fields. *Phys. Lett.* **1964**, *12*, 1136–1139. [CrossRef]
101. Higgs, P.W. Broken symmetries and the masses of gauge bosons. *Phys. Rev. Lett.* **1964**, *13*, 508–509. [CrossRef]
102. Feyerabend, P.K. *Against Method: Outline of an Anarchistic Theory of Knowledge*; NLB: Singapore, 1970.
103. Kuhn, T.S. *The Structure of Scientific Revolutions*; Univesrity of Chicago Press: Chicago, IL, USA, 1970.
104. Armour, J.A.L.; Davison, A.; Spector, T.D.; McManus, I.C. Genome-wide association study of handedness excludes simple genetic models. *Heredity* **2014**, *112*, 221–225. [CrossRef]
105. Morgan, M.J.; Corballis, M.C. On the biological basis of human laterality: II. The mechanisms of inheritance. *Behav. Brain Sci.* **1978**, *2*, 270–278. [CrossRef]
106. Jordan, H.E. Hereditary lefthandedness, with a note on twinning. (Study III.). *J. Genet.* **1914**, *4*, 67–81. [CrossRef]
107. Chamberlain, H.D. The inheritance of left-handedness. *J. Hered.* **1928**, *19*, 557–559. [CrossRef]
108. Rife, D.C. Handedness, with special reference to twins. *Genetics* **1940**, *25*, 178–186. [CrossRef]
109. Trankell, A. Aspects of genetics in psychology. *Am. J. Hum. Genet.* **1955**, *7*, 264–276. [PubMed]
110. Annett, M. A model of the inheritance of handedness and cerebral dominance. *Nature* **1964**, *204*, 59–60. [CrossRef] [PubMed]
111. Levy, J. A review of evidence for a genetic component in the determination of handedness. *Behav. Genet.* **1976**, *6*, 429–453. [CrossRef]
112. Beaton, A.A.; Corballis, M.; McManus, C. Obituary for Dr. Marian Annett, 1931–2018. *Laterality* **2020**, *25*, 405–412. [CrossRef]
113. Darwin, C. *A Monograph on the Sub-Class Cirripedia with Figures of All the Species: The Balanidae (Ir Sessile Cirripedes), The Verrucidae, etc etc etc*; The Ray Society: London, UK, 1854. Available online: http://darwin-online.org.uk/converted/pdf/1854_Balanidae_F339.2.pdf (accessed on 15 February 2022).
114. Darwin, C. *The Variation of Animals and Plants under Domestication*; In Two Volumes; John Murray: London, UK, 1868. Available online: http://darwin-online.org.uk/converted/pdf/1868_Variation_F877.2.pdf (accessed on 15 February 2022).
115. Palmer, A.R.; Strobeck, C. Fluctuating asymmetry: Measurement, analysis, patterns. *Annu. Rev. Ecol. Syst.* **1986**, *17*, 391–421. [CrossRef]
116. Polak, M. *Developmental Instability: Causes and Consequences*; Oxford University Press on Demand: Oxford, UK, 2003.
117. Palmer, A.R.; Strobeck, C. Fluctuating asymmetry analyses revisited. In *Developmental Instability (DI): Causes and Consequences*; Polak, M., Ed.; Oxford University Press: Oxford, UK, 2003; pp. 279–319.
118. Van Valen, L. A study of fluctuating asymmetry. *Evolution* **1962**, *16*, 125–142. [CrossRef]
119. Ludwig, W. *Das Rechts-Links-Problem im Tierreich und beim Menschen*; Springer: Berlin/Heidelberg, Germany, 1932.
120. Palmer, A.R.; Strobeck, C. Fluctuating asymmetry as a measure of developmental stability: Implications of non-normal distributions and power of statistical tests. *Acta Zool. Fenn.* **1992**, *191*, 13.

121. Graham, J.H. Nature, Nurture, and Noise: Developmental Instability, Fluctuating Asymmetry, and the Causes of Phenotypic Variation. *Symmetry* **2021**, *13*, 1204. [CrossRef]
122. Graham, J.H.; Özener, B. Fluctuating asymmetry of human populations: A review. *Symmetry* **2016**, *8*, 154. [CrossRef]
123. Graham, J.H.; Raz, S.; Hel-Or, H.; Nevo, E. Fluctuating asymmetry: Methods, theory, and applications. *Symmetry* **2010**, *2*, 466–540. [CrossRef]
124. Layton, W.M. Random determination of a developmental process. *J. Hered.* **1976**, *67*, 336–338. [CrossRef] [PubMed]
125. Afzelius, B.A. A human syndrome caused by immotile cilia. *Science* **1976**, *193*, 317–319. [CrossRef] [PubMed]
126. Collins, R.L. On the inheritance of handedness. I: Laterality in inbred mice. *J. Hered.* **1968**, *59*, 9–12. [CrossRef] [PubMed]
127. Collins, R.L. On the inheritance of handedness. II: Selection for sinistrality in mice. *J. Hered.* **1969**, *60*, 117–119. [CrossRef]
128. Joober, R.; Karama, S. Randomness and nondeterminism: From genes to free will with implications for psychiatry. *J. Psychiatry Neurosci.* **2021**, *46*, E500. [CrossRef]
129. Rolls, E.T.; Deco, G. *The Noisy Brain: Stochastic Dynamics as a Principle of Brain Function*; Oxford University Press: Oxford, UK, 2010.
130. Linneweber, G.A.; Andriatsilavo, M.; Dutta, S.B.; Bengochea, M.; Hellbruegge, L.; Liu, G.; Ejsmont, R.K.; Straw, A.D.; Wernet, M.; Hiesinger, P.R. A neurodevelopmental origin of behavioral individuality in the Drosophila visual system. *Science* **2020**, *367*, 1112–1119. [CrossRef]
131. Newman, H.H.; Patterson, J.T. A case of normal identical quadruplets in the nine-banded armadillo, and its bearing on the problems of identical twins and of sex determination. *Biol. Bull.* **1909**, *17*, 181–187. [CrossRef]
132. Ballouz, S.; Pena, M.T.; Knight, F.M.; Adams, L.B.; Gillis, J.A. The transcriptional legacy of developmental stochasticity. *bioRxiv* **2019**. [CrossRef]
133. Begley, S. Scientists Aren't Sure Why Identical Twins Differ. Armadillo Quads Offer an Answer. *Statnews* **2019**. Available online: https://www.statnews.com/2019/2012/2020/armadillo-quadruplets-genetic-determinism/ (accessed on 20 December 2021).
134. Melnick, M.; Shields, E. Allelic restriction: A biologic alternative to multifactorial threshold inheritance. *Lancet* **1976**, *307*, 176–179. [CrossRef]
135. Dempster, E.L.; Pidsley, R.; Schalkwyk, L.C.; Owens, S.; Georgiades, A.; Kane, F.; Kalidindi, S.; Picchioni, M.; Kravariti, E.; Toulopoulou, T. Disease-associated epigenetic changes in monozygotic twins discordant for schizophrenia and bipolar disorder. *Hum. Mol. Genet.* **2011**, *20*, 4786–4796. [CrossRef] [PubMed]
136. Gillis, J. Twitter. 12 December 2019. Available online: https://twitter.com/JesseAGillis/status/1205251818013560837 (accessed on 15 February 2022).
137. Cavalli-Sforza, L.L.; Bodmer, W.F. *The Genetics of Human Populations*; W. H. Freeman: San Francisco, CA, USA, 1971.
138. Papadatou-Pastou, M.; Ntolka, E.; Schmitz, J.; Martin, M.; Munafò, M.R.; Ocklenburg, S.; Paracchini, S. Human handedness: A meta-analysis. *Psychol. Bull.* **2020**, *146*, 481. [CrossRef] [PubMed]
139. McManus, I.C. Familial sinistrality: The utility of calculating exact genotype probabilities for individuals. *Cortex* **1995**, *31*, 3–24. [CrossRef]
140. McManus, I.C. Handedness in twins: A critical review. *Neuropsychologia* **1980**, *18*, 347–355. [CrossRef]
141. Sicotte, N.L.; Woods, R.P.; Mazziotta, J.C. Handedness in Twins: A Meta-analysis. *Laterality* **1999**, *4*, 265–286. [CrossRef]
142. Medland, S.E.; Duffy, D.L.; Wright, M.J.; Geffen, G.M.; Martin, N.G. Handedness in twins: Joint analysis of data from 35 samples. *Twin Res. Hum. Genet.* **2006**, *9*, 46–53. [CrossRef] [PubMed]
143. Medland, S.E.; Duffy, D.L.; Wright, M.J.; Geffen, G.M.; Hay, D.A.; Levy, F.; van Beisjterveldt, C.E.M.; Willemsen, G.; Townsend, G.C.; White, V.; et al. Genetic influences on handedness: Data from 25,732 Australian and Dutch twin families. *Neuropsychologia* **2009**, *47*, 330–337. [CrossRef]
144. Pfeifer, L.S.; Schmitz, J.; Papadatou-Pastou, M.; Peterburs, J.; Paracchini, S.; Ocklenburg, S. Handedness in Twins: Meta-Analyses. *BMC Psychol.* **2022**, *10*, 11. [CrossRef] [PubMed]
145. Danforth, C.H. Resemblance and difference in twins: Twins that look and act alike attract attention first, while dissimilar ones are apt to be overlooked. *J. Hered.* **1919**, *10*, 399–409. [CrossRef]
146. Lauterbach, C.E. Studies in twin resemblance. *Genetics* **1925**, *10*, 525–568. [CrossRef]
147. Rife, D.C. Genetic studies of monozygotic twins. III: Mirror-imaging. *J. Hered.* **1933**, *24*, 443–446. [CrossRef]
148. Zazzo, R. *Les Jumeaux: Le Couple et la Personne. Tome Premier: L'individuation Somatique*; Presses Universitaires de France: Paris, France, 1960.
149. Newman, H.H.; Freeman, F.N.; Holzinger, K.J. *Twins: A Study of Heredity and Environment*; University of Chicago Press: Chicago, IL, USA, 1937.
150. Brimicombe, M. Notes and Queries: Mirror-Image Twins. *The Guardian*, 20 April 2000; 16.
151. Baker-Cohen, K.F. Visceral and vascular transposition in fishes, and a comparison with similar anomalies in man. *Am. J. Anat.* **1961**, *109*, 37–55. [CrossRef] [PubMed]
152. Orlebeke, J.F.; Knol, D.L.; Koopmans, J.R.; Boomsma, D.I.; Bleker, O.P. Left-handedness in twins: Genes or environment? *Cortex* **1996**, *32*, 479–490. [CrossRef]
153. James, W.H.; Orlebeke, J.F. Determinants of handedness in twins. *Laterality Asymmetries Body Brain Cogn.* **2002**, *7*, 301–307. [CrossRef]
154. Nagylaki, T.; Levy, J. 'The sound of one paw clapping' isn't sound. *Behav. Genet.* **1973**, *3*, 279–292. [CrossRef] [PubMed]
155. Wilson, P.T.; Jones, H.E. Left-handedness in twins. *Genetics* **1932**, *17*, 560–571. [CrossRef] [PubMed]

156. McManus, I.C. Handedness and birth stress. *Psychol. Med.* **1981**, *11*, 485–496. [CrossRef]
157. Bakan, P.; Dibb, G.; Reed, P. Handedness and birth stress. *Neuropsychologia* **1973**, *11*, 363–366. [CrossRef]
158. Searleman, A.; Porac, C.; Coren, S. Relationship between birth order, birth stress, and lateral preferences: A critical review. *Psychol. Bull* **1989**, *105*, 397–408. [CrossRef]
159. Marlow, N.; Roberts, B.L.; Cooke, R.W.I. Laterality and prematurity. *Arch. Dis. Child* **1989**, *64*, 1713–1716. [CrossRef]
160. Morley, R.; Leeson-Payne, C.; Lister, G.; Lucas, A. *Hand Preference at 7.5 to 8 Years in Children Weighing under 1850 Grams at Birth*; Iowa State University: Ames, IA, USA, 1993.
161. Heikkilä, K.; Vuoksimaa, E.; Saari-Kemppainen, A.; Kaprio, J.; Rose, R.J.; Haukka, J.; Pitkäniemi, J.; Iivanainen, M. Higher prevalence of left-handedness in twins? Not after controlling birth time confounders. *Twin Res. Hum. Genet.* **2015**, *18*, 526–532. [CrossRef] [PubMed]
162. McGlone, J. Sex differences in human brain organization: A critical survey. *Behav. Brain Sci.* **1980**, *3*, 215–227. [CrossRef]
163. McGlone, J.; Kertesz, A. Sex differences in cerebral processing of visuo-spatial tasks. *Cortex* **1973**, *9*, 313–320. [CrossRef]
164. Seddon, B.; McManus, I.C. The Incidence of Left-Handedness: A Meta-Analysis [1993]. *PsyArXiv* **2019**. [CrossRef]
165. McManus, I.C. The inheritance of left-handedness. In *Biological Asymmetry and Handedness (Ciba Foundation Symposium 162)*; Bock, G.R., Marsh, J., Eds.; Wiley: Chichester, UK, 1991; pp. 251–281.
166. Papadatou-Pastou, M.; Martin, M.; Munafò, M.R.; Jones, G.V. Sex differences in left-handedness: A meta-analysis of 144 studies. *Psychol. Bull.* **2008**, *134*, 677–699. [CrossRef] [PubMed]
167. Jones, G.V.; Martin, M. Language Dominance, Handedness, and Sex: Recessive X-Linkage Theory and Test. *Cortex* **2010**, *46*, 781–786. [CrossRef] [PubMed]
168. McManus, I.C. Precisely wrong? The problems with the Jones and Martin genetic model of sex differences in handedness and language lateralisation. *Cortex* **2010**, *46*, 700–702. [CrossRef] [PubMed]
169. Schmitz, J.; Zheng, M.; Lui, K.F.; McBride, C.; Ho, C.S.-H.; Paracchini, S. Quantitative multidimensional phenotypes improve genetic analysis of laterality traits. *Transl. Psychiatry* **2022**. [CrossRef]
170. Shaywitz, B.A.; Shaywitz, S.E.; Pugh, K.R.; Constable, R.T.; Skudlarski, P.; Fulbright, R.K.; Bronen, R.A.; Fletcher, J.M.; Shankweiler, D.P.; Katz, L.; et al. Sex differences in the functional organization of the brain for language. *Nature* **1995**, *373*, 607–609. [CrossRef] [PubMed]
171. Sommer, I.E.C.; Kahn, R.S. Sex differences in handedness and language lateralisation. In *Language Lateralisation and Psychosis*; Sommer, I., Khan, R.S., Eds.; Cambridge University Press: Cambridge, UK, 2009; pp. 101–117.
172. Levy, J. Lateral specialization of the human brain: Behavioral manifestations and possible evolutionary basis. In *The Biology of Behavior*; Kiger, J.A., Ed.; Oregon State University Press: Corvallis, OR, USA, 1972.
173. Levy, J.; Reid, M. Variations in cerebral organization as a function of handedness, hand posture in writing, and sex. *J. Exp. Psychol.* **1978**, *107*, 119–144. [CrossRef]
174. Hirnstein, M.; Hugdahl, K.; Hausmann, M. Cognitive sex differences and hemispheric asymmetry: A critical review of 40 years of research. *Laterality Asymmetries Body Brain Cogn.* **2019**, *24*, 204–252. [CrossRef] [PubMed]
175. Marr, D. *Vision: A Computational Investigation into the Human Representation and Processing of Visual Information*; W. H. Freeman: New York, NY, USA, 1982.
176. Rolls, E.T. David Marr's Vision: Floreat computational neuroscience. *Brain* **2011**, *134*, 913–916. [CrossRef]
177. Chen, C.-F.; Foley, J.; Tang, P.-C.; Li, A.; Jiang, T.X.; Wu, P.; Widelitz, R.B.; Chuong, C.M. Development, regeneration, and evolution of feathers. *Annu. Rev. Anim. Biosci.* **2015**, *3*, 169–195. [CrossRef] [PubMed]
178. Miller, G. *The Mating Mind: How Sexual Choice Shaped the Evolution of Human Nature*; William Heinemann: London, UK, 2000.
179. McManus, I.C. *Right Hand, Left Hand: The Origins of Asymmetry in Brains, Bodies, Atoms and Cultures*; Weidenfeld and Nicolson: London, UK; Harvard University Press: Cambridge, MA, USA, 2002.
180. Kincaid, P. *The Rule of the Road: An International Guide to History and Practice*; Greenwood Press: New York, NY, USA, 1986.
181. Weyl, H. *Symmetry*; Princeton University Press: Princeton, NJ, USA, 1952.
182. Mitchell, K.J.; Levin, S. The genetics of brain wiring: From molecule to mind. *PLoS Biol.* **2007**, *5*, e113. [CrossRef]
183. Ferrell, J.E., Jr. Bistability, bifurcations, and Waddington's epigenetic landscape. *Curr. Biol.* **2012**, *22*, R458–R466. [CrossRef]
184. Rabajante, J.F.; Babierra, A.L. Branching and oscillations in the epigenetic landscape of cell-fate determination. *Prog. Biophys. Mol. Biol.* **2015**, *117*, 240–249. [CrossRef]
185. Fruchart, M.; Hanai, R.; Littlewood, P.B.; Vitelli, V. Non-reciprocal phase transitions. *Nature* **2021**, *592*, 363–369. [CrossRef]
186. McRobie, A. *The Seduction of Curves: The Lines of Beauty that Connect Mathematics, Art, and the Nude*; Princeton University Press: Princeton, NJ, USA, 2017.
187. McManus, I.C.; Murray, B.; Doyle, K.; Baron-Cohen, S. Handedness in childhood autism shows a dissociation of skill and preference. *Cortex* **1992**, *28*, 373–381. [CrossRef]
188. Cornish, K.M.; McManus, I.C. Hand preference and hand skill in children with autism. *J. Autism Dev. Disord.* **1996**, *26*, 597–609. [CrossRef]
189. Tapley, S.M.; Bryden, M.P. A group test for the assessment of performance between the hands. *Neuropsychologia* **1985**, *23*, 215–221. [CrossRef]
190. McManus, I.C.; Van Horn, J.D.; Bryden, P.J. The Tapley and Bryden test of performance differences between the hands: The original data, newer data, and the relation to pegboard and other tasks. *Laterality* **2016**, *21*, 396. [CrossRef] [PubMed]

191. Annett, M. The growth of manual preference and speed. *Br. J. Psychol.* **1970**, *61*, 545–558. [CrossRef] [PubMed]
192. Ocklenburg, S.; Gunturkun, O. *The Lateralized Brain: The Neuroscience and Evolution of Hemispheric Asymmetries*; Academic Press: Cambridge, MA, USA, 2017.
193. McManus, I.C.; Buckens, G.; Harris, N.; Flint, A.; Ng, H.L.A.; Vovou, F. Faking handedness: Individual differences in ability to fake handedness, social cognitions of the handedness of others, and a forensic application using Bayes' theorem. *Laterality* **2018**, *23*, 67–100. [CrossRef] [PubMed]
194. McManus, I.C. Right- and left-hand skill: Failure of the right shift model. *Br. J. Psychol.* **1985**, *76*, 1–16. [CrossRef]
195. Steele, J. Handedness in past human populations: Skeletal markers. *Laterality* **2000**, *5*, 193–220. [CrossRef]
196. Satz, P. Pathological left-handedness: An explanatory model. *Cortex* **1972**, *8*, 121–135. [CrossRef]
197. McManus, I.C. The interpretation of laterality. *Cortex* **1983**, *19*, 187–214. [CrossRef]
198. Everitt, B.S.; Hand, D.J. *Finite Mixture Distributions*; Chapman and Hall: London, UK, 1991.
199. McLachlan, G.J.; Lee, S.X.; Rathnayake, S.I. Finite mixture models. *Annu. Rev. Stat. Appl.* **2019**, *6*, 355–378. [CrossRef]
200. Stasinopoulos, M.D.; Rigby, R.A.; Heller, G.Z.; Voudouris, V.; De Bastiani, F. *Flexible Regression and Smoothing: Using GAMLSS in R*; CRC Press: Boca Raton, FL, USA, 2017.
201. Boker, S.M.; Neale, M.C.; Maes, H.H.; Wilde, M.J.; Spiegel, M.; Brick, T.R.; Estabrook, R.; Bates, T.C.; Mehta, P.; von Oertzen, T. OpenMx user guide. *Release* **2012**, *1*, 1919.
202. Neale, M.C.; Hunter, M.D.; Pritikin, J.N.; Zahery, M.; Brick, T.R.; Kirkpatrick, R.M.; Estabrook, R.; Bates, T.C.; Maes, H.H.; Boker, S.M. OpenMx 2.0: Extended structural equation and statistical modeling. *Psychometrika* **2016**, *81*, 535–549. [CrossRef]
203. McManus, I.C. The power of a procedure for detecting mixture distributions in laterality data. *Cortex* **1984**, *20*, 421–426. [CrossRef]
204. Gordon, H.W. Cerebral organization in bilinguals: I. Lateralization. *Brain Lang.* **1980**, *9*, 255–268. [CrossRef]
205. Sulpizio, S.; Del Maschio, N.; Fedeli, D.; Abutalebi, J. Bilingual language processing: A meta-analysis of functional neuroimaging studies. *Neurosci. Biobehav. Rev.* **2020**, *108*, 834–853. [CrossRef] [PubMed]
206. Corballis, M.C. Some difficulties with difficulty. *Psychol. Rep.* **1968**, *22*, 15–22. [CrossRef] [PubMed]
207. R Core Team. *R: A Language and Environment for Statistical Computing*; R Foundation for Statistical Computing: Vienna, Austria, 2020. Available online: https://www.R-project.org/ (accessed on 15 February 2022).
208. Brumer, I.; De Vita, E.; Ashmore, J.; Jarosz, J.; Borri, M. Implementation of clinically relevant and robust fMRI-based language lateralization: Choosing the laterality index calculation method. *PLoS ONE* **2020**, *15*, e0230129. [CrossRef]
209. Wilke, M.; Lidzba, K. LI-tool: A new toolbox to assess lateralization in functional MR-data. *J. Neurosci. Methods* **2007**, *163*, 128–136. [CrossRef] [PubMed]
210. Seghier, M.L.; Kherif, F.; Josse, G.; Price, C.J. Regional and hemispheric determinants of language laterality: Implications for preoperative fMRI. *Hum. Brain Mapp.* **2011**, *32*, 1602–1614. [CrossRef]
211. Deppe, M.; Knecht, S.; Henningsen, H.; Ringelstein, E.-B. AVERAGE: A Windows program for automated analysis of event related cerebral blood flow. *J. Neurosci. Methods* **1997**, *75*, 147–154. [CrossRef]
212. Bishop, D.V.M.; Bates, T.C. Heritability of language laterality assessed by functional transcranial Doppler ultrasound: A twin study. *Wellcome Open Res.* **2020**, *4*, 161. [CrossRef]
213. Chlebus, P.; Mikl, M.; Brázdil, M.; Pažourková, M.; Krupa, P.; Rektor, I. fMRI evaluation of hemispheric language dominance using various methods of laterality index calculation. *Exp. Brain Res.* **2007**, *179*, 365–374. [CrossRef]
214. Wegrzyn, M.; Mertens, M.; Bien, C.G.; Woermann, F.G.; Labudda, K. Quantifying the confidence in fMRI-based language lateralisation through laterality index deconstruction. *Front. Neurol.* **2019**, *10*, 655. [CrossRef] [PubMed]
215. Kurthen, M.; Helmstaedter, C.; Linke, D.; Hufnagel, A.; Elger, C.; Schramm, J. Quantitative and qualitative evaluation of patterns of cerebral language dominance: An amobarbital study. *Brain Lang.* **1994**, *46*, 536–564. [CrossRef] [PubMed]
216. Neville, A.C. *Animal Asymmetry*; Edward Arnold: London, UK, 1976.
217. Palmer, A.R. Animal asymmetry. *Curr. Biol.* **2009**, *19*, R473–R477. [CrossRef] [PubMed]
218. Watson, T. An account of some cases of transposition observed in the human body. *Lond. Med. Gaz.* **1836**, *18*, 393–403.
219. Torgersen, J. Situs inversus, asymmetry and twinning. *Am. J. Hum. Genet.* **1950**, *2*, 361–370.
220. Siewert, A.K. Ueber einen Fall von Bronchiectasie bei einem Patienten mit Situs inversus viscerum. *Berl. Klin. Wochenschr.* **1904**, *41*, 139–141.
221. Kartagener, M. Zur Pathogenese der Bronchiektasien. I. Mitteilung: Bronchiektasien bei Situs viscerum inversus. *Beiträge Zur Klin. Und Erforsch. Der Tuberk. Und Der Lungenkrankenheiten* **1933**, *83*, 489–501. [CrossRef]
222. Afzelius, B.A.; Mossberg, B. Immotile cilia. *Thorax* **1980**, *35*, 401. [CrossRef]
223. Berdon, W.E.; McManus, I.C.; Afzelius, B.A. More on Kartagener's syndrome and the contributions of Afzelius and A.K. Siewert. *Pediatr. Radiol.* **2004**, *34*, 585–586. [CrossRef]
224. Berdon, W.E.; Willi, U. Situs inversus, bronchiectasis, and sinusitis and its relation to immotile cilia: History of the diseases and their discoverers-Manes Kartagener and Bjorn Afzelius. *Pediatr. Radiol.* **2004**, *34*, 38–42. [CrossRef]
225. Fuchs, J.L.; Schwark, H.D. Neuronal primary cilia: A review. *Cell Biol. Int.* **2004**, *28*, 111–118. [CrossRef] [PubMed]
226. Tihen, J.A.; Charles, D.R.; Sippel, T.D. Inherited visceral inversion in mice. *J. Hered.* **1948**, *39*, 29–31. [CrossRef] [PubMed]
227. Hummel, K.P.; Chapman, D.B. Visceral inversion and associated anomalies in the mouse. *J. Hered.* **1959**, *50*, 9–13. [CrossRef]
228. Afzelius, B.A. Inheritance of randomness. *Med. Hypotheses* **1996**, *47*, 23–26. [CrossRef]

229. Supp, D.M.; Witte, D.P.; Potter, S.S.; Brueckner, M. Mutation of an axonemal dynein affects left-right asymmetry in inversus viscerum mice. *Nature* **1997**, *389*, 963–966. [CrossRef]
230. Levin, M.; Johnson, R.L.; Stern, C.D.; Kuehn, M.; Tabin, C. A molecular pathway determining left-right asymmetry in chick embryogenesis. *Cell* **1995**, *82*, 803–814. [CrossRef]
231. Nonaka, S.; Tanaka, Y.; Okada, Y.; Takeda, S.; Harada, A.; Kanai, Y.; Kido, M.; Hirokawa, N. Randomisation of left-right asymmetry due to loss of nodal cilia generating leftward flow of extraembryonic fluid in mice lacking KIF3B motor protein. *Cell* **1998**, *95*, 829–837. [CrossRef]
232. Okada, Y.; Nonaka, S.; Tanaka, Y.; Saijoh, Y.; Hamada, H.; Hirokawa, N. Abnormal nodal flow precedes situs inversus in *iv* and *inv* mice. *Mol. Cell* **1999**, *4*, 459–468. [CrossRef]
233. Okada, Y.; Takeda, S.; Tanaka, Y.; Izpisua Belmonte, J.C.; Hirokawa, N. Mechanism of nodal flow: A conserved symmetry breaking event in left-right axis determination. *Cell* **2005**, *121*, 633–644. [CrossRef]
234. Levin, M.; Palmer, A.R. Left–right patterning from the inside out: Widespread evidence for intracellular control. *Bioessays* **2007**, *29*, 271–287. [CrossRef]
235. Levin, M.; Klar, A.J.; Ramsdell, A.F. Introduction to provocative questions in left–right asymmetry. *Philos. Trans. R. Soc. B Biol. Sci.* **2016**, *371*, 20150399. [CrossRef] [PubMed]
236. Meeks, M.; Walne, A.; Spiden, S.; Simpson, H.; Mussaffi, G.H.; Hamam, H.D.; Fehaid, E.L.; Cheehab, M.; Al Dabbagh, M.; Polak, C.S.; et al. A locus for primary ciliary dyskinesia maps to chromosome 19q. *J. Med. Genet.* **2000**, *37*, 241–244. [CrossRef] [PubMed]
237. Yano, J.; Valentine, M.S.; Van Houten, J.L. Novel insights into the development and function of cilia using the advantages of the Paramecium cell and its many cilia. *Cells* **2015**, *4*, 297–314. [CrossRef] [PubMed]
238. Brown, J.M.; Witman, G.B. Cilia and diseases. *Bioscience* **2014**, *64*, 1126–1137. [CrossRef] [PubMed]
239. Lovera, M.; Lüders, J. The ciliary impact of nonciliary gene mutations. *Trends Cell Biol.* **2021**, *31*, 876–887. [CrossRef] [PubMed]
240. Reiter, J.F.; Leroux, M.R. Genes and molecular pathways underpinning ciliopathies. *Nat. Rev. Mol. Cell Biol.* **2017**, *18*, 533–547. [CrossRef]
241. Legendre, M.; Zaragosi, L.-E.; Mitchison, H.M. Motile cilia and airway disease. *Semin. Cell Dev. Biol.* **2021**, *110*, 19–33. [CrossRef]
242. Lucas, J.S.; Davis, S.D.; Omran, H.; Shoemark, A. Primary ciliary dyskinesia in the genomics age. *Lancet Respir. Med.* **2020**, *8*, 202–216. [CrossRef]
243. McManus, I.C.; Martin, N.; Stubbings, G.F.; Chung, E.M.K.; Mitchison, H.M. Handedness and *situs inversus* in primary ciliary dyskinesia. *Proc. R. Soc. Lond. Ser. B* **2004**, *271*, 2579–2582. [CrossRef]
244. McManus, I.C.; Mitchison, H.M.; Chung, E.M.K.; Stubbings, G.F.; Martin, N. Primary Ciliary Dyskinesia (Siewert's /Kartagener's Syndrome): Respiratory symptoms and psycho-social impact. *BMC Pulm. Med.* **2003**, *3*, 4. Available online: www.biomedcentral.com/1471-2466/1473/1474/abstract (accessed on 15 February 2022). [CrossRef]
245. Whalley, S.; McManus, I.C. Living with primary ciliary dyskinesia: A prospective qualitative study of knowledge sharing, symptom concealment, embarrassment, mistrust, and stigma. *BMC Pulm. Med.* **2006**, *6*, 25. Available online: www.biomedcentral.com/1471-2466/1476/1425 (accessed on 15 February 2022). [CrossRef] [PubMed]
246. McManus, I.C.; Stubbings, G.F.; Martin, N. Stigmatisation, physical illness and mental health in primary ciliary dyskinesia. *J. Health Psychol.* **2006**, *11*, 467–482. [CrossRef] [PubMed]
247. Ryzhkova, A.I.; Sazonova, M.A.; Sinyov, V.V.; Galitsyna, E.V.; Chicheva, M.M.; Melnichenko, A.A.; Grechko, A.V.; Postnov, A.Y.; Orekhov, A.N.; Shkurat, T.P. Mitochondrial diseases caused by mtDNA mutations: A mini-review. *Ther. Clin. Risk Manag.* **2018**, *14*, 1933. [CrossRef] [PubMed]
248. Duman, D.; Tekin, M. Autosomal recessive nonsyndromic deafness genes: A review. *Front. Biosci.* **2012**, *17*, 2213. [CrossRef]
249. Nance, W.E. The genetics of deafness. *Mental Retard. Dev. Disabil. Res. Rev.* **2003**, *9*, 109–119. [CrossRef]
250. Mahner, M.; Kary, M. What exactly are genomes, genotypes and phenotypes? And what about phenomes? *J. Theor. Biol.* **1997**, *186*, 55–63. [CrossRef]
251. John, B.; Lewis, K.R. Chromosome Variability and Geographic Distribution in Insects: Chromosome rather than gene variations provide the key to differences among populations. *Science* **1966**, *152*, 711–721. [CrossRef]
252. Gottesman, I.I.; Shields, J. Genetic theorizing and schizophrenia. *Br. J. Psychiatry* **1973**, *122*, 15–30. [CrossRef]
253. Gottesman, I.I.; Gould, T.D. The endophenotype concept in psychiatry: Etymology and strategic intentions. *Am. J. Psychiatry* **2003**, *160*, 636–645. [CrossRef]
254. Glahn, D.C.; Knowles, E.E.; McKay, D.R.; Sprooten, E.; Raventós, H.; Blangero, J.; Gottesman, I.I.; Almasy, L. Arguments for the sake of endophenotypes: Examining common misconceptions about the use of endophenotypes in psychiatric genetics. *Am. J. Med. Genet. Pt. B Neuropsychiatr. Genet.* **2014**, *165*, 122–130. [CrossRef]
255. Kendler, K.S.; Neale, M.C. Endophenotype: A conceptual analysis. *Mol. Psychiatry* **2010**, *15*, 789–797. [CrossRef] [PubMed]
256. Walters, J.T.R.; Owen, M.J. Endophenotypes in psychiatric genetics. *Mol. Psychiatry* **2007**, *12*, 886–890. [CrossRef] [PubMed]
257. Lenzenweger, M.F. Endophenotype, intermediate phenotype, biomarker: Definitions, concept comparisons, clarifications. *Depress. Anxiety* **2013**, *30*, 185–189. [CrossRef] [PubMed]
258. Forero, D.A.; López-León, S.; González-Giraldo, Y.; Dries, D.R.; Pereira-Morales, A.J.; Jiménez, K.M.; Franco-Restrepo, J.E. APOE gene and neuropsychiatric disorders and endophenotypes: A comprehensive review. *Am. J. Med. Genet. Pt. B Neuropsychiatr. Genet.* **2018**, *177*, 126–142. [CrossRef]

259. Allen, A.J.; Griss, M.E.; Folley, B.S.; Hawkins, K.A.; Pearlson, G.D. Endophenotypes in schizophrenia: A selective review. *Schizophr. Res.* **2009**, *109*, 24–37. [CrossRef]
260. Hirnstein, M.; Hugdahl, K. Excess of non-right-handedness in schizophrenia: Meta-analysis of gender effects and potential biases in handedness assessment. *Br. J. Psychiatry* **2014**, *205*, 260–267. [CrossRef]
261. Ribolsi, M.; Lisi, G.; Di Lorenzo, G.; Koch, G.; Oliveri, M.; Magni, V.; Pezzarossa, B.; Saya, A.; Rociola, G.; Rubino, I.A. Perceptual pseudoneglect in schizophrenia: Candidate endophenotype and the role of the right parietal cortex. *Schizophr. Bull.* **2013**, *39*, 601–607. [CrossRef]
262. Rodriguez, A.; Kaakinen, M.; Moilanen, I.; Taanila, A.; McGough, J.J.; Loo, S.; Järvelin, M.-R. Mixed-handedness is linked to mental health problems in children and adolescents. *Pediatrics* **2010**, *125*, e340–e348. [CrossRef]
263. Bishop, D.V. Cerebral asymmetry and language development: Cause, correlate, or consequence? *Science* **2013**, *340*, 6138. [CrossRef]
264. Tenconi, E.; Santonastaso, P.; Degortes, D.; Bosello, R.; Titton, F.; Mapelli, D.; Favaro, A. Set-shifting abilities, central coherence, and handedness in anorexia nervosa patients, their unaffected siblings and healthy controls: Exploring putative endophenotypes. *World J. Biol. Psychiatry* **2010**, *11*, 813–823. [CrossRef]
265. Preslar, J.; Kushner, H.I.; Marino, L.; Pearce, B. Autism, lateralisation, and handedness: A review of the literature and meta-analysis. *Laterality* **2014**, *19*, 64–95. [CrossRef] [PubMed]
266. Ocklenburg, S.; Garland, A.; Ströckens, F.; Uber Reinert, A. Investigating the neural architecture of handedness. *Front. Psychol.* **2015**, *6*, 148. [CrossRef] [PubMed]
267. Guadalupe, T.; Willems, R.M.; Zwiers, M.P.; Arias Vasquez, A.; Hoogman, M.; Hagoort, P.; Fernandez, G.; Buitelaar, J.; Franke, B.; Fisher, S.E. Differences in cerebral cortical anatomy of left-and right-handers. *Front. Psychol.* **2014**, *5*, 261. [CrossRef]
268. Davison, A.; Thomas, P. Internet 'shellebrity'reflects on origin of rare mirror-image snails. *Biol. Lett.* **2020**, *16*, 20200110. [CrossRef] [PubMed]
269. Davison, A.; McDowell, G.S.; Holden, J.M.; Johnson, H.F.; Koutsovoulos, G.D.; Liu, M.M.; Hulpiau, P.; Van Roy, F.; Wade, C.M.; Banerjee, R. Formin is associated with left-right asymmetry in the pond snail and the frog. *Curr. Biol.* **2016**, *26*, 654–660. [CrossRef]
270. Annett, M. Handedness in the children of two left-handed parents. *Br. J. Psychol.* **1974**, *65*, 129–131. [CrossRef]
271. Annett, M. Family handedness in three generations predicted by the right shift theory. *Ann. Hum. Genet.* **1979**, *42*, 479–491. [CrossRef]
272. Klar, A.J.S. A Single Locus, *RGHT*, Species Preference for Hand Utilization in Humans. *Cold Spring Harb. Symp. Quant. Biol.* **1996**, *61*, 59–65.
273. McManus, I.C. Lifeline. *Lancet* **1997**, *349*, 1922.
274. Medland, S.E.; Lindgren, M.; Magi, R.; Neale, B.M.; Albrecht, E.; Esko, T.; Dvans, D.M.; Hottenga, J.J.; Ikram, M.A.; Mangino, M.; et al. *Meta-Analysis of GWAS for Handedness: Results from the ENGAGE Consortium*; Abstract, 2009 meeting of the American Society for Human Genetics; American Society for Human Genetics: Rockville, MD, USA, 2009.
275. Eriksson, N.; Macpherson, J.M.; Tung, J.Y.; Hon, L.S.; Naughton, B.; Saxonov, S.; Avey, L.; Wojcicki, A.; Pe'er, I.; Mountain, J. Web-based, participant-driven studies yield novel genetic associations for common traits. *PLoS Genet.* **2010**, *6*, e1000993. [CrossRef]
276. De Kovel, C.G.; Francks, C. The molecular genetics of hand preference revisited. *Sci. Rep.* **2019**, *9*, 5986. [CrossRef] [PubMed]
277. Harden, K.P. *The Genetic Lottery: Why DNA Matters for Social Equality*; Princeton University Press: Princeton, NJ, USA, 2021.
278. Cho, S. Is handedness exogenously determined? Counterevidence from South Korea. *Econ. Hum. Biol.* **2021**, *2021*, 101072. [CrossRef] [PubMed]
279. Sha, Z.; Pepe, A.; Schijven, D.; Castillo, A.C.; Roe, J.M.; Westerhausen, R.; Joliot, M.; Fisher, S.E.; Crivello, F.; Francks, C. Left-handedness and its genetic influences are associated with structural asymmetries mapped across the cerebral cortex in 31,864 individuals. *bioRxiv* **2021**. [CrossRef]
280. Gilbert, A.N.; Wysocki, C.J. Hand preference and age in the United States. *Neuropsychologia* **1992**, *30*, 601–608. [CrossRef]
281. Grimshaw, J.M.; Russell, I.T. Effect of clinical guidelines of medical practice: A sysematic review of rigorous evaluations. *Lancet* **1993**, *342*, 1317–1322. [CrossRef]
282. Grimshaw, G.M.; Bryden, M.P. Are there meaningful handedness subtypes? *J. Clin. Exp. Neuropsychol.* **1995**, *1*, 367.
283. McManus, I.C.; Porac, C.; Bryden, M.P.; Boucher, R. Eye dominance, writing hand and throwing hand. *Laterality* **1999**, *4*, 173–192. [CrossRef]
284. Plomin, R. *Blueprint: How DNA Makes Us Who We Are*; Mit Press: Cambridge, MA, USA, 2019.
285. Ma, Y.; Zhou, X. Genetic prediction of complex traits with polygenic scores: A statistical review. *Trends Genet.* **2021**, *37*, 995–1011. [CrossRef]
286. Timpson, N.J.; Greenwood, C.M.; Soranzo, N.; Lawson, D.J.; Richards, J.B. Genetic architecture: The shape of the genetic contribution to human traits and disease. *Nat. Rev. Genet.* **2018**, *19*, 110–124. [CrossRef]
287. Pocock, G.; Richards, C.D.; Richards, D.A. *Human Physiology*; Oxford University Press: Oxford, UK, 2013.
288. Steck, A.K.; Winter, W.E. Review on monogenic diabetes. *Curr. Opin. Endocrinol. Diabetes Obes.* **2011**, *18*, 252–258. [CrossRef]
289. Mitchell, K.J. What is complex about complex disorders? *Genome Biol.* **2012**, *13*, 238. [CrossRef] [PubMed]
290. Mak, C.M.; Lee, H.-C.H.; Chan, A.Y.-W.; Lam, C.-W. Inborn errors of metabolism and expanded newborn screening: Review and update. *Crit. Rev. Clin. Lab. Sci.* **2013**, *50*, 142–162. [CrossRef] [PubMed]
291. Garrod, A.E. *Inborn Errors of Metabolism*; H. Frowde and Hodder & Stoughton: London, UK, 1909.

292. Blau, N.; Shen, N.; Carducci, C. Molecular genetics and diagnosis of phenylketonuria: State of the art. *Expert Rev. Mol. Diagn.* **2014**, *14*, 655–671. [CrossRef] [PubMed]
293. Wade, K.; Lam, B.; Melvin, A.; Pan, W.; Corbin, L.; Hughes, D.; Rainbow, K.; Chen, J.; Duckett, K.; Liu, X. Frequency and phenotypic expression in childhood and early adulthood of loss of function mutations in the Melanocortin 4 Receptor in a UK birth cohort. *Nat. Med.* **2021**, *27*, 1088–1096. [CrossRef] [PubMed]
294. Wood, A.R.; Esko, T.; Yang, J.; Vedantam, S.; Pers, T.H.; Gustafsson, S.; Chu, A.Y.; Estrada, K.; Kutalik, Z.; Amin, N. Defining the role of common variation in the genomic and biological architecture of adult human height. *Nat. Genet.* **2014**, *46*, 1173–1186. [CrossRef]
295. Marouli, E.; Graff, M.; Medina-Gomez, C.; Lo, K.S.; Wood, A.R.; Kjaer, T.R.; Fine, R.S.; Lu, Y.; Schurmann, C.; Highland, H.M. Rare and low-frequency coding variants alter human adult height. *Nature* **2017**, *542*, 186–190. [CrossRef]
296. Boyle, E.A.; Li, Y.I.; Pritchard, J.K. The omnigenic model: Response from the authors. *J. Psychiatry Brain Sci.* **2017**, *2*, 1–3.
297. Boyle, E.A.; Li, Y.I.; Pritchard, J.K. An expanded view of complex traits: From polygenic to omnigenic. *Cell* **2017**, *169*, 1177–1186. [CrossRef]
298. Backman, J.D.; Li, A.H.; Marcketta, A.; Sun, D.; Mbatchou, J.; Kessler, M.D.; Benner, C.; Liu, D.; Locke, A.E.; Balasubramanian, S. Exome sequencing and analysis of 454,787 UK Biobank participants. *Nature* **2021**, *599*, 628–634. [CrossRef]
299. Cutting, G.R. Cystic fibrosis genetics: From molecular understanding to clinical application. *Nat. Rev. Genet.* **2015**, *16*, 45–56. [CrossRef]
300. Simcoe, M.; Valdes, A.; Liu, F.; Furlotte, N.A.; Evans, D.M.; Hemani, G.; Ring, S.M.; Smith, G.D.; Duffy, D.L.; Zhu, G. Genome-wide association study in almost 195,000 individuals identifies 50 previously unidentified genetic loci for eye color. *Sci. Adv.* **2021**, *7*, eabd1239. [CrossRef] [PubMed]
301. Li, Z.; Chen, J.; Yu, H.; He, L.; Xu, Y.; Zhang, D.; Yi, Q.; Li, C.; Li, X.; Shen, J. Genome-wide association analysis identifies 30 new susceptibility loci for schizophrenia. *Nat. Genet.* **2017**, *49*, 1576–1583. [CrossRef] [PubMed]
302. Häberling, I.S.; Badzakova-Trajkov, G.; Corballis, M.C. Asymmetries of the arcuate fasciculus in monozygotic twins: Genetic and nongenetic influences. *PLoS ONE* **2013**, *8*, e52315. [CrossRef] [PubMed]
303. Badzakova-Trajkov, G.; Häberling, I.S.; Corballis, M.C. Cerebral asymmetries in monozygotic twins: An fMRI study. *Neuropsychologia* **2010**, *48*, 3086–3093. [CrossRef] [PubMed]
304. Lux, S.; Keller, S.; Mackay, C.; Ebers, G.; Marshall, J.C.; Cherkas, L.; Rezaie, R.; Roberts, N.; Fink, G.R.; Gurd, J.M. Crossed cerebral lateralization for verbal and visuo-spatial function in a pair of handedness discordant monozygotic twins: MRI and fMRI brain imaging. *J. Anat.* **2008**, *212*, 235–248. [CrossRef] [PubMed]
305. Gurd, J.; Cowell, P.; Lux, S.; Rezai, R.; Cherkas, L.; Ebers, G. fMRI and corpus callosum relationships in monozygotic twins discordant for handedness. *Brain Struct. Funct.* **2013**, *218*, 491–509. [CrossRef]
306. Gurd, J.M.; Cowell, P.E. Discordant cerebral lateralisation for verbal fluency is not an artefact of attention: Evidence from MzHd twins. *Brain Struct. Funct.* **2015**, *220*, 59–69. [CrossRef]
307. Cowell, P.; Gurd, J. Handedness and the corpus callosum: A review and further analyses of discordant twins. *Neuroscience* **2018**, *388*, 57–68. [CrossRef]
308. Sommer, I.E.C.; Ramsey, N.F.; Bouma, A.; Kahn, R.S. Cerebral mirror-imaging in a monozygotic twin. *Lancet* **1999**, *354*, 1445–1446. [CrossRef]
309. Sommer, I.E.C.; Kahn, R.S. Language lateralization and handedness in twins: An argument against a genetic basis. In *Language lateralization and psychosis*; Sommer, I.E.C., Kahn, R.S., Eds.; Cambridge University Press: Cambridge, UK, 2009; pp. 87–100.
310. Ooki, S. An overview of human handedness in twins. *Front. Psychol.* **2014**, *5*, 10. [CrossRef]
311. Bycroft, C.; Freeman, C.; Petkova, D.; Band, G.; Elliott, L.T.; Sharp, K.; Motyer, A.; Vukcevic, D.; Delaneau, O.; O'Connell, J. The UK Biobank resource with deep phenotyping and genomic data. *Nature* **2018**, *562*, 203–209. [CrossRef] [PubMed]
312. McManus, I.C.; Mascie-Taylor, C.G.N. Hand-clasping and arm-folding: A review and a genetic model. *Ann. Hum. Biol.* **1979**, *6*, 527–558. [CrossRef] [PubMed]
313. Simpson, E.H. The interpretation of interaction in contingency tables. *J. R. Statist. Soc. Ser. B* **1951**, *13*, 241. [CrossRef]
314. Hernán, M.A.; Clayton, D.; Keiding, N. The Simpson's paradox unraveled. *Int. J. Epidemiol.* **2011**, *40*, 780–785. [CrossRef] [PubMed]
315. Liu, H.; Stufflebeam, S.M.; Sepulcre, J.; Hedden, T.; Buckner, R.L. Evidence from intrinsic activity that asymmetry of the human brain is controlled by multiple factors. *Proc. Natl. Acad. Sci. USA* **2009**, *106*, 20499–20503. [CrossRef] [PubMed]
316. Vigneau, M.; Beaucousin, V.; Hervé, P.-Y.; Duffau, H.; Crivello, F.; Houde, O.; Mazoyer, B.; Tzourio-Mazoyer, N. Meta-analyzing left hemisphere language areas: Phonology, semantics, and sentence processing. *Neuroimage* **2006**, *30*, 1414–1432. [CrossRef] [PubMed]
317. Vigneau, M.; Beaucousin, V.; Hervé, P.-Y.; Jobard, G.; Petit, L.; Crivello, F.; Mellet, E.; Zago, L.; Mazoyer, B.; Tzourio-Mazoyer, N. What is right-hemisphere contribution to phonological, lexico-semantic, and sentence processing?: Insights from a meta-analysis. *Neuroimage* **2011**, *54*, 577–593. [CrossRef] [PubMed]
318. Palmer, A.R. Symmetry breaking and the evolution of development. *Science* **2004**, *306*, 828–833. [CrossRef]
319. Kaul, D.; Papadatou-Pastou, M.; Learmonth, G. A meta-analysis of line bisection and landmark task performance in children. *Neuropsychol. Rev.* **2021**. [CrossRef]

320. Learmonth, G.; Papadatou-Pastou, M. A meta-analysis of line bisection and landmark task performance in older adults. *Neuropsychol. Rev.* **2021**. [CrossRef]
321. Häberling, I.S.; Corballis, P.M.; Corballis, M.C. Language, gesture, and handedness: Evidence for independent lateralized networks. *Cortex* **2016**, *82*, 72–85. [CrossRef]
322. Bernal, B.; Ardila, A. Bilateral representation of language: A critical review and analysis of some unusual cases. *J. Neuroling.* **2014**, *28*, 63–80. [CrossRef]
323. Rasmussen, T.; Milner, B. The role of early left-brain injury in determining lateralization of cerebral speech functions. *Ann. N. Y. Acad. Sci* **1977**, *299*, 355–369. [CrossRef] [PubMed]
324. Loring, D.W.; Meador, K.J.; Lee, G.P.; Murro, A.M.; Smith, J.R.; Flanigin, H.F.; Gallagher, B.B.; King, D.W. Cerebral language lateralization: Evidence from intracarotid amobarbital testing. *Neuropsychologia* **1990**, *28*, 831–838. [CrossRef]
325. Risse, G.L.; Gates, J.R.; Fangman, M.C. A reconsideration of bilateral language representation based on the intracarotid amobarbital procedure. *Brain Cogn.* **1997**, *33*, 118–132. [CrossRef] [PubMed]
326. Möddel, G.; Lineweaver, T.; Schuele, S.U.; Reinholz, J.; Loddenkemper, T. Atypical language lateralization in epilepsy patients. *Epilepsia* **2009**, *50*, 1505–1516. [CrossRef]
327. Bauer, P.R.; Reitsma, J.B.; Houweling, B.M.; Ferrier, C.H.; Ramsey, N.F. Can fMRI safely replace the Wada test for preoperative assessment of language lateralisation? A meta-analysis and systematic review. *J. Neurol. Neurosurg. Psychiatry* **2014**, *85*, 581–588. [CrossRef]
328. Janecek, J.K.; Swanson, S.J.; Sabsevitz, D.S.; Hammeke, T.A.; Raghavan, M.; Mueller, W.; Binder, J.R. Naming outcome prediction in patients with discordant Wada and fMRI language lateralization. *Epilepsy Behav.* **2013**, *27*, 399–403. [CrossRef]
329. Gardner, W.J. Injection of procaine in to the brain to locate the speech area in left-handed persons. *Arch. Neurol. Psychiatry* **1941**, *46*, 1035–1038. [CrossRef]
330. Wada, J. and T. Rasmussen. Intracarotid injection of sodium amytal for the lateralization of cerebral speech dominance. Experimental and clinical observations. *J. Neurosurg.* **1960**, *17*, 266–282. [CrossRef]
331. Snyder, P.J.; Harris, L.J. The intracarotid amobarbital procedure: An historical perspective. *Brain Cogn.* **1997**, *33*, 18–32. [CrossRef] [PubMed]
332. Branch, C.; Milner, B.; Rasmussen, T. Intracarotid sodium amytal for the lateralization of cerebral speech dominance: Observations in 123 patients. *J. Neurosurg.* **1964**, *21*, 399–405. [CrossRef] [PubMed]
333. Dym, R.J.; Burns, J.; Freeman, K.; Lipton, M.L. Is functional MR imaging assessment of hemispheric language dominance as good as the Wada test?: A meta-analysis. *Radiology* **2011**, *261*, 446–455. [CrossRef] [PubMed]
334. Mariën, P.; Paghera, B.; De Deyn, P.P.; Vignolo, L.A. Adult crossed aphasia in dextrals revisited. *Cortex* **2004**, *40*, 41–74. [CrossRef]
335. Subirana, A. Handedness and cerebral dominance. In *Handbook of Clinical Neurology*; Vinken, P.J., Bruyn, G.W., Eds.; North-Holland: Amsterdam, The Netherlands, 1969; Volume 4, pp. 248–272.
336. Hécaen, H.; Albert, M.L. *Human Neuropsychology*; John Wiley: New York, NY, USA, 1978.
337. Gloning, I.; Gloning, K.; Haub, G.; Quatember, R. Comparison of verbal behavior in right-handed and non right-handed patients with anatomically verified lesion of one hemisphere. *Cortex* **1969**, *5*, 43–52. [CrossRef]
338. Subirana, A. The prognosis in aphasia in relation to cerebral dominance and handedness. *Brain* **1958**, *81*, 415–425. [CrossRef]
339. Goodglass, H., Geschwind, N. Language disorders (aphasia). In *Handbook of Perception, Language and Speech*; Carterette, E.C., Friedman, M.P., Eds.; Academic Press: New York, NY, USA, 1976; Volume 7.
340. Hicks, R.E.; Kinsbourne, M. Human handedness. *Asymmetr. Funct. Brain* **1978**, 523–549.
341. Zangwill, O.L. *Cerebral Dominance and Its Relation to Psychological Function*; Oliver and Boyd: Edinburgh, UK, 1960.
342. von Monakow, C. Über den gegenwaertigen stand der frage nach der lokalisation im grosshirn. In *Ergebnisse der Physiologie, Band I, Abt II*; Asher, L., Spiro, K., Eds.; Bergman: Wiesbaden, Germany, 1902.
343. Riese, W. Cerebral dominance: Its origin, its history and its nature. *Clio Med.* **1970**, *5*, 319–326.
344. Finger, S.; Koehler, P.J.; Jagella, C. The Monakow concept of diaschisis: Origins and perspectives. *Arch. Neurol.* **2004**, *61*, 283–288. [CrossRef]
345. Carrera, E.; Tononi, G. Diaschisis: Past, present, future. *Brain* **2014**, *137*, 2408–2422. [CrossRef]
346. Engelhardt, E.; Gomes, M.D.M. *Shock, Diaschisis and Von Monakow*; SciELO Brasil: Sao Paulo, Brasil, 2013.
347. Basso, A.; Gardelli, M.; Grassi, M.P.; Mariotti, M. The role of the right hemisphere in recovery from aphasia. Two case studies. *Cortex* **1989**, *25*, 555–566. [CrossRef]
348. Anglade, C.; Thiel, A.; Ansaldo, A.I. The complementary role of the cerebral hemispheres in recovery from aphasia after stroke: A critical review of literature. *Brain Injury* **2014**, *28*, 138–145. [CrossRef] [PubMed]
349. Cadilhac, D.A.; Kim, J.; Lannin, N.A.; Kapral, M.K.; Schwamm, L.H.; Dennis, M.S.; Norrving, B.; Meretoja, A. National stroke registries for monitoring and improving the quality of hospital care: A systematic review. *Int. J. Stroke* **2016**, *11*, 28–40. [CrossRef] [PubMed]
350. Croquelois, A.; Bogousslavsky, J. Stroke aphasia: 1500 consecutive cases. *Cerebrovasc. Dis.* **2011**, *31*, 392–399. [CrossRef] [PubMed]
351. Hoffmann, M. Higher cortical function deficits after stroke: An analysis of 1000 patients from a dedicated cognitive stroke registry. *Neurorehab. Neural Repair* **2001**, *15*, 113–127. [CrossRef]
352. Strauss, E.; Satz, P.; Wada, J. An examination of the crowding hypothesis in epileptic patients who have undergone the carotid amytal test. *Neuropsychologia* **1990**, *28*, 1221–1227. [CrossRef]

353. Danguecan, A.N.; Smith, M.L. Re-examining the crowding hypothesis in pediatric epilepsy. *Epilepsy Behav.* **2019**, *94*, 281–287. [CrossRef]
354. Lust, J.; Geuze, R.; Groothuis, A.; Bouma, A. Functional cerebral lateralization and dual-task efficiency—Testing the function of human brain lateralization using fTCD. *Behav. Brain Res.* **2011**, *217*, 293–301. [CrossRef]
355. Luders, E.; Thompson, P.M.; Toga, A.W. The development of the corpus callosum in the healthy human brain. *J. Neurosci.* **2010**, *30*, 10985–10990. [CrossRef]
356. Tzourio-Mazoyer, N. Intra-and inter-hemispheric connectivity supporting hemispheric specialization. In *Micro-, Meso- and Macro-Connectomics of the Brain*; Kennedy, H., Van Essen, D.C., Christen, Y., Eds.; Spinger International: Cham, Switzerland, 2016; pp. 129–146.
357. Schüz, A.; Braitenberg, V. The human cortical white matter: Quantitative aspects of cortico-cortical long-range connectivity. In *Cortical Areas: Unity and Diversity, Conceptual Advances in Brain Research*; Schüz, A., Miller, R., Eds.; Taylor and Francis: London, UK, 2002; pp. 377–386.
358. Geschwind, N. Disconnexion syndromes in animals and man: Part I. *Brain* **1965**, *88*, 237–294. [CrossRef]
359. Geschwind, N. Disconnexion syndromes in animals and man: Part II. *Brain* **1965**, *88*, 585. [CrossRef] [PubMed]
360. Mesulam, M.-M. Fifty years of disconnexion syndromes and the Geschwind legacy. *Brain* **2015**, *138*, 2791–2799. [CrossRef] [PubMed]
361. Catani, M.; Fytche, D.H. The rises and falls of disconnection syndromes. *Brain* **2005**, *128*, 2224–2239. [CrossRef] [PubMed]
362. Papadatou-Pastou, M. Using meta-analysis for the study of handedness: Four examples. In Proceedings of the Workshop—Handedness and Language, Wassenaar, The Netherlands, 27–30 November 2013.
363. Aggleton, J.P.; Kentridge, R.W.; Good, J.M.M. Handedness and musical ability: A study of professional orchestral players, composers, and choir members. *Psychol. Music* **1994**, *22*, 148–156. [CrossRef]
364. Peters, M. Are there more lefthanders among the mathematically gifted? Benbow's evidence is weak. *Can. J. Psychol.* **1992**.
365. Benbow, C.P. Possible biological correlates of precocious mathematical reasoning ability. *Trends Neurosci.* **1987**, *10*, 17–20. [CrossRef]
366. Winner, E.; von Karolyi, C.; Malinsky, D.; French, L.; Seliger, S.; Ross, E.; Weber, E. Dyslexia and visual-spatial talents: Compensation vs deficit model. *Brain Lang.* **2001**, *76*, 81–110. [CrossRef] [PubMed]
367. Von Karolyi, C.; Winner, E. Dyslexia and visual spatial talents: Are they connected? In *Students with Both Gifts and Learning Disabilities*; Newman, T.M., Sternberg, R.J., Eds.; Springer: New York, NY, USA, 2004; pp. 95–118.
368. Chamberlain, R.; Brunswick, N.; Siev, S.; McManus, I.C. Meta-analytic findings reveal lower means but higher variances in visuospatial ability in dyslexia. *Br. J. Psychol.* **2018**, *109*, 897–916. [CrossRef]
369. Gotestam, K.O. Lefthandedness among students of architecture and music. *Percept. Motor Skills* **1990**, *70*, 1323–1327. [CrossRef]
370. Peterson, J.M.; Lansky, L.M. Left-handedness among architects: Partial replication and some new data. *Percept. Motor Skills* **1977**, *45*, 1216–1218. [CrossRef]
371. O'Kusky, J.; Strauss, E.; Kosaka, B.; Wada, J.; Li, D.; Druhan, M.; Petrie, J. The corpus callosum is larger with right-hemisphere cerebral speech dominance. *Ann. Neurol.* **1988**, *24*, 379–383. [CrossRef]
372. Beaton, A.A. The relation of planum temporale asymmetry and morphology of the corpus callosum to handedness, gender and dyslexia: A review of the evidence. *Brain Lang.* **1997**, *60*, 255–322. [CrossRef] [PubMed]
373. Driesen, N.R.; Raz, N. The influence of sex, age, and handedness on corpus callosum morphology: A meta-analysis. *Psychobiology* **1995**, *23*, 240–247. [CrossRef]
374. Luders, E.; Cherbuin, N.; Thompson, P.M.; Gutman, B.; Anstey, K.J.; Sachdev, P.; Toga, A.W. When more is less: Associations between corpus callosum size and handedness lateralization. *Neuroimage* **2010**, *52*, 43–49. [CrossRef] [PubMed]
375. Bernard, F.; Zemmoura, I.; Ter Minassian, A.; Lemée, J.-M.; Menei, P. Anatomical variability of the arcuate fasciculus: A systematical review. *Surg. Radiol. Anat.* **2019**, *41*, 889–900. [CrossRef] [PubMed]
376. Catani, M.; Mesulam, M. The arcuate fasciculus and the disconnection theme in language and aphasia: History and current state. *Cortex* **2008**, *44*, 953–961. [CrossRef]
377. Toga, A.W.; Clark, K.A.; Thompson, P.M.; Shattuck, D.W.; Van Horn, J.D. Mapping the human connectome. *Neurosurgery* **2012**, *71*, 1–5. [CrossRef]
378. Van Essen, D.C.; Smith, S.M.; Barch, D.M.; Behrens, T.E.; Yacoub, E.; Ugurbil, K.; Consortium, W.-M.H. The WU-Minn human connectome project: An overview. *Neuroimage* **2013**, *80*, 62–79. [CrossRef]
379. Jahanshad, N.; Lee, A.D.; Barysheva, M.; McMahon, K.L.; de Zubicaray, G.I.; Martin, N.G.; Wright, M.J.; Toga, A.W.; Thompson, P.M. Genetic influences on brain asymmetry: A DTI study of 374 twins and siblings. *Neuroimage* **2010**, *52*, 455–469. [CrossRef]
380. Willems, R.M.; Van der Haegen, L.; Fisher, S.E.; Francks, C. On the other hand: Including left-handers in cognitive neuroscience and neurogenetics. *Nat. Rev. Neurosci.* **2014**, *15*, 193–201. [CrossRef]
381. Bailey, L.M.; McMillan, L.E.; Newman, A.J. A sinister subject: Quantifying handedness-based recruitment biases in current neuroimaging research. *Eur. J. Neurosci.* **2020**, *51*, 1642–1656. [CrossRef] [PubMed]
382. Westerhausen, R.; Huster, R.J.; Kreuder, F.; Wittling, W.; Schweiger, E. Corticospinal tract asymmetries at the level of the internal capsule: Is there an association with handedness? *Neuroimage* **2007**, *37*, 379–386. [CrossRef]

383. Propper, R.E.; O'Donnell, L.J.; Whalen, S.; Tie, Y.; Norton, I.H.; Suarez, R.O.; Zollei, L.; Radmanesh, A.; Golby, A.J. A combined fMRI and DTI examination of functional language lateralization and arcuate fasciculus structure: Effects of degree versus direction of hand preference. *Brain Cogn.* **2010**, *73*, 85–92. [CrossRef] [PubMed]
384. McManus, I.C.; Moore, J.; Freegard, M.; Rawles, R. Science in the making: Right Hand, Left Hand: III: The incidence of left-handedness. *Laterality* **2010**, *15*, 186–208. [CrossRef] [PubMed]
385. Coren, S.; Porac, C. Fifty centuries of right-handedness: The historical record. *Science* **1977**, *198*, 631–632. [CrossRef] [PubMed]
386. Faurie, C.; Raymond, M. Handedness frequency over more than 10,000 years. *Proc. R. Soc. Lond. Ser. B* **2004**, *271*, S43–S45. [CrossRef] [PubMed]
387. Frayer, D.W.; Lozano, M.; Bermúdez de Castro, J.M.; Carbonell, E.; Arsuaga, J.L.; Radovcic, J.; Fiore, I.; Bondioli, L. More than 500,000 years of right-handedness in Europe. *Laterality* **2012**, *17*, 51–69. [CrossRef]
388. Pitts, M.; Roberts, M. *Fairweather Eden: Life in Britain Half a Million Years Ago as Revealed by the Excavations at Boxgrove*; Century: London, UK, 1997.
389. Toth, N. Archaeological evidence for preferential right handedness in the lower and middle Pleistocene and its possible implications. *J. Hum. Evol.* **1985**, *14*, 607–614. [CrossRef]
390. McManus, I.C. Handedness, cerebral lateralization and the evolution of language. In *The Descent of Mind: Psychological Perspectives on Hominid Evolution*; Corballis, M.C., Lea, S.E.G., Eds.; Oxford University Press: Oxford, UK, 1999; pp. 194–217.
391. Hopkins, W.D. Comparative and familial analysis of handedness in great apes. *Psychol. Bull.* **2006**, *132*, 538–559. [CrossRef]
392. Hori, M. Frequency-dependent natural selection in the handedness of scale-eating cichlid fish. *Science* **1993**, *260*, 216–219. [CrossRef]
393. Faurie, C.; Raymond, M. Handedness, homicide and negative frequency-dependent selection. *Proc. R. Soc.* **2004**, *272*, 25–28. [CrossRef]
394. Groothuis, T.G.G.; McManus, I.C.; Schaafsma, S.M.; Geuze, R.H. The fighting hypothesis in combat: How well does the highting hypothesis explain human left-handed minorities? *Ann. N. Y. Acad. Sci.* **2013**, *1288*, 100–109. [CrossRef] [PubMed]
395. Richardson, T.; Gilman, R.T. Left-handedness is associated with greater fighting success in humans. *Sci. Rep.* **2019**, *9*, 15402. [CrossRef] [PubMed]
396. Schaafsma, S.M.; Geuze, R.H.; Riedstra, B.; Schiefenhövel, W.; Bouma, A.; Groothuis, T.G. Handedness in a nonindustrial society challenges the fighting hypothesis as an evolutionary explanation for left-handedness. *Evol. Hum. Behav.* **2012**, *33*, 94–99. [CrossRef]
397. Graw, J.; Brackmann, H.-H.; Oldenburg, J.; Schneppenheim, R.; Spannagl, M.; Schwaab, R. Haemophilia A: From mutation analysis to new therapies. *Nat. Rev. Genet.* **2005**, *6*, 488–501. [CrossRef] [PubMed]
398. Johnson, T.; Barton, N. Theoretical models of selection and mutation on quantitative traits. *Philos. Trans. R. Soc. B Biol. Sci.* **2005**, *360*, 1411–1425. [CrossRef]
399. Barton, N.H. Pleiotropic models of quantitative variation. *Genetics* **1990**, *124*, 773–782. [CrossRef]
400. Genomes Project Consortium. A map of human genome variation from population scale sequencing. *Nature* **2010**, *467*, 1061. [CrossRef]
401. Curtis, D. Analysis of rare coding variants in 200,000 exome-sequenced subjects reveals novel genetic risk factors for type 2 diabetes. *Diabetes/Metab. Res. Rev.* **2021**, *38*, e3482. [CrossRef]
402. Hill, W.D.; Arslan, R.C.; Xia, C.; Luciano, M.; Amador, C.; Navarro, P.; Hayward, C.; Nagy, R.; Porteous, D.J.; McIntosh, A.M.; et al. Genomic analysis of family data reveals additional genetic effects on intelligence and personality. *Mol. Psychiatry* **2018**, *23*, 2347–2362. [CrossRef]
403. Penke, L.; Denissen, J.J.; Miller, G.F. The evolutionary genetics of personality. *Eur. J. Pers.* **2007**, *21*, 549–587. [CrossRef]
404. Wells, K.L.; Hadad, Y.; Ben-Avraham, D.; Hillel, J.; Cahaner, A.; Headon, D.J. Genome-wide SNP scan of pooled DNA reveals nonsense mutation in FGF20 in the scaleless line of featherless chickens. *BMC Genom.* **2012**, *13*, 257. [CrossRef] [PubMed]
405. Abbott, U.; Asmundson, V. Scaleless, an inherited ectodermal defect in the domestic fowl. *J. Hered.* **1957**, *48*, 63–70. [CrossRef]
406. Terrill, R.S.; Shultz, A.J. On the Multifunctionality of Feathers and the Evolution of BIrds. *EcoEvoRxiv* **2021**. [CrossRef]
407. Prum, R.O.; Berv, J.S.; Dornburg, A.; Field, D.J.; Townsend, J.P.; Lemmon, E.M.; Lemmon, A.R. A comprehensive phylogeny of birds (Aves) using targeted next-generation DNA sequencing. *Nature* **2015**, *526*, 569–573. [CrossRef] [PubMed]
408. Zhou, Z. The origin and early evolution of birds: Discoveries, disputes, and perspectives from fossil evidence. *Naturwissenschaften* **2004**, *91*, 455–471. [CrossRef] [PubMed]
409. Rogers, L.J.; Vallortigara, G.; Andrew, R.J. *Divided Brains: The Biology and Behaviour of Brain Asymmetries*; Cambridge University Press: Cambridge, UK, 2013.
410. Babcock, L.E. Trilobite malformations and the fossil record of behavioral asymmetry. *J. Paleontol.* **1993**, *67*, 217–229. [CrossRef]
411. Palmer, A.R. From symmetry to asymmetry: Phylogenetic patterns of asymmetry variation in animals and their evolutionary significance. *Proc. Natl. Acad. Sci. USA* **1996**, *93*, 14279–14286. [CrossRef]
412. Vallortigara, G.; Rogers, L.J.; Bisazza, A. Possible evolutionary origins of cognitive brain lateralization. *Brain Res. Rev.* **1999**, *30*, 164–175. [CrossRef]
413. Giljov, A.; Karenina, K.; Ingram, J.; Malschichev, Y. Parallel Emergence of True Handedness in the Evolution of Marsupials and Placentals. *Curr. Biol.* **2015**, *25*, 1878–1884. [CrossRef]

414. Meguerditchian, A.; Vauclair, J.; Hopkins, W.D. On the origins of human handedness and language: A comparative review of hand preferences for bimanual coordinated actions and gestural communication in nonhuman primates. *Dev. Psychobiol.* **2013**, *55*, 637–650. [CrossRef]
415. Caspar, K.R.; Pallasdies, F.; Mader, L.; Sartorelli, H.; Begall, S. The evolution and biological correlates of hand preferences in anthropoid primates. *bioRxiv* **2021**. [CrossRef]
416. Waddington, C.H. Canalization of development and the inheritance of acquired characters. *Nature* **1942**, *150*, 563–565. [CrossRef]
417. Siegal, M.L.; Bergman, A. Waddington's canalization revisited: Developmental stability and evolution. *Proc. Natl. Acad. Sci. USA* **2002**, *99*, 10528–10532. [CrossRef] [PubMed]
418. Kilner, P.J.; Yang, G.-Z.; Wilkes, A.J.; Mohiaddin, R.H.; Firmin, D.N.; Yacoub, M.H. Asymmetric redirection of flow through the heart. *Nature* **2000**, *404*, 759–761. [CrossRef] [PubMed]
419. Concha, M.L.; Wilson, S.W. Asymmetry in the epithalamus of vertebrates. *J. Anat.* **2001**, *199*, 63–84. [CrossRef] [PubMed]
420. Gutiérrez-Ibáñez, C.; Reddon, A.R.; Kreuzer, M.B.; Wylie, D.R.; Hurd, P.L. Variation in asymmetry of the habenular nucleus correlates with behavioural asymmetry in a cichlid fish. *Behav. Brain Res.* **2011**, *221*, 189–196. [CrossRef] [PubMed]
421. Ahumada-Galleguillos, P.; Lemus, C.G.; Díaz, E.; Osorio-Reich, M.; Härtel, S.; Concha, M.L. Directional asymmetry in the volume of the human habenula. *Brain Struct. Funct.* **2017**, *222*, 1087–1092. [CrossRef]
422. Lawson, R.P.; Drevets, W.C.; Roiser, J.P. Defining the habenula in human neuroimaging studies. *Neuroimage* **2013**, *64*, 722–727. [CrossRef]
423. Namboodiri, V.M.K.; Rodriguez-Romaguera, J.; Stuber, G.D. The habenula. *Curr. Biol.* **2016**, *26*, R873–R877. [CrossRef]
424. Lawson, R.; Nord, C.; Seymour, B.; Thomas, D.; Dayan, P.; Pilling, S.; Roiser, J. Disrupted habenula function in major depression. *Mol. Psychiatry* **2017**, *22*, 202–208. [CrossRef] [PubMed]

Review

Recent Advances in Handedness Genetics

Silvia Paracchini 

School of Medicine, University of St Andrews, North Haugh, St Andrews KY16 9TF, UK; sp58@st-andrews.ac.uk

Abstract: Around the world, about 10% people prefer using their left-hand. What leads to this fixed proportion across populations and what determines left versus right preference at an individual level is far from being established. Genetic studies are a tool to answer these questions. Analysis in twins and family show that about 25% of handedness variance is due to genetics. In spite of very large cohorts, only a small fraction of this genetic component can be pinpoint to specific genes. Some of the genetic associations identified so far provide evidence for shared biology contributing to both handedness and cerebral asymmetries. In addition, they demonstrate that handedness is a highly polygenic trait. Typically, handedness is measured as the preferred hand for writing. This is a very convenient measure, especially to reach large sample sizes, but quantitative measures might capture different handedness dimensions and be better suited for genetic analyses. This paper reviews the latest findings from molecular genetic studies as well as the implications of using different ways of assessing handedness.

Keywords: handedness; neurodevelopment; GWAS; heritability; quantitative trait; polygenic scores

Citation: Paracchini, S. Recent Advances in Handedness Genetics. *Symmetry* **2021**, *13*, 1792. https://doi.org/10.3390/sym13101792

Academic Editor: David A. Becker

Received: 29 August 2021
Accepted: 23 September 2021
Published: 26 September 2021

1. Is Handedness a Genetic Trait?

Before embarking in the search for the genetics factors of any traits, the most fundamental question is whether a trait is influenced by a genetic component. More specifically, we are asking whether the variability observed in the population for that particular trait is influenced by genetics. Several observations confirm that a genetic component contributes to handedness.

Most people can readily say whether they are left- or right-handed, especially for highly skilled task like writing with a pen. Of course, it is possible to learn writing with the nonpreferred hand, but at least at the beginning, that would feel an un-natural act. Based on these observations we can state that it is in our nature to have a preferred hand for writing, which is the right hand for most people. Probably because of the minority status, left-handers were stigmatised throughout history and cultures. In fact, it is quite common to hear of left-handers being forced to use their right hand for some tasks such as writing. Instead, the reverse, i.e., forcing right-handers to use the left hand, is very unlikely. This phenomenon is well-documented in the UK Biobank data showing that the prevalence of left-handedness increases in younger participants probably because of stronger stigma in older generations [1]. A recent meta-analysis, confirmed the same historical trend and that left-handedness tends to converge to around 10% across populations [2]. Although left-handedness prevalence tends to remain low in some countries, e.g., China, this seems to be a cultural effect. For example, a 1980s survey reported that less than 1% Chinese students are left-handed [3]. A more recent study, reported a higher prevalence of left-handedness (6%) in a Chinese cohort living in Hong Kong, possibly as a result of the westernisation of this region [4]. Therefore, left-handedness, not only is a minority status, but appears to be fixed to a constant frequency. This fixed prevalence is suggestive of evolutionary forces maintaining the ratio of 1 left- to 9 right-handers possibly through genetic mechanisms. This scenario could be explained by a frequency dependent selection process where the minor trait has an advantage but only until it remains at low prevalence in the population [5–7]. A cost is clearly associated to left-handedness, else we would observe it at a 50% frequency in populations.

The link between handedness and language is another indicator of the biological nature of handedness. Although both hemispheres are engaged during language tasks, for the majority of people, hemispheric dominance resides in the left side. Right hemisphere dominance for language is rare and observed preferentially in left-handers [8] (see also Corballis [9] and Vingerhoets et al. [10] in this issue for details on functional and anatomical brain asymmetries). This link is weak but suggests some common pathways control the establishment of brain asymmetries and contribute to both language and handedness.

Family and twin studies provide the most compelling case in support of genetics, indicating that at least one quarter of handedness variance is determined by genetic factors [11]. However, the remaining 75% are not necessarily influenced by nongenetic or environmental factors. For example, intrinsic variability linked to developmental processes might explain a large of portion of the remaining variability across people, as argued by Kevin Mitchell [12] and, more recently, by Chris McManus [13], as part of a discussion setting the vision for the future of laterality research [14,15]. The idea is that, while the general developmental stages of an individual are directed by biological processes tightly regulated by our genes, a random component allows fluctuations from the general plan. Such fluctuations, which are actually part of the biological plan itself, could play an important role in determining an individual's characteristics, including handedness. Under this view, the actual genetic component of handedness is expected to be much higher than what (~25%) predicted by twin studies. McManus' prediction is that very few environment factors are likely to play any significant role in establishing the direction of hand preference.

2. How to Measure Handedness

Having established a firm and conspicuous genetic component underlying a trait, the next question is how best to measure the phenotype for genetic analyses. Handedness appears to be a very straightforward phenotype, with most people being able to define themselves as either left- or right-handed, typically on the basis of their preferred hand for writing. The majority of individuals also carry out other tasks preferentially with the same hand they used for writing, either the left or the right one. However, a minority, defined as mixed-handed, prefer using different hands for different activities (e.g., writing with the right but throwing a ball with the left hand) and a small group, or ambidextrous, has no clear hand preference between the two hands. In total, mixed-handed and ambidextrous individuals are about 9% of the population, a group almost as big as the left-handers [2]. Tools like the Edinburgh Handedness Inventory (EHI) and Annett's questionnaire [16,17], which record the preferred hands for a dozen of activities or items, allows identifying these individuals. While most people will answer "right" or "left" for all items, there will be a group without consistent preferences. Instead, one task alone, e.g., hand preference for writing, cannot identify this group. A third possibility is to measure handedness as relative hand skills by assessing how better one individual performs with one hand versus the other. This approach leads to continuous measures, or laterality quotients (LQ; Figure 1). The pegboard task, which records the time taken to move pegs in a row of holes [17], is a commonly used tool to derive such scores. A key question is whether different handedness measures, which require significant time or resources to be collected in large cohorts, offer any specific advantage for genetic studies over the self-reported measure of hand preference for writing [18].

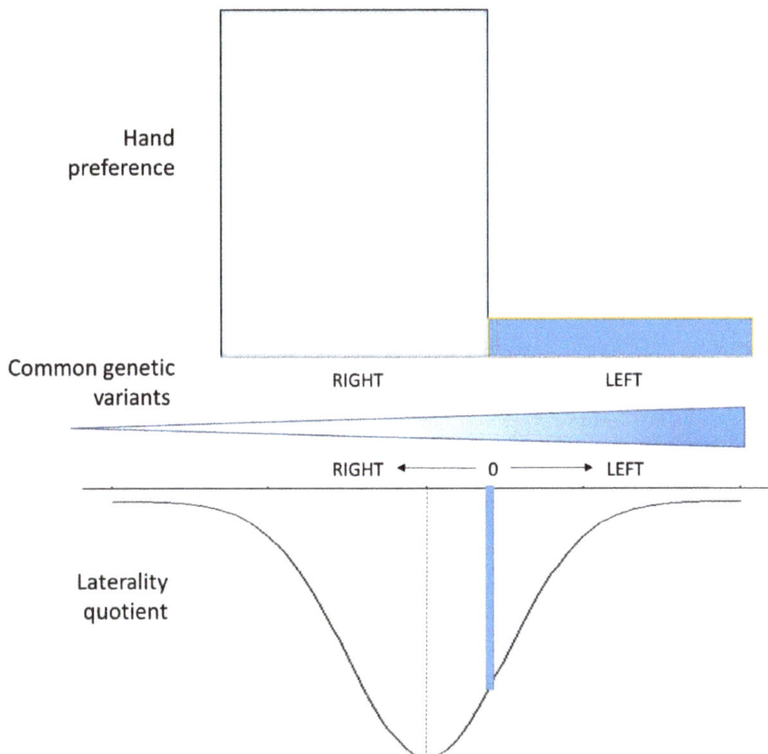

Figure 1. Polygenic model for handedness. Handedness is typically measured as hand preference (top bars). But it can also be measured along a continuum using laterality quotient (bottom curve, shown upside-down for convenience). Hand preference leads to two categories: right and left distributed in a 9:1 ratio. Laterality quotients (LQ) assess relative hand skills and how much an individual is lateralised in addition to a left v right direction. A value of zero (0, aligned along blue line) indicates equal ability with both hands and separates left and right handers for that particular skill. Different LQ identify a general left v right component, but do not correlate perfectly with hand preference. The chance of being left-handed increases with accumulation of multiple genetic variants represented by gradient in middle of figure. Poor correlation across handedness measures suggest that different pools of common variants contribute to different measures, although we expect some overlaps. For example, different genetic studies reported associations with different set of genes with cytoskeletal functions. Although hand preference is a convenient measure, which can be easily collected in very large cohorts, LQ might be better suited to identify genetics underlying handedness.

A starting point to address this issue is to examine how different measures correlate with each other and whether different types of assessment can be used interchangeably. This can be done in population-based cohorts that include thousands of participants characterised with multiple handedness measures. For example, participants of the Avon Longitudinal Study of Parents and Children (ALSPAC) cohort [19] were assessed with hand preference at different time points, handedness questionnaires, and multiple motor tasks, which can be used to derive LQ (N up to 8000). Thanks to these data, we showed that different laterality measures are poorly correlated with one another [20] and, beyond capturing a general left/right component, they tap in different laterality dimensions. Moderate correlation (0.42) for handedness measures derived from the EHI and the pegboard task was also reported in 205 twin pairs recruited in Hong Kong. Remarkably, both measures presented similar heritability estimates at around 20%, but the low correlation suggest that most likely they are underpinned by different genetic factors [4].

The UK Biobank [21] with its multilayers of biological, genetic, clinical, and behavioural data for half million study participants is revolutionising different lines of research, including in the field of laterality, as we will discuss later (see also Corballis in this issue [9]). However, in this cohort the handedness assessment is limited to the self-reported preference for writing which leads to three categories: "right hand", "left hand" and "both hands". These data present some peculiarities. The rate of ambidexterity is reported at ~1.5% in the population, which is higher than expected. In fact, individuals who can write equally well with both hands are extremely rare. The heritability estimates for left-handedness and ambidexterity are also puzzling. At behavioural level, the identification of siblings (N = 20,277 pairs) and other relatives (N = 49,788 pairs) led to a heritability (h^2) estimate of 12% for left-handedness [22]. This sample size was too small to derive a reliable estimate for ambidexterity with the same approach. Instead, genome-wide molecular data showed that common genetic variants, or single nucleotide polymorphisms (SNPs), capture up to 6% and 15% of the heritability (${h_g}^2$) for left-handedness and ambidexterity, respectively. The higher ${h_g}^2$ observed for ambidexterity is a potentially exciting finding, but it also revealed some bias. Thanks to the molecular data, it is possible to test the genetic correlation across different traits. This analysis revealed that ambidexterity did not show genetic correlation with left-handedness or other neurodevelopmental, neural, and psychiatric traits as expected and as observed for left-handedness [22]. Instead, ambidexterity showed genetic correlation with the risk of being injured. A possibility is that the ambidexterity measure (reported as "being able to write with both hands") may be a consequence of injuries that force the use of the nonpreferred hand. However, to further complicate the situation a very recent genetic study for dyslexia—a neurodevelopmental phenotype—found a significant genetic correlation with the UK Biobank ambidexterity measure (but not with left-handedness) in over a million individuals derived from the 23andMe database [23]. The results of these studies show that even extremely large samples are not sufficient to disentangle patterns of associations between various binary traits and emphasise the importance of the quality of the phenotypes used for genetic studies.

More detailed handedness assessments were possible in smaller cohorts. The ALSPAC cohort is exceptional for the richness of measures collected over three decades. In addition to multiple measures, it also offers the advantage of a family structure design with both behavioural and genetic data available in parents and children. Taking advantage of these features, we were able to derive and compare heritability estimates across different handedness measures [24]. We found that ${h_g}^2$ for left-handedness, as a categorical measure, was 8%—a slightly higher but comparable figure to the 6% observed in the UK Biobank. When transforming the categorical phenotypes in quantitative scores, the ${h_g}^2$ for measures of hand preference derived from the summary of EHI scores was 21% (this transformation was achieved by regressing out effects of sex, age, and the first two principal components for ancestry as described by Verhoef et al. [25]). This 21% figure is similar to the estimates derived from behavioural analysis in twins. The same analysis for individual items showed variability across activities. For example, the highest heritability estimate (42%) was observed for the "hand used to cut" item. The same item presented the higher heritability (32%) also in a Japanese study [26]. These data both support the benefit of using quantitative phenotypic transformations and indicate that individual, rather than summary or composite measures, might be a more powerful tool to capture genetic factors underlying handedness. The same conclusion was reached by a study in a Mexican sample [27] and support the idea that different handedness measures capture different components of handedness. A key feature of quantitative phenotypes is that they distinguish both the poorly and the extremely lateralised individuals in addition to the left- and right-handedness direction (Figure 1). Therefore, if genetic factors contribute to the degree of lateralization rather than the direction of handedness, such effect will not be captured by individual measures of hand preference. When we applied the same phenotypic transformations to laterality measures other than handedness (i.e., foot and eye preference), we found that the heritability of foot preference was 28%—higher than what observed for handedness—but was negligible for eye preference. This finding, in agreement with behavioural data from a previous study [28], suggests that other laterality measures beyond hand preference have the potential to lead to significant genetic discoveries.

However, the ideal scenario of having multiple handedness measures in large cohorts remain challenging. The "preferred hand for writing" is a very convenient way to assess handedness because it can simply be a box ticking as part of larger studies. For example, large cohorts primarily designed for studying the genetics of various diseases, can then be reanalyse for the genetics of hand preference at no extra cost. However, a hand preference measure might not capture genetic factors contributing to different aspects of handedness. In particular, it is not an ideal way to identify mixed-handed or poorly lateralised individuals. Currently, large scale collection on LQ measures is challenging and requires significant resource. For example, it would require dedicated personnel to collect, record and entry the data. As we move towards increasing digitalisation of every aspect of our lives, online platforms could offer a viable route for the collection of laterality data in large populations.

3. There Is No Handedness Gene

Having firmly established that a large generic genetic component underlies handedness, the next question is what specific genes determine whether one individual is right- or left-handed? Currently, our best answer is "many and not one in particular". It is now universally accepted that there is no single gene or single allele determining left handedness, contrary to what predicted by the theories proposed in the 80s [29,30]. However, it is important to recognise the values of these theories, which fitted with the data available at the time and played key roles in driving research efforts in the field. Thanks to recent advances in genomic technologies, we are now appreciating the highly polygenic nature of neurodevelopmental traits and of common human diseases. In fact, such complexity is much higher than it was anticipated only 10 years ago [31]. Genomic technologies include both genotyping of known variants, or SNPs, used for genome-wide association studies (GWAS) and resequencing to discover rare or *de novo* variants. Both technologies have generated data for hundreds of thousands of individuals and, considering that, as we just discussed, hand preference for writing is a very straightforward variable to be collected, a large amount of data for gene mapping are available. The fact that no specific individual gene with large effect was identified yet unequivocally excludes the possibility that there could be one single genetic factor causing left-handedness. Instead, an increasing number of genes with small effect sizes are being found in parallel with the analysis of increasingly large cohorts, confirming the polygenic nature of this trait (Table 1).

4. Resequencing the Genome

The ability to resequence the human genome at affordable cost allows the identification of genetic variants characterised with large effect sizes on the phenotype. Such effect tends to be disruptive and reduce the fitness of an individual. The large effect size is usually due to changes in the coding sequence that in turn alter the function of the corresponding protein. Whole exome sequencing (WES) technology targets specifically the coding regions (~2% of the entire genome) and offers an efficient way to discover such variants. Whole genome sequencing (WGS) instead covers the entire genome. Compared to that of WES, WGS is more expensive and poses the challenge of handling and interpreting a very large amount of data. It is estimated that each of us carry thousands of rare variants and up to 100 *de novo* mutations that are not inherited from our parents. Dissecting which ones might be relevant for the phenotype under investigation is not straightforward. WES has the advantages of being cheaper in terms of data generation and more straightforward in terms of data handling cost compared to that of WGS. The downside is that WES cannot detect potential functional mutations located in regions far from genes, e.g., regulatory enhancers, and is not ideal for the identification or larger insertions and duplications. While sample size remains a key factor for the interpretability of sequencing studies, the selection of participants is also vital. In general, the severity of the phenotype could be a good indicator of the presence of causative rare variants. For example, WES studies are a powerful tool for the identification of mutations causing undiagnosed severe neurodevelopmental delays [32,33]. However, these are examples of clearly severe and debilitating phenotypes that cannot be compared directly with a left-handedness status, even in the case of "strong" or "extreme" left hand preference assessed with the EHI and LQ measures. A strategy that mirror this selection criteria was adopted by a WGS study that focussed on individuals with *situs inversus*, a left-right reversal of the visceral organs, and who presented an elevated rate left-handedness [34]. Although in a few cases mutations in genes known to contribute to laterality defects were detected, no obvious genetic causes were identified for five individuals with *situs inversus*, three of which were also left-handers. These data suggest that even extreme asymmetric phenotypes are not a specific category cause by single variants with large effects.

Another strategy for the selection of individuals in sequencing studies is to focus on families presenting a clear inheritance pattern suggestive of a mutation that co-segregates with the phenotype. This approach was successful in mapping genes underlying different traits and diseases including language-related disorders. Although the discovery of the *FOXP2* gene goes back to the pre-genomic era, it was due to the observation of a severe speech and language disorder in multiple members of a large multigenerational family, consistently with the presence of a dominant mutation [35]. Other mutations contributing to language impairment were identified either through the analysis of large families [36], as well as in individual cases selected for severity [37]. Such approaches demonstrate the power of sequencing studies in detecting single causative genetic factors also in the context of highly polygenic traits like language impairment. This scenario shows that polygenic traits can result from single mutations, however these mutations are likely to occur in different genes, and therefore, are difficult to detect. Causation can be inferred when the same variant or different variants in the same gene are observed in multiple individuals. Very few sequencing studies, conducted specifically to map genes causing left-handedness, were conducted so far. Two separate WES studies sequenced members of families that practiced consanguineous marriage and presented an overrepresentation of non-right-handed individuals. The assumption of the sequencing studies was that left-handers in these families would carry a causative genetic variant [38,39]. Neither of the studies found any compelling evidence that this was the case. While a causative mutation located outside the regions covered by WES cannot be completely ruled out, the most likely interpretation of these negative findings is to add support to the polygenic nature of handedness. Given the limited number of sequencing studies, we cannot reach definitive conclusions, and the identification of single mutations directly causing left-handedness remains a possibility. However, considering the evidence collected so far, we expect this scenario to be an exception rather than the rule.

5. Handedness GWAS

Increasingly large GWAS for handedness measures led to a growing number of statistically significant genetic associations (Table 1).

The most recent GWAS, and the largest to date (N = 1,766,671) conducted by Cuellar-Partida et al., for a categorical definition of handedness confirmed the highly polygenic nature of handedness [22]. Such an impressive sample size was reached by analysing study participants from the UK Biobank, 23andMe (https://www.23andme.com, accessed on 25 August 2021), and the International Handedness Consortium. The study identified 48 statistically significant associations, of which 41 were associations with left-handedness, and 7 with ambidexterity. In addition to the detection of these associations, what the study did not find is equally compelling. Firstly, there was no single genetic factor associated with a large effect and, second, it is clear that many other genetic factors, beyond these 48 associations, remain to be identified. Overall, these observations confirm the highly polygenic nature of handedness, which is expected to implicate a much larger pool of genes than the ~40 genes predicted by McManus and colleagues in 2013 [40]. Cuellar–Partida et al.

conducted multiple analyses in addition to individual marker-traits associations providing new insights into the biological pathways contributing to handedness [22]. They observed that the genetic correlation between left-handedness and ambidexterity was very low. As discussed above, this is probably explained by a bias introduced by the self-reported measures, which are do not capture genuine ambidexterity. Instead, tissue- and pathway-enrichment found that genetic associations for left-handedness (but not ambidexterity) suggested, as expected, a role of the central nervous system. In particular, left-handedness was associated with genes involved in the activity or formation of microtubules, including *MAP2, TUBB, TUBB3, NDRG1, TUBB4A, TUBA1B, BUB3,* and *TTC28*. Microtubules are major components of the cytoskeleton and are essential for many processes, such as cell division, cell motility, intracellular transport, and maintenance of cell shape. Increasing evidence is supporting the role of microtubules in neurodevelopment and neurodevelopmental disorders [41,42]. Given the association between handedness and some psychiatric conditions, e.g., schizophrenia [43], Cuellar–Partida and colleagues suggested that microtubule-mediated processes could mediate the link between asymmetries and disorders. Microtubules were proposed as a key element to explain this complex link by Wiberg and colleagues in an earlier GWAS conducted in a subset of the UK Biobank individuals (N~400,000) [44]. The findings from our previous GWAS for a LQ derived from the pegboard task, and conducted in a much smaller sample (N = 728), proposed that shared biological pathways contributing to the establishment of left/right anatomical differences would also contribute to handedness and brain asymmetries [45]. Specifically, we suggested cilia-mediated processes as one of these biological pathways [46–48]. Cilia are microtubule-based cellular structures with sensory and motility function. During early development, cilia are critical in pattering the left/right axis determination and mutations in genes controlling cilia formation and function lead to laterality defects (See Vingerhoets et al. in this issue for a detailed explanations of this biological pathways [10]). The specific marker-trait associations from our study did not replicate in the larger GWAS for categorical measures of handedness, and it is possible that the lack of replication is due to the limited power of the original study, which led to false positives. Alternatively, it is possible that the different results are explained by the use of a quantitative LQ versus categorical phenotypes. As discussed earlier, the LQ measure different handedness dimensions better suited to capture the underlying genetic component. Beyond the individual associations, microtubules functions and formation (e.g., cilia and cytoskeleton dynamics) are unifying themes across the different studies. Together, the molecular genetics studies support the polygenic nature of handedness (Figure 1), disproving the single-gene theories and suggesting a scenario more in line with the liability threshold model [49]. This model proposed that binary traits are the results of multiple factors, each contributing a small effect, and normally distributed in the population. A threshold along the liability distribution determines the status of an individual for one of the two trait categories.

6. Genetics, Handedness and Brain Asymmetries

The first link between handedness and brain imaging genetics was suggested by Wiberg and colleagues in their GWAS [44]. One of the top associations with handedness, i.e., the rs199512 SNP located in the *WNT3* gene, was also associated with measures of white matter structural connectivity in brain regions involved in language, including the tracts linking Broca's and temporoparietal junction areas. A limitation of this analysis was the use of a single marker. Instead, a feature of large GWAS is that they allow the generation of polygenic risk scores (PRS), which capture the cumulative effect of associated variants [50]. It is then possible to test whether PRS for a particular trait derived from large GWAS (the training sample) influence other traits in separate samples which can be small in size (the target sample). For example, PRSs for educational attainment were among the first to become available [51] and were derived from increasingly large cohorts of up to 1.1 million individuals [52]. PRSs for educational attainment were tested for association with different cognitive, behavioural, and clinical traits, and were shown to account for about 2.1% of the variance in measures of reading abilities and dyslexia [53–55].

Under this principle, PRS for categorical measures of handedness, derived from a subset of UK Biobank participants (N = 331,037) [56], were tested by Ocklenburg and colleagues in a cohort of N = 296 participants [57]. They found that the PRS for hand preference were associated with LQ, showing the potential advantages of quantitative measures of handedness to capture genetic effects in samples of a modest size. Instead, no associations were detected with the brain measures selected for this study that focussed specifically on asymmetries in grey matter macrostructures.

Mapping genetic variants to functional and anatomical brain data is extremely challenging because of the large number of tests required by these analyses and high heterogeneity of the methods used in different studies. The ENIGMA (Enhancing Neuro-Imaging Genetics through Meta-Analysis; http://enigma.ini.usc.edu/) (accessed on 25 August 2021) consortium provides a platform to address these challenges and includes a working group focussed on brain laterality.

A number of studies looked for brain markers that could correlate with handedness measures. Brain imaging data in the UK Biobank provided evidence for associations between handedness and the overall anatomical hemispheric twist, or "torque" [58], and differences in functional connectivity in the language-associated regions in both hemispheres [44]. The next question is to ask whether associations between handedness and brain asymmetries could be mediated by shared genetics. In a very recent study, Sha and colleagues assessed the relationship between handedness and cortical asymmetries by generating asymmetry maps for cortical thickness and surface area in 28,802 right-handed and 3062 left-handed UK Biobank participants [59]. They found several regions that differed between left- and right-handers, consistent with a shift of neuronal resources to the hemisphere controlling the dominant hand. This means a general less leftward/more rightward shift for left-handers, who have a right hemisphere dominance for the preferred hand. Next, the same study derived PRS for handedness in an independent training sample of individuals from the UK Biobank to be tested in the target sample of individuals selected for the initial brain imaging analysis. As expected, the PRS were associated with handedness in the target sample. However, the handedness PRS also showed associations with cortical surface area asymmetries that differed between left- and right-handers. Specifically, PRS increasing the chances of left-handedness were associated with increased average rightward asymmetry in the fusiform cluster and decreased average leftward asymmetry in the anterior insula clusters. Tubulin-associated genes featured among the genes associated with cortical asymmetries. This is not surprising considering that these types of genes were enriched in the associations with handedness.

Table 1. GWAS for handedness measures.

Reference	N Participants	Cohorts	Handedness Phenotype	N Associated Genes
Eriksson et al. 2010 [60]	9126	23andMe	Handedness questionnaire	none
Scerri et al. 2011 [61]	744	Dyslexia cohorts and ALSPAC	LQ from pegboard task	1
Brandler et al. 2013 [45]	728 + 2666	Dyslexia cohorts and ALSPAC	LQ from pegboard task	1
Wiberg et al. 2019 [44]	~400,000	UK Biobank	Hand preference	4
De Kovel et al. 2019 [56]	331,037	UK Biobank	Hand preference	3
Cuellar-Partida et al. 2021 [22]	1,766,671	UK Biobank, 23andMe, International Handedness Consortium	Hand preference	48

These studies illustrate the challenges of conducting these types of analyses, which require large samples and rigorous methodology. Resources like the UK Biobank are a real gamechanger for this field. The large sample size allows detecting subtle effects of genes associated to complex phenotypes. These findings are the initial step to start disentangling at molecular level the relationship between handedness and cerebral asymmetries.

7. Conclusions

The two critical elements for the success of genetic studies are the sample size and the quality of the phenotype. Resources such as the UK Biobank demonstrate how large sample sizes allow the detection of subtle effects, as well as linking different types of data collected in relatively homogeneous ways across many individuals. Such studies led to the identification of specific genes associated to hand preference, implicating specific biological pathways, such as the function and formation of microtubules, to be relevant to both handedness and cerebral asymmetries. These discoveries relied on the use of the preferred hand for writing as handedness phenotype. This is a very convenient measure for the collection of large-scale data. However, these discoveries explain only a tiny fraction of the genetics contributing to handedness, and many more genes remain to be identified. While even larger samples characterised with hand preference measures will probably lead to the discovery of additional genes, the use of different types of handedness measures could provide another valid route for gene discovery. The modest correlation across handedness measures indicates that each of them captures a distinct dimension of handedness. Some of these measures also present heritability estimates that are higher than those observed for categorical measures of hand preference, and therefore, are more suited for genetic studies. In an ideal scenario, multiple handedness measures collected in large samples are likely to lead to novel breakthroughs. With the increased level of digitalisation and online testing [62], these types of datasets are becoming a more likely and extremely exciting possibility. For now, one of the key advances in the field is a new appreciation for the complexity that underlies handedness, a trait apparently very simple at both the behavioural and molecular level.

Funding: This work received no external funding. SP is supported by the Royal Society [UF150663]. The APC was funded by The University of St Andrews.

Conflicts of Interest: The author declares no conflict of interest.

References

1. De Kovel, C.G.F.; Carrión-Castillo, A.; Francks, C. A large-scale population study of early life factors influencing left-handedness. *Sci. Rep.* **2019**, *9*, 584. [CrossRef]
2. Papadatou-Pastou, M.; Ntolka, E.; Schmitz, J.; Martin, M.; Munafò, M.R.; Ocklenburg, S.; Paracchini, S. Human handedness: A meta-analysis. *Psychol. Bull.* **2020**, *146*, 481–524. [CrossRef] [PubMed]
3. HI, K. Why are there (almost) no left-handers in China? *Endeavour* **2013**, *37*, 71–81.
4. Zheng, M.; McBride, C.; Ho, C.S.-H.; Chan, J.K.-C.; Choy, K.W.; Paracchini, S. Prevalence and heritability of handedness in a Hong Kong Chinese twin and singleton sample. *BMC Psychol.* **2020**, *8*, 37. [CrossRef] [PubMed]
5. Ghirlanda, S.; Vallortigara, G. The evolution of brain lateralization: A game-theoretical analysis of population structure. *Proc. R. Soc. B Biol. Sci.* **2004**, *271*, 853–857. [CrossRef] [PubMed]
6. Faurie, C.; Raymond, M. The fighting hypothesis as an evolutionary explanation for the handedness polymorphism in humans: Where are we? *Ann. N. Y. Acad. Sci.* **2013**, *1288*, 110–113. [CrossRef]
7. Faurie, C.; Raymond, M.; Uomini, N. Origins, Development, and Persistence of Laterality in Humans. In *Laterality in Sports;* Academic Press: Cambridge, MA, USA, 2016; pp. 11–30. [CrossRef]
8. Mazoyer, B.; Zago, L.; Jobard, G.; Crivello, F.; Joliot, M.; Perchey, G.; Mellet, E.; Petit, L.; Tzourio-Mazoyer, N. Gaussian Mixture Modeling of Hemispheric Lateralization for Language in a Large Sample of Healthy Individuals Balanced for Handedness. *PLoS ONE* **2014**, *9*, e101165. [CrossRef]
9. Corballis, M. How Asymmetries Evolved: Hearts, Brains, and Molecules. *Symmetry* **2021**, *13*, 914. [CrossRef]
10. Vingerhoets, G.; Gerrits, R.; Verhelst, H. Atypical Brain Asymmetry in Human Situs Inversus: Gut Feeling or Real Evidence? *Symmetry* **2021**, *13*, 695. [CrossRef]
11. Medland, S.E.; Duffy, D.L.; Wright, M.; Geffen, G.M.; Hay, D.A.; Levy, F.; Van-Beijsterveldt, C.E.; Willemsen, G.; Townsend, G.C.; White, V.; et al. Genetic influences on handedness: Data from 25,732 Australian and Dutch twin families. *Neuropsychologia* **2009**, *47*, 330–337. [CrossRef]
12. Mitchell, K.J. *Innate: How the Wiring of Our Brains Shapes Who We Are. Innate: How the Wiring of Our Brains Shapes Who We Are;* Princeton University Press: Princeton, NJ, USA, 2020.
13. McManus, C. Is any but a tiny fraction of handedness variance likely to be due to the external environment? *Laterality* **2021**, *26*, 310–314. [CrossRef]

14. Ocklenburg, S.; Berretz, G.; Packheiser, J.; Friedrich, P. Laterality 2020: Entering the next decade. *Laterality* **2021**, *26*, 265–297. [CrossRef]
15. Ocklenburg, S.; Berretz, G.; Packheiser, J.; Friedrich, P. Laterality 2020: Response to the article commentaries. *Laterality* **2021**, *26*, 348–357. [CrossRef]
16. Oldfield, R.C. The assessment and analysis of handedness: The Edinburgh inventory. *Neuropsychologia* **1971**, *9*, 97–113. [CrossRef]
17. Annett, M. A classification of hand preference by association analysis. *Br. J. Psychol.* **1970**, *61*, 303–321. [CrossRef]
18. Paracchini, S.; Scerri, T. Genetics of human handedness and laterality. In *Lateralized Brain Functions*; Rogers, L., Vallortigara, G., Eds.; Springer: Berlin/Heidelberg, Germany, 2017; pp. 523–552.
19. Boyd, A.; Golding, J.; Macleod, J.; Lawlor, D.A.; Fraser, A.; Henderson, J.; Molloy, L.; Ness, A.; Ring, S.; Davey Smith, G. Cohort Profile: The 'children of the 90s'—The index offspring of the Avon Longitudinal Study of Parents and Children. *Int. J. Epidemiol.* **2013**, *42*, 111–127. [CrossRef]
20. Castillo, C.B.; Lynch, A.G.; Paracchini, S. Different laterality indexes are poorly correlated with one another but consistently show the tendency of males and females to be more left- and right-lateralized, respectively. *R. Soc. Open Sci.* **2020**, *7*, 191700. [CrossRef]
21. Bycroft, C.; Freeman, C.; Petkova, D.; Band, G.; Elliott, L.T.; Sharp, K.; Motyer, A.; Vukcevic, D.; Delaneau, O.; O'Connell, J.; et al. The UK Biobank resource with deep phenotyping and genomic data. *Nat. Cell Biol.* **2018**, *562*, 203–209. [CrossRef] [PubMed]
22. Cuellar-Partida, G.; Tung, J.Y.; Eriksson, N.; Albrecht, E.; Aliev, F.; Andreassen, O.A.; Barroso, I.; Beckmann, J.S.; Boks, M.P.; Boomsma, D.I.; et al. Genome-wide association study identifies 48 common genetic variants associated with handedness. *Nat. Hum. Behav.* **2021**, *5*, 59–70. [CrossRef] [PubMed]
23. Doust, C.; Fontanillas, P.; Eising, E.; Gordon, S.D.; Wang, Z.; Alagöz, G.; Molz, B.; St Pourcain, B.; Francks, C.; Marioni, R.E.; et al. Discovery of 42 Genome-Wide Significant Loci Associated with Dyslexia. *medRxiv* **2021**. [CrossRef]
24. Schmitz, J.; Zheng, M.; Lui, K.F.H.; McBride, C.; Ho, C.S.-H.; Paracchini, S. Quantitative multidimensional phenotypes improve genetic analysis of laterality traits. *bioRxiv* **2021**. [CrossRef]
25. Verhoef, E.; Shapland, C.Y.; Fisher, S.E.; Dale, P.S.; Pourcain, B.S. The developmental origins of genetic factors influencing language and literacy: Associations with early-childhood vocabulary. *J. Child Psychol. Psychiatry* **2021**, *62*, 728–738. [CrossRef]
26. Suzuki, K.; Ando, J. Genetic and environmental structure of individual differences in hand, foot, and ear preferences: A twin study. *Laterality* **2013**, *19*, 113–128. [CrossRef]
27. Warren, D.M.; Stern, M.; Duggirala, R.; Dyer, T.D.; Almasy, L. Heritability and linkage analysis of hand, foot, and eye preference in Mexican Americans. *Laterality* **2006**, *11*, 508–524. [CrossRef]
28. Packheiser, J.; Schmitz, J.; Berretz, G.; Carey, D.P.; Paracchini, S.; Papadatou-Pastou, M.; Ocklenburg, S. Four meta-analyses across 164 studies on atypical footedness prevalence and its relation to handedness. *Sci. Rep.* **2020**, *10*, 14501. [CrossRef]
29. Annett, M. *Left, Right, Hand and Brain: The Right Shift Theory*; Psychology Press: London, UK, 1985.
30. McManus, I. Right- and left-hand skill: Failure of the right shift model. *Br. J. Psychol.* **1985**, *76*, 1–16. [CrossRef]
31. Claussnitzer, M.; Cho, J.H.; Collins, R.; Cox, N.J.; Dermitzakis, E.T.; Hurles, M.E.; Kathiresan, S.; Kenny, E.E.; Lindgren, C.M.; MacArthur, D.G.; et al. A brief history of human disease genetics. *Nat. Cell Biol.* **2020**, *577*, 179–189. [CrossRef] [PubMed]
32. Wright, C.F.; Fitzgerald, T.W.; Jones, W.D.; Clayton, S.; McRae, J.F.; van Kogelenberg, M.; King, D.A.; Ambridge, K.; Barrett, D.M.; Bayzetinova, T.; et al. Genetic diagnosis of developmental disorders in the DDD study: A scalable analysis of genome-wide research data. *Lancet* **2015**, *385*, 1305–1314. [CrossRef]
33. Diquigiovanni, C.; Bergamini, C.; Diaz, R.; Liparulo, I.; Bianco, F.; Masin, L.; Baldassarro, V.A.; Rizzardi, N.; Tranchina, A.; Buscherini, F.; et al. A novel mutation in SPART gene causes a severe neurodevelopmental delay due to mitochondrial dysfunction with complex I impairments and altered pyruvate metabolism. *FASEB J.* **2019**, *33*, 11284–11302. [CrossRef] [PubMed]
34. Postema, M.C.; Carrion-Castillo, A.; Fisher, S.E.; Vingerhoets, G.; Francks, C. The genetics of situs inversus without primary ciliary dyskinesia. *Sci. Rep.* **2020**, *10*, 3677. [CrossRef] [PubMed]
35. Lai, C.S.L.; Fisher, S.; Hurst, J.A.; Vargha-Khadem, F.; Monaco, A. A forkhead-domain gene is mutated in a severe speech and language disorder. *Nat. Cell Biol.* **2001**, *413*, 519–523. [CrossRef]
36. Martinelli, A.; Rice, M.L.; Talcott, J.B.; Diaz, R.; Smith, S.; Raza, M.H.; Snowling, M.J.; Hulme, C.; Stein, J.; Hayiou-Thomas, M.E.; et al. A rare missense variant in the ATP2C2 gene is associated with language impairment and related measures. *Hum. Mol. Genet.* **2021**, *30*, 1160–1171. [CrossRef]
37. Ceroni, F.; Simpson, N.H.; Francks, C.; Baird, G.; Conti-Ramsden, G.; Clark, A.; Bolton, P.F.; Hennessy, E.R.; Donnelly, P.; Bentley, D.R.; et al. Homozygous microdeletion of exon 5 in ZNF277 in a girl with specific language impairment. *Eur. J. Hum. Genet.* **2014**, *22*, 1165–1171. [CrossRef]
38. Ocklenburg, S.; Barutçuoğlu, C.; Özgören, A.Ö.; Özgören, M.; Erdal, E.; Moser, D.; Schmitz, J.; Kumsta, R.; Güntürkün, O. The Genetics of Asymmetry: Whole Exome Sequencing in a Consanguineous Turkish Family with an Overrepresentation of Left-Handedness. *Symmetry* **2017**, *9*, 66. [CrossRef]
39. Kavaklioglu, T.; Ajmal, M.; Hameed, A.; Francks, C. Whole exome sequencing for handedness in a large and highly consanguineous family. *Neuropsychologia* **2016**, *93*, 342–349. [CrossRef] [PubMed]
40. McManus, I.C.; Davison, A.; Armour, J. Multilocus genetic models of handedness closely resemble single-locus models in explaining family data and are compatible with genome-wide association studies. *Ann. N. Y. Acad. Sci.* **2013**, *1288*, 48–58. [CrossRef]

41. Del Castillo, U.; Norkett, R.; Gelfand, V.I. Unconventional Roles of Cytoskeletal Mitotic Machinery in Neurodevelopment. *Trends Cell Biol.* **2019**, *29*, 901–911. [CrossRef] [PubMed]
42. Lasser, M.; Tiber, J.; Lowery, L.A. The Role of the Microtubule Cytoskeleton in Neurodevelopmental Disorders. *Front. Cell. Neurosci.* **2018**, *12*, 165. [CrossRef]
43. Hirnstein, M.; Hugdahl, K. Excess of non-right-handedness in schizophrenia: Meta-analysis of gender effects and potential biases in handedness assessment. *Br. J. Psychiatry* **2014**, *205*, 260–267. [CrossRef] [PubMed]
44. Wiberg, A.; Ng, M.; Al Omran, Y.; Alfaro-Almagro, F.; McCarthy, P.; Marchini, J.; Bennett, D.; Smith, S.; Douaud, G.; Furniss, D. Handedness, language areas and neuropsychiatric diseases: Insights from brain imaging and genetics. *Brain* **2019**, *142*, 2938–2947. [CrossRef]
45. Brandler, W.M.; Morris, A.P.; Evans, D.M.; Scerri, T.S.; Kemp, J.P.; Timpson, N.J.; Pourcain, B.S.; Smith, G.D.; Ring, S.M.; Stein, J.; et al. Common Variants in Left/Right Asymmetry Genes and Pathways Are Associated with Relative Hand Skill. *PLoS Genet.* **2013**, *9*, e1003751. [CrossRef] [PubMed]
46. Brandler, W.M.; Paracchini, S. The genetic relationship between handedness and neurodevelopmental disorders. *Trends Mol. Med.* **2014**, *20*, 83–90. [CrossRef] [PubMed]
47. Shore, R.; Covill, L.; Pettigrew, K.; Brandler, W.M.; Diaz, R.; Xu, Y.; Tello, J.A.; Talcott, J.B.; Newbury, D.F.; Stein, J.; et al. The handedness-associated PCSK6 locus spans an intronic promoter regulating novel transcripts. *Hum. Mol. Genet.* **2016**, *25*, 1771–1779. [CrossRef] [PubMed]
48. Paracchini, S.; Diaz, R.; Stein, J. Advances in Dyslexia Genetics—New Insights into the Role of Brain Asymmetries. *Adv. Genet.* **2016**, *96*, 53–97. [CrossRef] [PubMed]
49. Pearson, K.; Lee, A. On the inheritance of characters not capable if exact quantitative measurement. *Philos. Trans. R. Soc. Lond. Ser. A Math. Phys. Sci.* **1901**, *195*, 79–150.
50. Wray, N.R.; Wijmenga, C.; Sullivan, P.F.; Yang, J.; Visscher, P.M. Common Disease Is More Complex than Implied by the Core Gene Omnigenic Model. *Cell* **2018**, *173*, 1573–1580. [CrossRef]
51. Rietveld, C.A.; Medland, S.E.; Derringer, J.; Yang, J.; Esko, T.; Martin, N.W.; Westra, H.-J.; Shakhbazov, K.; Abdellaoui, A.; Agrawal, A.; et al. GWAS of 126,559 Individuals Identifies Genetic Variants Associated with Educational Attainment. *Science* **2013**, *340*, 1467–1471. [CrossRef]
52. Lee, J.J.; Wedow, R.; Okbay, A.; Kong, E.; Maghzian, O.; Zacher, M.; Nguyen-Viet, T.A.; Bowers, P.; Sidorenko, J.; Linnér, R.K.; et al. Gene discovery and polygenic prediction from a genome-wide association study of educational attainment in 1.1 million individuals. *Nat. Genet.* **2018**, *50*, 1112–1121. [CrossRef]
53. Selzam, S.; Dale, P.S.; Wagner, R.K.; DeFries, J.C.; Cederlöf, M.; O'Reilly, P.F.; Krapohl, E.; Plomin, R. Genome-Wide Polygenic Scores Predict Reading Performance throughout the School Years. *Sci. Stud. Read.* **2017**, *21*, 334–349. [CrossRef]
54. Gialluisi, A.; Andlauer, T.F.M.; Mirza-Schreiber, N.; Moll, K.; Becker, J.; Hoffmann, P.; Ludwig, K.U.; Czamara, D.; Pourcain, B.S.; Brandler, W.; et al. Genome-wide association scan identifies new variants associated with a cognitive predictor of dyslexia. *Transl. Psychiatry* **2019**, *9*, 77. [CrossRef]
55. Gialluisi, A.; Andlauer, T.F.M.; Mirza-Schreiber, N.; Moll, K.; Becker, J.; Hoffmann, P.; Ludwig, K.U.; Czamara, D.; Pourcain, B.S.; Honbolygó, F.; et al. Genome-wide association study reveals new insights into the heritability and genetic correlates of developmental dyslexia. *Mol. Psychiatry* **2020**. [CrossRef]
56. De Kovel, C.G.F.; Francks, C. The molecular genetics of hand preference revisited. *Sci. Rep.* **2019**, *91*, 5986. [CrossRef]
57. Ocklenburg, S.; Metzen, D.; Schlüter, C.; Fraenz, C.; Arning, L.; Streit, F.; Güntürkün, O.; Kumsta, R.; Genç, E. Polygenic scores for handedness and their association with asymmetries in brain structure. *Anat. Embryol.* **2021**, *1*, 1–13. [CrossRef]
58. Kong, X.-Z.; Postema, M.; Schijven, D.; Castillo, A.C.; Pepe, A.; Crivello, F.; Joliot, M.; Mazoyer, B.; Fisher, S.E.; Francks, C. Large-Scale Phenomic and Genomic Analysis of Brain Asymmetrical Skew. *Cereb. Cortex* **2021**, *31*, 4151–4168. [CrossRef] [PubMed]
59. Sha, Z.; Pepe, A.; Schijven, D.; Carrion Castillo, A.; Roe, J.M.; Westerhausen, R.; Joliot, M.; Fisher, S.E.; Crivello, F.; Francks, C. Left-handedness and its genetic influences are associated with structural asymmetries mapped across the cerebral cortex in 31,864 individuals. *bioRxiv* **2021**. [CrossRef]
60. Eriksson, N.; MacPherson, J.M.; Tung, J.Y.; Hon, L.S.; Naughton, B.; Saxonov, S.; Avey, L.; Wojcicki, A.; Pe'Er, I.; Mountain, J. Web-Based, Participant-Driven Studies Yield Novel Genetic Associations for Common Traits. *PLoS Genet.* **2010**, *6*, e1000993. [CrossRef] [PubMed]
61. Scerri, T.S.; Brandler, W.M.; Paracchini, S.; Morris, A.P.; Ring, S.M.; Richardson, A.J.; Talcott, J.B.; Stein, J.; Monaco, A.P. PCSK6 is associated with handedness in individuals with dyslexia. *Hum. Mol. Genet.* **2010**, *20*, 608–614. [CrossRef] [PubMed]
62. Parker, A.J.; Woodhead, Z.V.J.; Thompson, P.A.; Bishop, D.V.M. Assessing the reliability of an online behavioural laterality battery: A pre-registered study. *Laterality* **2020**, *26*, 359–397. [CrossRef]

Review

Atypical Brain Asymmetry in Human Situs Inversus: Gut Feeling or Real Evidence?

Guy Vingerhoets *, **Robin Gerrits** and **Helena Verhelst**

Department of Experimental Psychology, Ghent University, 9000 Ghent, Belgium; robin.gerrits@ugent.be (R.G.); helena.verhelst@ugent.be (H.V.)
* Correspondence: guy.vingerhoets@ugent.be

Abstract: The alignment of visceral and brain asymmetry observed in some vertebrate species raises the question of whether this association also exists in humans. While the visceral and brain systems may have developed asymmetry for different reasons, basic visceral left–right differentiation mechanisms could have been duplicated to establish brain asymmetry. We describe the main phenotypical anomalies and the general mechanism of left–right differentiation of vertebrate visceral and brain laterality. Next, we systematically review the available human studies that explored the prevalence of atypical behavioral and brain asymmetry in visceral situs anomalies, which almost exclusively involved participants with the mirrored visceral organization (situs inversus). The data show no direct link between human visceral and brain functional laterality as most participants with situs inversus show the typical population bias for handedness and brain functional asymmetry, although an increased prevalence of functional crowding may be present. At the same time, several independent studies present evidence for a possible relation between situs inversus and the gross morphological asymmetry of the brain torque with potential differences between subtypes of situs inversus with ciliary and non-ciliary etiologies.

Keywords: situs inversus; heterotaxy; brain asymmetry; visceral asymmetry; vertebrate asymmetry; human laterality; left-right differentiation; brain torque; ciliopathy

Citation: Vingerhoets, G.; Gerrits, R.; Verhelst, H. Atypical Brain Asymmetry in Human Situs Inversus: Gut Feeling or Real Evidence?. *Symmetry* **2021**, *13*, 695. https://doi.org/10.3390/sym13040695

Academic Editor: Sebastian Ocklenburg

Received: 24 February 2021
Accepted: 14 April 2021
Published: 16 April 2021

A glossary of terms is available at the end of this paper

1. Introduction

Vertebrates' visceral and central nervous systems demonstrate a strikingly asymmetric organization with a strong population bias toward a prototypical left–right configuration [1,2]. As both systems serve fundamentally different biological functions, it seems plausible to assume that the reasons behind their asymmetry may be entirely different and that their left–right differentiation evolved independently. While this may be true, it does not preclude the possibility that basic mechanisms for establishing left–right differentiation of the viscera have been reused to establish central nervous system laterality and that there may be a link between both manifestations of asymmetry. The strong population bias in visceral and brain asymmetry makes it difficult to determine whether they develop independently or related. Research turned to atypical conditions of visceral laterality to investigate possible relationships. Animal studies showed that some species like newts and zebrafish appear to align their brain and visceral asymmetry, mediated by *nodal*-related events [3–5]. In the frequent-situs-inversus (fsi) line of zebrafish, visceral reversal is accompanied by neuroanatomical reversals in the diencephalon, particularly epithalamic nuclei, which are believed to be involved in the functional lateralization of the vertebrate central nervous system [6]. In line with this claim, diencephalic reversals of fsi zebrafish correlate with the reversal of some (but not all) lateralized behavioral responses [7]. Do we anticipate a similar association in humans?

We will approach this outstanding question by describing the phenotypes and development of left–right asymmetry of the visceral system and the central nervous system and discuss the possible links between their mechanisms of left–right differentiation. Ultimately, a valid test for the hypothesis of an association between human visceral and neural asymmetry is to investigate the prevalence of atypical brain asymmetry in participants with visceral situs anomalies. In a systematic review, we discuss the studies that provide empirical evidence on behavioral, brain functional, and brain structural asymmetries in participants with situs anomalies. However, first, we will briefly explain the relevance of studies on asymmetry for the evolution of development.

Fluctuating asymmetry, directional asymmetry, and antisymmetry constitute three observable types of asymmetry within a population. Fluctuating asymmetry is the amount of deviation from perfect bilateral symmetry, and it manifests as small differences between the left and the right sides due to random errors in individual development. Fluctuating asymmetry is caused by genetic or environmental stress and is taken to measure developmental instability reflecting the level of stress in populations or of individual quality [8]. Directional asymmetry refers to the phenomenon that most individuals in a population are asymmetrical in the same direction, whereas in antisymmetry, dextral and sinistral forms are equally present within a species [9]. The typical asymmetrical position of the internal organs in vertebrates is an example of directional asymmetry, and the equal number of male fiddler crabs with a larger left or right claw is the prototypical example of antisymmetry. The latter two types of asymmetry have been proposed as informative traits to investigate evolution mechanisms as they are easy to define, easy to compare, and have evolved multiple times independently [9]. Differences in the heritability of antisymmetry (absent) and directional symmetry (present) contribute to understanding the evolutionary origin of novel forms, and it has been posited that directional asymmetry appears to have evolved through genetic assimilation (phenotype precedes genotype) almost as frequently as through conventional mutation-mode (genotype precedes phenotype) [9]. Comparing asymmetry patterns across species is relevant to investigate the evolutionary history of gene-expression patterns and anatomical asymmetries. The nodal signaling cascade, which takes a central place in vertebrate asymmetry, provides an important example of cascade capture and trait canalization [9]. In fact, a comparison of the key nodal cascade genes in lower chordates and vertebrates surprisingly suggests that the ancestral target of the nodal cascade might have been brain asymmetry [9].

2. Left–Right Asymmetry of the Visceral System

2.1. Phenotypes of Situs Viscerum

Like all vertebrates, humans establish left–right asymmetry of the thoracic and abdominal organ position during embryogenesis [1,10]. The position (*situs*, Latin) of the internal organs (*viscera*, Latin) in the human body shows a strong population bias toward an asymmetric organization with the heart's apex and aorta, bi-lobed lung, stomach and spleen on the left side of the body midline, and the heart's vena cava, most of the liver and the tri-lobed lung on the right side [11]. This typical configuration is called situs *solitus* (from Latin, meaning habitual), presents in about 99.99% of the human population and is taken to reflect optimal packing and transfer of body fluids [11]. Anomalies of this arrangement span a wide range of laterality defects whose classification remains without general consensus, thus hampering pathological, genetic, and epidemiological research [12,13]. As etiological and morphological boundaries between atypical manifestations of visceral situs remain to be settled, there is general agreement on the main two phenotypic subgroups of situs anomalies; the complete or partial reversal of the typical condition termed situs *inversus* (from Latin, meaning inverted), and the mirroring of either the typical left or right visceral configuration, called *heterotaxy* (from Greek *heteros*: other, different and *taxis*: arrangement) (Figure 1). As a rule, situs inversus and heterotaxy occur in different families, but occasionally they present in the same (often consanguineous) family [14,15]. Epidemiological studies estimate the prevalence of human visceral laterality defects between 1/5000 and 1/11,000 live births [12,16,17].

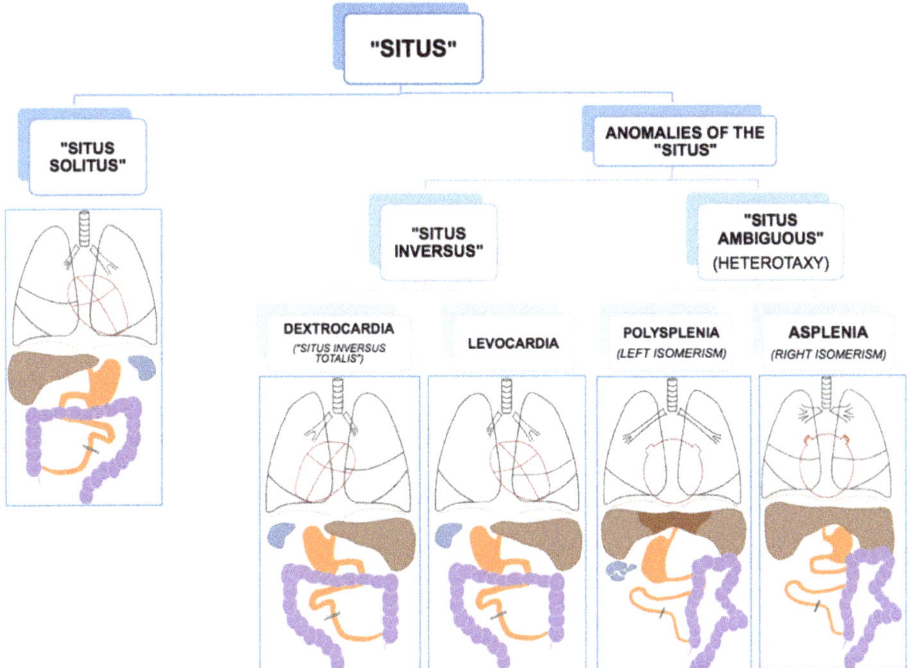

Figure 1. Anatomical dispositions of the viscera in the different types of "situs" (radiological convention). Reprinted with kind permission by Dr. Francisco Barqueros Escuer, https://dx.doi.org/10.26044/ecr2019/C-2735 (accessed on 1 October 2020).

2.2. *Situs Inversus*

Complete reversal of the standard visceral arrangement with the heart now in a right-sided position (*dextrocardia*) is referred to as *situs inversus totalis*. Prevalence reports vary widely and have been estimated between 1/6000–1/33,000 live births [12,17,18]. The condition itself is not associated with adverse medical complications as complete mirroring through the midsagittal plane of organs, blood vessels, nerves, and lymphatics do not interfere with their morphology nor positional relationships [11,19]. People with situs inversus totalis can live perfectly healthy lives, and medical problems may arise only in case of organ transplantation/donation or atypical symptom lateralization (for example, in appendicitis). Because of its limited clinical repercussions, situs inversus totalis is believed to be underdiagnosed. Nevertheless, structural malformations, such as congenital heart disease, may occur more frequently in situs inversus than in situs solitus [20,21]. In rare cases (1/2,000,000), situs inversus is not complete, and the heart is in its usual position (*levocardia*), while the other organs are in reversed position. Isolated levocardia is often associated with severe cardiovascular malformations because of the heart's unusual position compared to the other organs and their connections [20]. In about a quarter of cases, situs inversus occurs as part of a congenital syndrome in which medical complications are more prominent [11]. One of these syndromes, primary ciliary dyskinesia, has elucidated the importance of tiny hair-like organelles (cilia) in the ontogenesis of visceral asymmetry and will be discussed in more detail below.

2.3. *Heterotaxy*

An entirely different type of situs anomaly is heterotaxy, also referred to as situs *ambiguus*, as the defect presents as a complete loss of left–right laterality in the arrangement of the visceral organs along the superior–inferior axis. In contrast to situs inversus, heterotaxy syndrome alters the structure of visceral organs, particularly the heart, including the attachment of the large blood vessels, with the major morbidity and mortality resulting from complex cardiovascular malformations [13,16,22]. Prevalence figures for heterotaxy are estimated at 1/8000–1/12,000 live births [12,16,17]. Although classic heterotaxy accounts for only 3% of all congenital heart defects, gene mutations causing heterotaxy are also known to result in isolated cardiovascular malformations with no other visceral abnormalities, suggesting that the real prevalence of genetic heterotaxy is probably higher [19,23]. Two general types of heterotaxy, called isomerism, are described, although their exact morphology and its resulting abnormalities vary from patient to patient [11,19]. In left isomerism, morphologically left structures present on both sides of the body in the same individual. In this case, atrial cavities are morphologically left, both lungs will be bi-lobar with long main bronchial branches, the spleen is present but consists of multiple small and poorly functioning parts (polysplenia). In right isomerism, the right-sided visceral configuration is copy-mirrored to the left resulting in morphologically right atrial cavities, two tri-lobar lungs with short main bronchi, and an absent spleen (asplenia). In both conditions, the morphologically altered liver lies across the midline of the body, and intestinal malrotation is a typical feature, as well as cardiac malformations, the latter being more severe and sometimes life-threatening in right isomerism.

2.4. Cause of Visceral Situs Anomalies

Situs viscerum anomalies are congenital conditions due to heterogeneous genetic mutations that impact left–right patterning in early embryogenesis [19]. Genes involved in left–right axis development have emerged from animal studies and reveal a complex genetic cascade of left–right differentiation prior to the appearance of morphological asymmetry [14]. Most situs anomalies occur due to sporadic mutations, and many different genetic factors or genes cause the condition among different people or families [24]. Environmental and stochastic influences may also play a role as in a substantial number of cases, no clear monogenetic basis for their condition can be found [25]. In some families, situs viscerum anomalies present with an autosomal dominant, autosomal recessive (most commonly), or even X-linked pattern of inheritance [11,19]. Situs anomalies may arise as a variable manifestation of a syndrome encompassing a broader spectrum of defects [11]. Situs inversus, for example, sometimes occurs in cystic renal disease, Bardet-Biedl syndrome, and retinitis pigmentosa [24]. The best-known example of syndromal situs inversus, however, is when situs inversus arises as a symptom of primary ciliary dyskinesia (PCD), accounting for about 20 to 25% of its cases [19,21,26]. Primary ciliary dyskinesia is a causally heterogeneous group of autosomal recessive disorders characterized by a defect in the motility of small hair-like organelles (cilia) that protrude from the cell surface into extracellular space and perform various transport-related functions in the human body [27,28]. Ciliary motility is important for moving fluids and particles over epithelial surfaces, and cilia play crucial roles in various signal transduction pathways. Motile ciliogenesis requires a complex genetic program, and mutations of involved genes have been associated with ciliopathies, including primary ciliary dyskinesia (*DNAH5*, *DNAH11*, *DNAI1*, …) [26,28–30]. Ciliopathies give rise to a complex spectrum of disease and developmental mutant phenotypes that can be organ-specific or have broadly pleiotropic effects [31]. The diagnosis of primary ciliary dyskinesia is commonly based on electron microscopy showing abnormalities in structure and function of dynein arms or outright absence of cilia [26]. Affected individuals (1/10,000 to 1/20,000 live births [30,32]) have chronic upper respiratory tract (sinusitis) and lower respiratory tract (bronchiectasis) infections as well as chronic ear infections (otitis media) due to defective mucociliary clearance [26,29,33]. Reduced male fertility caused by decreased sperm motility, variable female infertility, and decreased sense of smell can also be part of the spectrum. About half of the patients with primary ciliary dyskinesia and associated sinusitis and bronchiectasis also have situs inversus (a triad of symptoms known as Kartagener syndrome [34]), while the other half is situs solitus [29]. Given the specificity of the ciliary mutation causing visceral inversion versus those causing respiratory problems, most but not all subgroups of the PCD syndrome will affect the genetic cascade induced by ciliary motion at the embryonic node (see below). Hence, the incidence of situs inversus in primary ciliary dyskinesia is estimated slightly less than the often reported 50%, and the Kartagener triad is expected in 1/22,000 live births [32]. Cardiac malformations suggestive of heterotaxy are found in 6–12% of individuals with primary ciliary dyskinesia [22,35], but it is generally believed that the condition is associated with a (near) randomization of left–right directionality rather than a loss of left–right specification [14]. The occurrence of a monozygotic twin pair with primary ciliary dyskinesia and with discordant visceral situs underlines the arbitrary nature of situs directionality in this condition [36].

3. Left–Right Visceral Development

3.1. Motile Cilia at the Primitive Node

The vertebrate left–right axis is established after developing its dorsal–ventral and anterior–posterior axes, and it is crucial for the correct positioning and morphogenesis of the internal organs [1,10]. The formation of the left–right axis involves several steps that have been investigated in several model organisms, such as the frog, zebrafish, chick, pig, and mouse (for a more detailed account, see [1,10,31,37–39]). While some genetic mechanisms are shared between vertebrates (like the expression of *nodal*, *lefty1*, *lefty2* and *pitx2*), other steps of the process seem to have diverged in evolution [10]. In fact, variation in the nodal cascade among vertebrates was said to resemble an hourglass, a conserved core set of genes listed above, with divergent genetic elements upstream and downstream that largely outnumber the shared core [9]. In most model organisms, symmetry breaking is established at the primitive node, a short-lived embryonic cavity filled with extracellular fluid that forms at the anterior tip of the primitive streak, a line of cells that establishes bilateral symmetry in the embryo, marks its future posterior side, and signals the beginning of gastrulation. Gastrulation is an important period in embryogenesis, which essentially consists of the differentiation of cells into an ectoderm, mesoderm, and endoderm layer. The left–right organizer or primitive node develops about 17 days postovulatory. The formation of the node coincides with the formation of motile cilia whose rotation produces a coordinated and unidirectional flow of the extracellular fluid that will induce symmetry breaking during gastrulation. It is important to point out that earlier asymmetries in the localization of some molecules have been established in some species and it has been claimed that asymmetries might exist perhaps as early as fertilization [37,40]. It is also important to note that not all species have fluid producing nodal cilia (absent in the chick and pig) yet show similarly strong population asymmetries of the viscera, which suggests that alternative cilia-independent symmetry breaking mechanisms at the node exist or that the cilia function as transmitters or amplifiers, but not initiators, of the asymmetrization [40]. In any case, in species with nodal cilia, such as the mouse, fish, and frog, (experimental) disruption of cilia functioning results in situs anomalies [41,42]. Reversal of flow in wild-type embryos results in L–R inversion, and introducing a leftward flow in mutants with ciliopathy restores typical L–R asymmetry [42,43]. While these experimental manipulations of ciliary flow are, of course, not possible in human embryos, the Kartagener syndrome clearly establishes humans as a species in which ciliary malfunction impacts visceral asymmetry. It may seem strange that a lack or impaired nodal flow caused by dysfunctional or absent cilia would result in L–R inversion instead of randomization, but models have been proposed to explain this [44].

3.2. Propagation of the Signal to the Lateral Plate Mesoderm and Organ Primordia

Due to their tilt and chiral nature, cilia that arise from nodal cells at the center of the nodal pit produce a clockwise (from tip to base) rotational motion that creates a leftward "nodal flow" towards the left periphery of the node [45] (Figure 2). Fluid flow is sensed by mechanosensory and/or chemosensory cilia in peripherally-located crown cells at the lateral ends of the pit [42]. These events cause intracellular Ca^{2+} levels to increase on the left side of the node, which results in asymmetries in gene expression and the establishment of a L–R axis [31]. The resulting asymmetric gene expression is then propagated to the lateral plate mesoderm—sheets of embryonic tissue at the peripheral left and right side of the embryo that will form the body wall and circulatory system—where a cascade of asymmetric left-sided gene expression is established (*nodal*, *lefty2*, *pitx2*). Several mechanisms have been proposed to explain the propagation of signaling from the node to the lateral plate mesoderm either directly by diffusion of *nodal* or by a cascade of signaling events via sonic hedgehog (*shh*) or bone morphogenetic protein (*bms*) that asymmetrically affect nodal expression [1]. In any case, the expression of *nodal* and the *lefty* genes (nodal antagonists) is transient and exclusively on the left side [1]. Finally, this asymmetric signaling is propagated from the lateral plate mesoderm to organ primordia for proper morphogenesis of the viscera to occur (*pitx2*). It is proposed that *nodal* acts as a determinant for leftness because cells that receive nodal signals will adopt left-side morphology, and those that lack nodal signals will adopt right-side morphology. In mutations in which *nodal* is bilaterally expressed in the lateral plate mesoderm, embryos will develop left isomerism, and in those that lack nodal signal on either side, embryos will develop right isomerism [1]. While heterotaxy may result from deficits in any of the above steps, they more often occur at one of the later stages. Situs inversus, on the other hand, is believed to originate from a more initial deficit in nodal flow caused by defectively operating cilia when the total direction of left–right asymmetry is determined. Animal models identified over 100 genes involved in left–right patterning, and more are to come [24]. Their mutations, in combination with reduced penetrance and variable expressivity, predict vast differences in phenotypical presentation of situs anomalies.

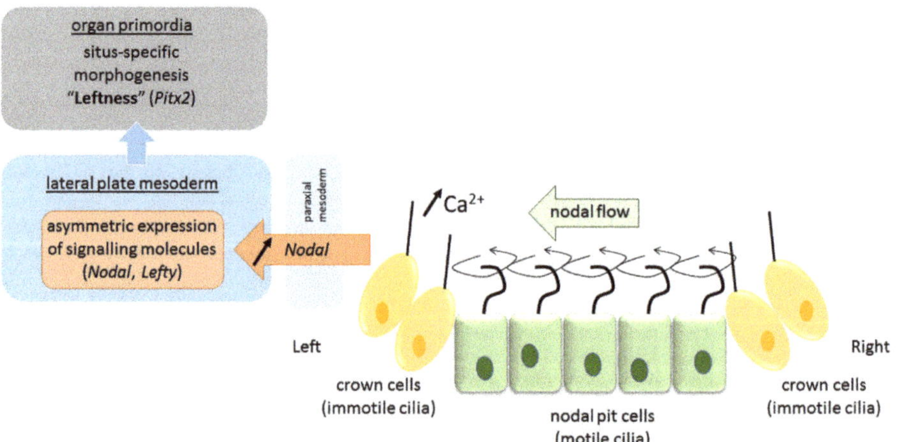

Figure 2. Pathway of visceral left–right determination in the vertebrate.

4. Left–Right Asymmetry of the Neurocognitive System

Like the visceral organs, our mental organs, by which we mean the biological substrates of cognitive functions, are asymmetrically represented in the brain. The advantages of hemispheric functional lateralization are explained in terms of improved parallel processing and the avoidance of useless duplications that saves neural space and evades competition between redundant control centers [46,47]. In addition to a bias favoring an asymmetric brain functional organization, there is also a bias toward a prototypical asymmetric configuration at the level of the population. Most humans have their left hemisphere in charge of language, manual dexterity (giving rise to handedness), and praxis (learned gestures), and the right hemisphere in control of spatial attention, face recognition, and prosody of speech [2]. The asymmetric arrangement gives rise to functional segregation between the left and right hemispheres. The existence of a population bias for exactly this configuration suggests that it may possess a biological advantage, but it remains to be explained why and how this would be the case. One possible way to look into this is by investigating alternative configurations of brain organization and explore their relationship with behavior.

4.1. Phenotypes of Brain Functional Organization

Recently, we have argued for the existence of three major categories in the phenotypes of functional brain segregation: typical, reversed typical, and atypical functional segregation [2]. Evidence for this distinction comes from studies investigating the asymmetry of more than one function in the same individuals. In random sample studies, this is achieved by investigating a random sample of the population [2]. The results of the available random sample studies are summarized in Table 1. Most studies probed two asymmetric functions. All used a language task as a typically left hemispheric function, and most used a spatial task to investigate right hemisphere dominance. Results reveal that most people show typical lateralization of the investigated functions and that a (substantial) minority of about 30% does not conform to this typical pattern (though many studies oversampled left-handers, which may have boosted this prevalence estimate). In about 20% of the participants, usually segregated functions were lateralized in the same hemisphere, a condition called crowding as the hemisphere is more crowded with functional representations. In about 10% of the participants, all investigated functions were lateralized in the atypical hemisphere resulting in a mirrored image of the prototypical functional segregation [2]. Evidence that this mirrored pattern of functional segregation extends beyond two atypically lateralized functions comes from selective sample research. In this type of investigation, participants are recruited based on the atypical lateralization of one function (usually language) to probe the lateralization of other functions. All these studies have been performed in left-handers as they are known to have a higher prevalence of atypical language dominance and revealed a concomitant reversal of the other investigated function [48–50]. In a recent study, five different lateralized functions were tested, and about 80% of the participants that had atypical language lateralization demonstrated complete or near complete reversal of all other functions as well [51]. In the remaining 20%, typical (or reversed typical) functional segregation was compromised more substantially, with two functions showing atypical lateralization, while the other three functions had conventional lateralization [51].

Table 1. Random sample studies that investigated more than one lateralized function in the same individuals.

Author and Year	LH Function	RH Function	N (#Sinistrals) *	Typical	Reversed	Crowded	Method **
Bryden et al., 1983 [52]	Language	Spatial dysfunction	270 (140)	72% (RH) 47% (LH)	12% (RH) 12% (LH)	16% (RH) 44% (LH)	Lesions
McNeely and Parlow, 2001 [53]	Language	Prosody	73 (7)	78%	22%	0%	Dichotic listening
Floël et al. 2005 [54]	Language	Spatial attention	75 (38)	95% (RH) 60% (LH)	2.5% (RH) 8% (LH)	2.5% (RH) 32% (LH)	fTCD
Whitehouse and Bischop, 2009 [55]	Language	Spatial memory	75 (30)	75%	0%	25%	fTCD
Badzakova-Trajkov et al. 2010 [56]	Language	Spatial attention Face processing	155 (48)	Majority	2%	Rest	fMRI
Rosch et al., 2012 [57]	Language	Spatial attention	20 (0)	65%	15%	20%	fTCD
Groen et al., 2012 [58]	Language	Spatial memory	60 (13)	58%	5%	32%	fTCD
Zago et al., 2016 [59]	Language	Spatial attention	293 (151)	80%	3.5%	15%	fMRI
Estimated proportion				±70%	±10%	±20%	

* Many "random-sample" studies included a proportionally higher number of left-handers to explore the effect of handedness; ** fTCD: functional transcranial Doppler ultrasonography; fMRI: functional magnetic resonance imaging.

4.2. Reversed Typical Functional Segregation

Together, these data confirm typical functional segregation in the majority of people, but they also show that alternative arrangements are not uncommon [2]. One alternative phenotype is a mirror reversal of typical functional segregation, which so far has been documented exclusively in left-handers [51]. Brain-wise, the reversed typical segregation phenotype is somewhat comparable with the visceral anomaly of situs inversus totalis, although its human population prevalence seems at least 100 times higher. Most random sample studies found no correlation between the laterality of different functions, suggesting that functions lateralize independently from other functions' laterality. Independent lateralization seems difficult to reconcile with a complete or near-complete reversal of five asymmetric functions in the same individual, let alone in 80% of a selective group. The odds that five independently lateralizing functions would each assume dominance in the atypical hemisphere in the same individual is extremely small. One way of reconciling independent lateralization and the observation of reversed typical functional segregation is achieved by assuming the existence of a generic blueprint of functional brain organization. Functions can develop their degree of lateralization more or less independently from other functions, but the origin of this process is seeded in a directional building plan that, on rare occasions, seems to have been flipped [2]. This assumption can explain the phenotype of the mirrored mind (mens inversus totalis, from mens, mentis (Latin) meaning mind) and at the same time allows for the independence of functional laterality indices. The assumption also predicts that the frequency by which functions occasionally deviate from the standard pattern (crowding) is not very different between the typical and reversed typical conditions as both mechanisms (independency of lateralization degree and reversal of the directional blueprint) are likely to be unrelated.

4.3. Atypical Functional Segregation

A second alternative phenotype groups conditions that show a more chaotic pattern of lateralization, as seen in individuals that have some functions showing typical and others showing atypical asymmetry. In these cases, the habitual functional segregation seems to be lost [2]. The visceral homolog of this phenotype category that we termed atypical functional segregation seems more akin to heterotaxy, where a loss of left–right asymmetry in the arrangement of the visceral organs is assumed, and that presents vast individual differences in organ displacement. While this comparison may seem farfetched at first, it has been raised before in the context of dissociated functional laterality [60], and there are more similarities between both conditions than meet the eye: variability of presentation, functional impact, and isomerism. As described above, the individual presentation of heterotaxy is very diverse, and the same gene mutation may cause severe heterotaxy affecting different organs in one individual and isolated cardiovascular malformation with no other visceral abnormalities in another. Similarly, atypical functional segregation can result from one or multiple functions deviating from the prototypical constellation [51]. Heterotaxy impacts the relationship between organs and is associated with more frequent and more severe medical problems than is situs inversus. Likewise, we reported evidence that healthy participants who show increased deviation from standard brain functional segregation perform significantly worse on a neuropsychological test battery compared to participants with typical or reversed typical segregation, suggesting that atypical functional segregation may be cognitively disadvantageous [51,61]. Finally, heterotaxy, at least theoretically, presents as two possible categories or isomerisms that copy-mirrors the left or right visceral morphology to both sides of the body. The brain functional homolog of this manifestation might be bilateral functional representation. Although bilateral functional representation has not been investigated at a multifunction level, it has received some attention at the single-function level. Research has shown that a small group of right and left-handers do not show clear-cut lateralization for language [62]. This group is said to have mixed or bilateral representation for language. Analyzing the left and right hemispheric activation patterns of these participants with a machine learning approach distinguished participants with a bilaterally dominant language representation from those with a bilaterally non-dominant pattern [63]. These findings are in line with observations from pre-surgical Wada-testing where some patients show speech arrest following sedation of either hemisphere, and other patients do not show speech arrest following sedation of either hemisphere [64–66].

In summary, alternative organizations of hemispheric functional segregation can be distinguished in two broad phenotypical categories that show at least some common properties with the main phenotypical subgroups of visceral anomalies. It remains to be determined whether these similarities are merely the product of the finite set of options imposed by our categorization or whether they reflect more fundamental principles that share a biological mechanism.

5. Left–Right Brain Development

5.1. Neurulation

The origin of brain symmetry breaking remains to be determined, but here too, an uneven distribution of molecules is believed to initiate left–right patterning [67]. During gastrulation and opposite to the primitive streak, the ectodermic tissue thickens and flattens to become the neural plate (about 19 days postovulatory). During that stage, the notochord appears below where the primitive streak and node used to be in the mesodermic tissue, and which will induce the start of neurulation. Neurulation is the process where the ectodermal neural plate folds into a neural tube (about 25 days postovulatory). The neural tube will later develop into the central nervous system (CNS). Primary cilia are involved in neurulation by neural tube patterning and closure through regulation of Sonic hedgehog signaling, and also in neural stem cell pool regulation, neural differentiation, and migration [68]. During neural tube development, its most ventral part, adjacent to the notochord, becomes the floor plate, and its dorsal part becomes the roof plate. The floor and roof plates, respectively, project ventralizing (*nodal*, *lefty*, *shh*) and dorsalizing (bone morphogenetic protein (*bms*) that suppress default neural differentiation and instead promotes epithelial growth) inductive signals to the developing neural tube, of which its most rostral part will develop into the forebrain (Figure 3). Asymmetric secretion of morphogens from the floor and roof plates to the left and right sides of the neural tube is believed to break the symmetry of neural patterning and induce asymmetric expression of downstream genes [67,69]. In addition to the floor and roof plates, the most rostral part of the neural tube has a third patterning center, the anterior neural ridge. The anterior neural ridge is a major organizing center that emits rostralizing signals essential for developing the secondary prosencephalon (that will form telencephalon, thalamus, hypothalamus, and epithalamus) [67,69]. It has been suggested that the asymmetric expression of morphogens secreted from this region could reflect asymmetrical topographic mapping of functional regions in the cortex [70,71].

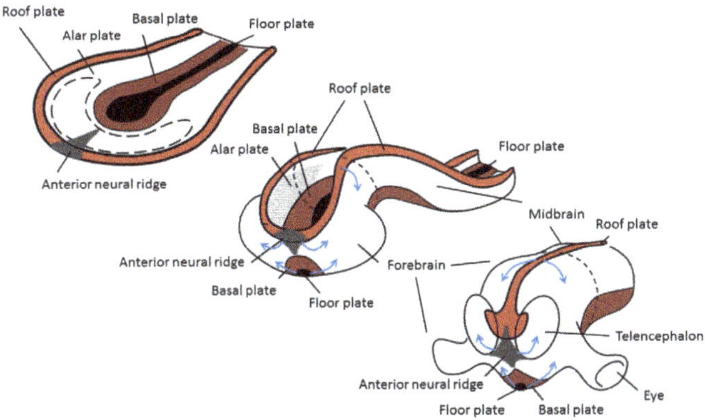

Figure 3. Changes during neurulation of the anterior neural section. Reprinted with permission from [72] and modified.

5.2. Asymmetric Development of the Central Nervous System

Empirical data on the asymmetry of gene expression in the left and right forebrains and midbrains of human embryos are available from 5 post-conception weeks onward [73]. By pooling data from voluntary medical abortions of healthy pregnancies and the Human Developmental Biology Resource (UK), the authors observed transcriptomic laterality in the anterior CNS regions of embryos between 5 and 14 weeks after conception. By joining the anterior CNS data with previous results of the midbrain and spinal cord regions of 4 to 8 week-old human embryos, the authors further reported evidence of age-dependent laterality of transcriptomic profiles for most structures indicating subtle differences in maturation rates between left and right CNS structures [73,74]. While both sides go through the same general developmental changes, one side appears to lead the other side at certain stages, and the laterality of the faster side is different from structure to structure. At 5 to 5.5 weeks post-conception, the spinal cord shows faster maturation on the left side than on the right, while the opposite pattern is observed for the midbrain and hindbrain [74]. By 7.5 weeks post-conception, the left choroid plexus, basal ganglia, diencephalon, and temporal cortex show faster maturation rates, but the rest of the cerebral cortex matures faster on the right side [73]. The observation of an early and differentiated pattern in the asymmetry of CNS structures with different functional destinations has led the authors to propose that brain asymmetry may be initiated/amplified at multiple locations [73]. For example, if faster maturation of the left spinal cord reflects observations of predominant right arm movements at 8 weeks post-conception (that is, prior to the innervation of the descending corticospinal tracts into the spinal cord), this could set the stage for the later cortical laterality of handedness, but would not necessarily influence the laterality of other functions or regions [73]. This suggestion is consistent with the weak correlations between the adult laterality of different brain functions like handedness and language [62] and with the results of gene ontology analysis that support the idea that handedness and language lateralization are ontogenetically independent phenotypes [75]. While subtle brain asymmetries in gene expression are already measurable at 5 weeks post-conception (i.e., approximately 7 weeks of gestational age), structural human fetal brain asymmetries become visible with current methods by the 11th week of gestational age for the choroid plexus [76], by the 16th week for the fetal cortex volume [77], by the 18th week for temporal lobe morphology [78], by the 20th week for sulcal folding [79,80], and by the 26th week for perisylvian hallmarks that have been associated with language [81]. The gap between genetic and morphological or functional brain asymmetries remains to be detailed [60].

6. Are Asymmetries of Visceral and Brain Development Related?

Visceral and neural patterning commence in close temporal proximity during the third and fourth week of human gestation, but it remains unclear whether the mechanisms that regulate visceral asymmetry also impact brain asymmetry. Asymmetric gene expression and the role of cilia seem potentially important factors for a link between visceral and brain manifestations of asymmetry.

Although *nodal* and *shh* pathways are also expressed during neurulation (cfr. floor plate induction), none of the reported 27 genes found to be differentially expressed in the left and right hemispheres of 12–14-week-old human fetal brains have known essential roles in visceral organ asymmetry [70]. Similar findings of lateralized gene expression with the more modern technique of transcriptomic profiling in post mortem temporal cortex from embryo to old age were reported, but here too, none of the reported genes have been associated with visceral anomalies [82]. On the other hand, relative hand skill in a cohort of individuals with a reading disability was associated with a variant in the gene *pcsk6*, an enzyme that cleaves *nodal* into an active form [83]. *Pcsk6* knockout mice display heterotaxy, and human variants of this gene are associated with heterotaxy and situs inversus as well, suggesting that handedness is at least in part controlled by genes that contribute to the determination of visceral asymmetry [83]. Human genes, like *GPC3*, associated (though not significant at a genome-wide threshold) with relative hand skill in the general population, cause situs anomalies when their orthologs are knocked out in mice [83,84].

Clinical evidence demonstrates the importance of cilia in human neurulation. Major ciliopathy-associated hereditary cerebral anomalies include neural tube defects, corpus callosum malformations, cerebellar hypoplasia, and hydrocephaly. Less severe neurological features, including cognitive deficits, autism spectrum disorders, and seizures, are also frequently observed in individuals with ciliopathies and hint at the possibility of more subtle cortical deficiencies [68]. Concerning laterality, genes most strongly associated with relative hand skill in a dyslexia cohort are involved in ciliogenesis, and their disruption in mice causes situs inversus [83,84]. In addition, cilia-related gene sets are more highly expressed in the right choroid plexus in the 7.5–13 post-conception age range [73]. The choroid plexus is also the first brain structure showing morphological asymmetry and is associated with the circulation of cerebrospinal fluid in the ventricles. Despite these observations, there is no clear evidence that cilia play a role in the initiation or propagation of central nervous system asymmetry [73].

7. Atypical Brain Asymmetry in Human Visceral Situs Anomalies

As the molecular regulation of brain asymmetry and its relationship with visceral lateralization remains to be elucidated, an alternative strategy of investigation is to look for evidence of atypical functional or structural brain asymmetry in people with situs anomalies. If the prevalence of behavioral, brain functional, or brain structural asymmetry differs between participants with typical and atypical visceral situs, then research would be better informed to explore more specific pathways of a possible link between human visceral and brain asymmetry. This approach is confronted with two major limitations: sample size and heterogeneous causality. As situs anomalies are inherently rare, it is extremely difficult to recruit many participants with atypical organ situs, especially if more intensive research protocols like neuroimaging are applied. In the absence of striking relations, small samples limit the statistical power to detect more subtle differences between typical and atypical groups in particular when only a subsample of participants shows a relation and others do not. This brings us to the second limitation of this approach, the heterogeneity of factors (genetic and other) that contribute to brain and visceral asymmetries. Different manifestations of situs anomalies have been associated with different genetic mutations, suggesting that genetic screening or at least a thorough description of the situs condition and family history should be used for categorization. Many gene mutations and combinations thereof have been associated with anomalies in visceral left–right patterning, and they are known to affect different steps and mechanisms of this complex process. It is plausible that some gene mutations bear no relation with brain asymmetrization, while others do. For example, in primary ciliary dyskinesia, the resulting randomization of organ situs is due to genetic mutations causing ciliary dysfunction. While cilia have a role in neurulation, it is unclear whether this includes lateralization of morphogens that induce brain asymmetry. Hence, a ciliopathy like primary cilia dyskinesia may not affect developing brain asymmetry at all. It is also possible that in people with situs inversus that have no primary ciliary dyskinesia, the origin of their situs anomaly is due to a temporary (or local) malfunction of nodal cilia or is caused by a different mechanism altogether. Even within the subgroup of situs inversus, etiological heterogeneity is substantial and extends beyond the role of genes. This was illustrated in a recent genome sequencing study of 15 cases with situs inversus totalis (SIT) [25]. The subgroup of six participants with primary ciliary dyskinesia (PCD) all presented with likely recessive PCD-associated mutations. Similar mutations were also detected in two of the non-PCD SIT participants, and in two other non-PCD SIT participants, recessive mutations in genes linked to situs inversus outside the context of PCD were found. In five of the nine non-PCD cases, however, no monogenic basis for their situs anomaly was found, which led the authors to consider early environmental or stochastic effects as possible causative factors.

8. Systematic Review

In March 2021, we performed a systematic literature search to address whether visceral situs anomalies have a different prevalence of brain and behavioral asymmetry [85]. The following platforms were searched: Web of Science (indexes: SCI-EXPANDED, SSCI, A&HCI, CPCI-S, CPCI-SSH, ESCI; range: 1972–2021), PubMed, and Google Scholar. In all cases, we searched for articles with the following strategy 1. Topic: situs inversus OR heterotaxy; 2. Topic: brain asymmetry OR brain functional asymmetry OR brain structural asymmetry OR behavio(u)ral asymmetry OR hemispheric dominance OR brain laterality; 3. #1 AND #2. We obtained a total of 79 records (WoS n = 63; PubMed n = 11; Google Scholar n = 5). Sixty-nine records were screened after the removal of duplicates. Records on animal research (n = 25), genetics (n = 17), and medical papers on comorbidities or laterality defects other than situs inversus or heterotaxy (n = 17) were excluded. Ten full-text articles were addressed for eligibility. In the references of these articles 7 further (mostly older) studies were identified that reported empirical data on the research question. While this manuscript was under revision, an additional paper on brain asymmetry in fetuses with laterality defects was accepted for publication and added to the review [86]. A total of 18 studies were included in the qualitative synthesis for this systematic review.

Many studies have not described the situs condition of their participants in detail, nor have they differentiated their already small samples of participants into separate categories or explored their genetic background. In the next section, we will summarize the behavioral, brain structural, and brain functional data on atypical asymmetry in participants with situs anomalies or, to be more precise, in participants with situs inversus as almost all brain and behavior-related research in this field has been performed within this subgroup.

8.1. Handedness in Situs Inversus

Already in 1836, Sir Thomas Watson remarked that individuals with situs inversus (SI) are no more left-handed than the rest of the population (reported in [87]). This observation was empirically confirmed in several remarkably large-scaled studies of the early to mid-20th century (Table 2) [88,89]. These early studies, however, suffer from poor behavioral assessment of handedness and poor etiological description of the situs anomaly. In addition, the reported prevalence of left-handedness around 6–7% is clearly lower than contemporary estimates of 10% [90], suggesting that cultural pressure against left-hand use and forced right-handedness may have underestimated natural left-hand preference in these cohorts. As a result, their findings might not provide a clear answer to the question at hand. Then follow two smaller studies based on hospital samples and reporting the low prevalence of non-right-handedness in 6 and 16 SI participants, respectively [91,92]. Unfortunately, very little information on recruitment and SI status or etiology was provided. In both studies the authors concluded that there was little evidence for a relationship between handedness and visceral position. Two later studies that focused on handedness and which recruited quite sizeable cohorts included PCD-related SI participants only. Both studies came to the conclusion that the prevalence of left-handedness in PCD-related SIT is no different from the rest of the population [32,93]. Given the reports of a possible genetic association between relative hand skill and ciliogenesis, typical handedness in PCD-related SIT may seem surprising and has been explained in terms of compensatory mechanism that allow the typical development of handedness to overrule the influence of ciliopathy [83,84]. For non-syndromal SI, the issue of handedness is less clear given the paucity or incompleteness of available data. Some information can be gathered from studies that investigated brain functional asymmetry in SI and which predominantly featured non-syndromal cases of SIT (Table 3). Together, these studies report on 22 sporadic cases that were explicitly reported to be free of PCD-symptoms [94–98]. Seventeen of these participants were right-handed, and 5 were left-handed (29% left-handedness). It needs to be remarked that all left-handers were reported by the same study in which 5 out of 9 non-PCD-related SIT had a left-hand preference (55% left-handedness) [98]. Interestingly, this study also recruited 6 PCD-related SIT cases, only one of which was a left-hander, a result that was in line with previous findings on hand preference in PCD-related SIT. Is the seemingly random hand preference in the non-PCD-related SIT participants of the Ghent-cohort an accidental finding? It may well be as none of the other studies even remotely suggested anything of the kind. Future research, preferably in a larger cohort of PCD and non-PCD-related SIT, is necessary to determine if the differential effect of situs inversus on handedness can be replicated. At the same time, the possibility that the etiology of the SI anomaly may differentially influence brain-related asymmetry underlines the importance of providing a detailed description of the SI participants' phenotype and, if possible, also take the genotype into account.

Table 2. Overview of handedness studies in situs inversus.

Author and Year	Sample Size	Situs Anomaly	Etiology	Source	Percent Left Handedness	Handedness Determination
Cockayne et al., 1938 [88]	115	Situs inversus	Not provided	From literature and the author's own series; most included were examined on account of illness or a congenital malformation	6.1% LH and 1.7% ambidexter	Unknown
Torgersen et al., 1950 [89]	Convenience sample of 160 from a total sample of 270	Situs inversus	Not provided	Mass X-ray photographs of the Norwegian population (200) + additional cases through hospitals and municipal health departments (70)	6.9% in SI compared to 3.5% in (715) siblings and 5.3% in (320) parents	Self-report obtained from the individual with SI
Everett et al., 1963, [91]	Convenience sample of 6 from a total sample of 10	Situs inversus totalis	Not provided	Hospital records search	0%	Self-report or hospital records
Matsumoto et al., 1997 [92]	16	Situs inversus	Not provided	Not reported	6.3%	Writing, drawing, throwing
McManus et al., 2004 [93]	46 PCD-SI and 42 PCD-SS, 334 controls	PCD-related situs inversus	PCD only	Family PCD-support group	15.2% in PCD-SI and 14.3 in PCD-SS, 8.1 in controls	Writing hand and laterality questionnaires
Afzelius and Stenram et al., 2006 [32]	Convenience sample of 112 from a total sample of 239 PCD patients (105 PCD-SI and 134 PCD-SS)	Situs inversus	PCD only	Obtained from contacting many Swedish clinicians	12.5% LH and 3.6 ambidexter in PCD, 14.3% LH in subgroup with PCD-SI	Self-report

Table 3. Overview of brain functional asymmetry studies in situs inversus.

Author and Year	Sample Size	Hand Preference *	Situs Anomaly	Etiology	Source	Method **	Function	Laterality ***
Woods et al., 1986 [94]	1 SI	RH	Situs inversus totalis	Sporadic case, no PCD-symptoms	Stroke patient	Lesion	Language	Aphasia following left stroke
Cohen et al., 1993 [99]	1 SI	RH	Heterotaxy (polysplenia)	Sporadic case	Stroke patient	Lesion	Language	Aphasia following right stroke (crossed aphasia)
Tanaka et al., 1999 [95]	9 SI and 24 controls	9 RH	Situs inversus totalis	8 sporadic cases, 1 PCD-SIT	Hospital search	DLT	Language	88.9% of SIT shows REA, 79.1% of controls shows REA
Kennedy et al., 1999 [96]	3 SI and 15 controls	3 RH (Annett handedness battery)	Situs inversus totalis	3 sporadic cases, normal general health	Hospital records search	fMRI	Language	100% of SIT show left-lateralized activation on each of two language tasks (word stem completion and semantics)
Ihara et al., 2010 [100]	3 SI and 11 controls	1 weak RH (EHI = 38), 1 weak LH (EHI = −20), 1 strong RH (EHI = 100)	Situs inversus totalis	Unknown, 1 SIT case has left temporal epilepsy	Unknown	MEG	Language	33% of SIT show left-lateralized activation during reading task, 66% show right-lateralized activation
Schuler et al., 2017 [97]	1 SIT and 1 control	RH (EHI = 100)	Situs inversus totalis	Sporadic case, no PCD-symptoms	Fetal screening	fMRI	Language	Typical left lateralization (LI = 0.48)
Vingerhoets et al., 2018 [61]	15 SIT and 15 controls	9 RH, 6 LH; 20% LH in PCD-SIT, 55% LH in non-PCD SIT, 40% LH overall (EHI)	Situs inversus totalis	9 sporadic cases, 6 PCD-related	Hospital records search	fMRI	Language	80% of SIT left lateralized (93% in controls)
							Praxis	73% of SIT left lateralized (87% in controls)
							Spatial attention	73% of SIT right lateralized (93% in controls)
							Face recognition	80% of SIT right lateralized (87% in controls)

* EHI: Edinburgh handedness inventory; ** DLT: dichotic listening test; fMRI: functional magnetic resonance imaging; MEG: magneto-encephalography; *** REA: right ear advantage.

8.2. Brain Functional Asymmetry in Situs Inversus

As mentioned in the previous paragraph, brain functional asymmetry was predominantly investigated in participants with a non-syndromal manifestation of SI (Table 3). The discussion starts in the late 1980s–early 1990s with the report of two right-handed stroke patients with visceral anomalies, one of which became aphasic following a left hemisphere cerebrovascular lesion [94] while the other, a patient with left isomerism heterotaxy, showed crossed-aphasia after a right hemisphere stroke [99]. More convincing evidence for typical language lateralization came from 9 SIT participants (only one with PCD-related SIT), who performed a dichotic listening paradigm and showed typical right ear advantage in all, but one case [95]. The advent of MRI research provided the opportunity of visualizing neural activation during cognitive tasks. A first fMRI study corroborated Tanaka's dichotic listening findings by showing typical left hemisphere lateralization for language in three non-syndromal SIT participants [96], but a decade later, a second fMRI study reported atypical right hemisphere lateralization for language in two out of three SIT cases [100]. Until now, all studies, including a longitudinal case study that used fMRI [97], had focused on language. Recently, research broadened to other lateralized functions, including praxis, spatial attention, and face recognition, in an fMRI study of 15 SIT participants, of which 6 had PCD-related SIT, and 9 had non-PCD-related SIT [61]. While 80% of this cohort had left hemisphere language dominance, suggesting generally typical language lateralization, a control group matched for handedness showed 93% leftward lateralization. The same trend was found for the three other tested functions that all showed more typical asymmetry in the matched controls compared to the SIT participants. The authors concluded that atypical functional segregation, that is, the likelihood that brain functional organization does not show the typical population pattern, is more frequent in SIT participants. No obvious difference in the level of deviation from typical functional segregation was observed between PCD and non-PCD-related SIT, but the small sample size limits proper statistical comparison. It can be argued that results on functional lateralization have been influenced by the unexpectedly high number of left-handers in this sample as left-handers have a higher prevalence of atypical functional lateralization [2], but atypical lateralization occurred equally frequently in the right-handed SIT participants. Together, the available data suggest that, while most people with SIT will show typical patterns of functional asymmetry, atypical lateralization of language and other asymmetric functions may be more frequent in SIT. It remains to be determined whether this is a general trend or associated with specific etiological characteristics.

8.3. Brain Structural Asymmetry in Situs Inversus

An overview of studies reporting on brain structural asymmetry in SI is provided in Table 4. If there is one consistent finding on brain asymmetry in SI, it is the observation that their cerebral torque is generally reversed than the typical human population bias. The cerebral or "Yakovlevian" torque is a gross anatomical and morphologically complex characteristic [101] that refers to an anti-clockwise twist of the brain about the ventral-dorsal axis. It is most often described in terms of its petalia, whereby the right frontal pole protrudes anteriorly to the right frontal pole, and the left occipital pole protrudes posteriorly to the right occipital pole. Typical petalia asymmetry is observed in 44% of modern human brains [102] and appears to be absent in non-human primates [101]. Reversed petalia were reported in 15 out of 23 SIT participants (65%), most of which were sporadic cases. Again, a possible distinction arises between syndromal and non-syndromal SIT as a recent study documented complete reversal of the petalia in 7 out of 9 non-PCD-related SIT participants (78%) and in none of the 6 PCD-related SIT participants [98]. It remains to be confirmed whether the reversed cerebral torque pairs with the reversal of intracranial vasculature and bony landmarks as suggested by one post-mortem study [103]. If it does, it would be an important argument for a link between different brain morphological asymmetries and a link between lateralized gradients of brain structural and visceral development.

Table 4. Overview of studies on brain structural asymmetry in situs inversus.

Author and Year	Sample Size	Hand Preference	Situs Anomaly	Etiology	Source	Method *	Cerebral Torque (Petalia)	Planum Temporale	Sylvian Fissure	Other **
Kennedy et al., 1999 [96]	3 SIT and 15 controls	RH (AHB = 1.24)	Situs inversus totalis	Sporadic cases, normal general health	Hospital records search	MRI	Reversed in all SIT	Volume L > R in 2 SIT and R > L in one SIT	Higher R > L in 2 SIT	
Tubbs et al., 2003 [103]	1 SIT	Unknown	Situs inversus totalis	Unknown, died from "natural causes"	Unknown	Autopsy	Reversed		Higher L > R	Reversed intracranial vasculature
Ihara et al., 2010 [100]	3 SIT and 11 controls	2 RH, 1 LH	Situs inversus totalis	Unknown, 1 SIT has temporal lobe epilepsy	Unknown	MRI	Reversed in all SIT	Volume L > R in 3 SIT		IFG volume L > R in 2 SIT, R > L in1 SIT
Leroy et al., 2015 [104]	6 SIT and 95 controls (48 RH)	5 RH, 1 LH	Situs inversus totalis	Cases from the Kennedy et al. and Ihara et al. studies	Cases from the Kennedy et al. and Ihara et al. studies	MRI				STS deeper on right (typical)
Schuler et al., 2017 [97]	1 SIT and 1 control	RH (EHI = 100)	Situs inversus totalis	Sporadic case, no PCD-symptoms	Fetal screening	MRI	Reversed in SIT	Absent asymmetry in SIT		STS deeper on right (typical)
Vingerhoets et al., 2018 [98] and Mannaert et al., 2019 [105]	15 SIT and 15 controls	9 RH, 6 LH	Situs inversus totalis	9 sporadic cases, 6 PCD-related	Hospital records search	MRI	Reversed in 78% of non-PCD SIT and in 0% of PCD SIT	Not different from controls	Same as controls	IFG volume same as controls; Heschl's gyrus and anterior insula volume same as controls; Arcuate fasciculus volume same as controls; Uncinate fasciculus lower volume in SIT than controls
Kienast et al., 2021 [86]	5 fetuses with ciliopathies and 21 fetuses with laterality defects; 26 age-matched controls	Not applicable	4 SIT; 8 dextrocardia; 4 situs ambiguus; 5 situs inversus abdominalis	Prenatal diagnosis reports	Hospital records search	MRI				Perisylvian fetal patterns and asymmetry indices do not differ between cases with laterality defects, ciliopathies, and normal controls

* MRI: magnetic resonance imaging; ** IFG: inferior frontal gyrus; STS: superior temporal sulcus.

No systematic reversals in other structural brain asymmetries have been reported in SI. Alleged language-related markers like the planum temporale, Sylvian fissure, inferior frontal gyrus, depth of the superior temporal sulcus, and the arcuate fasciculus show the same variability and directional bias as the general population. The available data are scarce, though, and the discovery of more subtle effects or between SIT-type differences awaits further research.

9. Discussion

The low prevalence and substantial phenotypical variability of human visceral laterality anomalies postpone consensus on clear classification criteria for subgroup determination. Still, two broad categories of anomalies are generally distinguished, *situs inversus* characterized by a complete or near-complete mirror reversal of typical visceral asymmetry, and *heterotaxy* described as a duplication of one of either asymmetric sides. Both phenotype categories are believed to result from different deficits in the complex developmental cascade of visceral left–right differentiation, but the exact causal implications for each step and each genetic mutation in that process remain to be elucidated. The same is true for the brain. While the prevalence of nonconventional brain organization is roughly 100 times more frequent than atypical visceral organization, it is more difficult to assess, and data are scarce. However, here too, two main categories of unconventional brain organization are advanced, *reversed functional segregation* presenting as a mirror image of the usual hemispheric task division, and *atypical functional segregation* characterized by functional crowding.

The substantial difference in the prevalence of atypical visceral and brain organization also brings the effect of evolutionary canalization to mind, the increased resistance of a trait to genetic and environmental perturbations over evolutionary time. Left-sided heart anatomy is a preserved trait in all living vertebrates, but the incidence of spontaneous reversal declines throughout vertebrate evolution from 5% in fish, 1–2% in amphibians, 0.1% in mammals, and 0.01% in humans [9]. Explanations for the evolutionary increase in canalization include increased predictability of symmetry breaking by cilia-controlled nodal flow or the more stable conditions of the placental environment [9]. Cladistic estimates of reversals in brain organization are not available, but the concept of evolutionary canalization may provide an interesting venue to explore the origin and timing of brain structural and functional asymmetries in humans by comparing prevalence measures of atypical laterality.

Apart from similarities in the overall appearance of the main phenotype subgroups of visceral and brain laterality anomalies, we should keep in mind that the visceral and neural systems serve fundamentally different biological functions and that the reasons for developing asymmetry in each system are likely to be dissimilar. Nevertheless, selfsame basic mechanisms for left–right differentiation may be employed by both systems to generate and/or propel asymmetry [9]. This possibility is hinted at by some mutant lines in vertebrate species that appear to align atypical visceral with atypical brain structural asymmetry and which also appears to impact their behavioral asymmetries [3,5]. One way to explore such a relation in humans is to investigate and compare the developmental cascades of visceral and brain laterality and scrutinize the molecular genetics underlying both mechanisms for biological links or similarities. The road toward asymmetry appears very complex and much of it, particularly concerning the brain, remains to be discovered.

An alternative way to explore possible relations lies in the direct comparison of phenotypes by investigating brain and behavioral asymmetries in individuals with situs anomalies. Delineation of atypical manifestations could provide molecular genetics with more specific targets to find associations between the developmental cascades of visceral and brain asymmetry. While this approach is hampered by the low prevalence of situs anomalies and the laborious assessment of brain asymmetries, several studies have contributed to this endeavor. However, samples are often small, and the range of phenotypes is restricted or poorly defined.

Most, if not all, studies on behavioral and brain asymmetry in situs anomalies focused on situs inversus (totalis). Probable reasons for this selective approach are the anticipation of more straightforward results and the better medical condition of participants with situs inversus compared to those with heterotaxy. In general, the studies appear to agree that situs inversus in humans is not inseparably associated with a reversal of brain and behavioral asymmetries as seen in some other species. On the contrary, most people with situs inversus seem to present with typical patterns of hemispheric specialization, although a higher prevalence of functional crowding in this group remains a possibility. At the same time, the findings hint at some more subtle effects that distinguish between types of situs inversus with different etiologies. More in particular, in PCD-related syndromal situs inversus, handedness and probably also brain torque reveal the same laterality bias as the general population. This finding can be taken to suggest that nodal ciliopathy and the eventually reversed subsequent molecular cascade that gives rise to visceral laterality has only little effect on hand preference and gross brain morphology. By contrast, situs inversus caused by non-ciliary, perhaps earlier, factors does seem to be accompanied by a reversal of the brain torque. This finding is reported by several independent studies from North America, Japan, and Europe and indeed hints at a possible relation between human visceral asymmetry and the asymmetrical shape of the brain organ. A possible venue to investigate a direct relation between both manifestations of directional asymmetry in humans would be to determine signed fluctuating asymmetry of the visceral and brain torque modules in a sample of humans, which do not necessarily need to have a visceral anomaly [8]. To corroborate and extend findings on brain asymmetry in visceral anomalies, future research should provide detailed phenotypical information of participants supplemented by genetic data if possible. Ideally, a consensus should be reached on core information to be reported that will allow open science and meta-analytic initiatives to gather larger samples of participants with situs anomalies and further understand possible interactions between human visceral and brain asymmetry.

Author Contributions: G.V. wrote the manuscript with support from R.G. and H.V. All authors have read and agreed to the published version of the manuscript.

Funding: This research received no external funding.

Institutional Review Board Statement: Not applicable (review).

Informed Consent Statement: Not applicable.

Data Availability Statement: Not applicable.

Conflicts of Interest: The authors declare no conflict of interest.

Glossary

Anterior neural ridge: The anterior neural ridge is a region in the neural plate and later neural tube, which secretes signaling molecules essential for developing the forebrain. **Antisymmetry**: Dextral and sinistral forms are equally present within a population. **Atypical hemispheric functional segregation**: Phenotype of hemispheric functional segregation in which the typical left–right segregation is lost due to one or more functions showing atypical dominance, while other functions do not. **Autosomal dominant disorder**: A pattern of inheritance in which an affected individual has one copy of a mutant gene and one normal gene on a pair of autosomal (one of the numbered, non-sex) chromosomes. **Autosomal recessive disorder**: A pattern of inheritance in which an affected individual requires two copies of a mutant gene on a pair of autosomal (one of the numbered, non-sex) chromosomes. **Behavioral asymmetry**: Left–right difference in behavior, like hand or foot preference, or the increased probability to retain words presented to the right ear versus those presented to the left ear. **Brain asymmetry**: Left–right differences in functional or structural (anatomical) characteristics between the two hemispheres. **Canalization (evolutionary canalization)**: Increased resistance of an established trait for genetic or environmental perturbations over evolutionary time. **Cascade capture**: The recruitment of genes or gene cascades for another duty. **Cilium/Cilia**: Small hair-like organelles that protrude from the larger cell body. Cilia can be motile or non-motile. Non-motile cilia serve as sensory organelles, much like a cellular antenna. Cells of the transient primitive node have singular motile cilia known as nodal cilia, critical for the establishment of left to right body asymmetry. **Ciliogenesis**: The building of the cell's cilium/cilia. Defects in ciliogenesis can lead to numerous human diseases related to non-functioning cilia (ciliopathies). **Ciliary motility**: The ability of some cilia types to produce motion by a molecular motor that drives its beating. Motile cilia have a function in the transport of fluids over the surface of cells. **Dextrocardia**: A rare congenital condition in which the heart's apex is located on the right side of the body. **Directional asymmetry**: Most individuals in a population are asymmetrical in the same direction (population bias). **Floor plate**: Located on the ventral midline of the embryonic neural tube, the floor plate is a glial structure that serves as an organizer to ventralize tissues in the embryo as well as to guide neuronal positioning and differentiation along the dorsoventral axis of the neural tube. **Fluctuating asymmetry**: The amount of deviation from perfect bilateral symmetry as reflected by small differences between the left and the right sides due to random errors in the individual development. **fMRI**: Functional magnetic resonance imaging is a non-invasive technique to measure and map changes in the brain's blood flow that coincide with brain activity. **Forebrain (prosencephalon)**: The rostral (forward-most) portion of the brain that will develop into the diencephalon (thalamus, hypothalamus, subthalamus, and epithalamus) and the telencephalon, which develops into the cerebrum. **fTCD**: Functional transcranial Doppler ultrasonography is a non-invasive technique to measure changes in the blood flow velocity of the basal segments of the cerebral arteries that coincide with brain activity. **Gastrulation**: A phase in early embryonic development during which the single-layered hollow sphere of cells (blastula) is reorganized into a multilayered structure (gastrula). By the end of gastrulation, the embryo has begun differentiation to establish distinct cell lineages and set up the basic axes of the body. **Genetic assimilation**: An alternative mechanism of variation (compared to mutations, in which genotype precedes phenotype) in which developmental plasticity creates novel phenotypes before heritable variation exists (phenotype precedes genotype). Genetic control over the new phenotype arises later through random mutations. **Genotype**: The particular type and arrangement of genes of an organism. **Hemispheric dominance**: The phenomenon that cognitive processes tend to be specialized to one side of the brain or the other, as demonstrated by aphasia following left hemisphere lesions and spatial neglect following right hemisphere lesions in most people. **Hemispheric functional segregation**: The division of labor in cognitive tasks between both hemispheres. In humans, hemispheric functional segregation shows a strong population bias toward prototypical segregation in which the left hemisphere is known to be dominant for language, fine motor control, and praxis (learned gestures), whereas the right hemisphere supports spatial attention, face recognition and prosody of speech. **Heterotaxy**: The loss of typical left–right laterality in the arrangement of the visceral organs along the superior–inferior axis, also referred to as situs ambiguus. **Kartagener syndrome**: A rare, autosomal recessive genetic ciliary disorder comprising the triad of situs inversus, chronic sinusitis, and bronchiectasis. **Lateral plate mesoderm**: A type of mesoderm that is found at the periphery of the embryo. ***Lefty***: A class of proteins related to the superfamily of growth factors that play a role in left–right asymmetry determination of organ systems during development. **Levocardia**: A condition where the heart is on the left (typical) side of the thoracic cavity. **Neural tube**: The embryonic precursor to the central nervous system, which is made up of the brain and spinal cord. **Neurulation**: The folding process in vertebrate embryos, which includes the transformation of the neural plate into the neural tube. ***Nodal***: A protein that is encoded by the human *NODAL* gene, which belongs to the transforming growth factor-beta superfamily. It is involved in cell differentiation in early embryogenesis, playing a key role in signal transfer from the primitive node, in the anterior primitive streak, to the lateral plate mesoderm. **Nodal flow**: The (leftward) movement of fluid at the primitive node caused by ciliary movement and taken to be a central process in symmetry breaking on the left–right axis. **Ortholog**: A homologous gene found in different species related by linear descent. **Phenotype**: The sum of an organism's observable characteristics or traits. ***Pitx2***: A protein that in humans is encoded by the *PITX2* gene. This protein acts as a transcription factor and is involved in developing the eye, tooth and abdominal organs. **Pleiotropy**: Occurs when one gene influences two or more seemingly unrelated phenotypic traits. Mutation in a pleiotropic gene may affect several traits simultaneously. **Primitive node**: The organizer for gastrulation in the vertebrate embryo. **Primary ciliary dyskinesia**: A rare, ciliopathic, genetically heterogeneous disorder that causes defects in the action of cilia lining the respiratory tract (lower and upper, sinuses, Eustachian tube, middle ear), fallopian tube, and flagellum of sperm cells. **Reversed typical hemispheric functional segregation**: Phenotype of hemispheric functional segregation in which the left–right laterality of functions is reversed than the typical organization seen in the population. While the habitual functional segregation is maintained, the phenotype is a mirror image of the usual functional brain organization. **Roof plate**: An embryonic organizing center consisting of specialized glial cells that occupy the dorsal midline of the vertebrate neural tube. The roof plate generates morphogenic signals along the length of the neuraxis, which control the specification and differentiation of dorsal neuronal cell types. ***Shh***: Sonic hedgehog (Shh) is a protein that, in humans, is encoded by the *SHH* gene. *Shh* plays a key role in developing many animals. In vertebrates, it is involved in organogenesis. **Signal-transducing pathway**: Signal transduction is the process by which a chemical or physical signal is transmitted through a cell as a series of molecular events, which ultimately results in a cellular response. The changes give rise to a chain of biochemical events known as a signaling pathway. **Situs ambiguus**: Medical term referring to a loss of the typical left–right positioning of thoracic and abdominal organs, also called heterotaxy. **Situs inversus (totalis)**: Medical term referring to a reversal of the typical position of thoracic and abdominal organs. **Situs solitus**: Medical term referring to the typical position of thoracic and abdominal organs. **Transcriptomics**: The study of the transcriptome—the complete set of RNA transcripts that are produced by the genome—using high-throughput methods, such as microarray analysis. **Typical hemispheric functional segregation**: Phenotype of hemispheric functional segregation that, due to a population bias, is most common in the human population. **Visceral asymmetry**: Refers to the asymmetry in left–right positioning of thoracic and abdominal organs.

References

1. Hamada, H.; Meno, C.; Watanabe, D.; Saijoh, Y. Establishment of vertebrate left-right asymmetry. *Nat. Rev. Genet.* **2002**, *3*, 103–113. [CrossRef]
2. Vingerhoets, G. Phenotypes in hemispheric functional segregation? Perspectives and challenges. *Phys. Life Rev.* **2019**. [CrossRef]
3. Long, S.; Ahmad, N.; Rebagliati, M. The zebrafish nodal-related gene southpaw is required for visceral and diencephalic left-right asymmetry. *Development* **2003**, *130*, 2303–2316. [CrossRef]
4. Concha, M.L.; Burdine, R.D.; Russell, C.; Schier, A.F.; Wilson, S.W. A nodal signaling pathway regulates the laterality of neuroanatomical asymmetries in the zebrafish forebrain. *Neuron* **2000**, *28*, 399–409. [CrossRef]
5. Wehrmaker, A. Right-Left Asymmetry and Situs Iversus in Triturus alpestris. *Wilhem Roux' Arch. Entwickl. Org.* **1969**, *163*, 1–32. [CrossRef] [PubMed]
6. Concha, M.L.; Wilson, S.W. Asymmetry in the epithalamus of vertebrates. *J. Anat.* **2001**, *199*, 63–84. [CrossRef] [PubMed]
7. Barth, K.A.; Miklosi, A.; Watkins, J.; Bianco, I.H.; Wilson, S.W.; Andrew, R.J. fsi zebrafish show concordant reversal of laterality of viscera, neuroanatomy, and a subset of behavioral responses. *Curr. Biol.* **2005**, *15*, 844–850. [CrossRef] [PubMed]
8. Klingenberg, C.P. Developmental instability as a research tool: Using patterns of fluctuating asymmetry to infer the developmental origins of morphological integration. In *Developmental Stability: Causes and Consequences*; Polak, M., Ed.; Oxford University Press: Oxford, UK, 2003; pp. 427–442.
9. Palmer, A.R. Symmetry breaking and the evolution of development. *Science* **2004**, *306*, 828–833. [CrossRef]
10. Nakamura, T.; Hamada, H. Left-right patterning: Conserved and divergent mechanisms. *Development* **2012**, *139*, 3257–3262. [CrossRef]
11. Aylsworth, A.S. Clinical aspects of defects in the determination of laterality. *Am. J. Med. Genet.* **2001**, *101*, 345–355. [CrossRef]

12. Lin, A.E.; Krikov, S.; Riehle-Colarusso, T.; Frias, J.L.; Belmont, J.; Anderka, M.; Geva, T.; Getz, K.D.; Botto, L.D.; Prevention, N.B.D. Laterality Defects in the National Birth Defects Prevention Study (1998–2007): Birth Prevalence and Descriptive Epidemiology. *Am. J. Med. Genet. Part A* **2014**, *164*, 2581–2591. [CrossRef]
13. Houyel, L.; Khoshnood, B.; Anderson, R.H.; Lelong, N.; Thieulin, A.C.; Goffinet, F.; Bonnet, D.; Grp, E.S. Population-based evaluation of a suggested anatomic and clinical classification of congenital heart defects based on the International Paediatric and Congenital Cardiac Code. *Orphanet J. Rare Dis.* **2011**, *6*. [CrossRef]
14. Casey, B. Two rights make a wrong: Human left-right malformations. *Hum. Mol. Genet.* **1998**, *7*, 1565–1571. [CrossRef]
15. Kosaki, K.; Casey, B. Genetics of human left-right axis malformations. *Semin. Cell Dev. Biol.* **1998**, *9*, 89–99. [CrossRef] [PubMed]
16. Lin, A.E.; Ticho, B.S.; Houde, K.; Westgate, M.N.; Holmes, L.B. Heterotaxy: Associated conditions and hospital-based prevalence in newborns. *Genet. Med.* **2000**, *2*, 157–172. [CrossRef] [PubMed]
17. Evans, W.N.; Acherman, R.J.; Restrepo, H. Heterotaxy in Southern Nevada: Prenatal Detection and Epidemiology. *Pediatric Cardiol.* **2015**, *36*, 930–934. [CrossRef]
18. Peeters, H.; Devriendt, K. Human laterality disorders. *Eur. J. Med. Genet.* **2006**, *49*, 349–362. [CrossRef] [PubMed]
19. Sutherland, M.J.; Ware, S.M. Disorders of Left-Right Asymmetry: Heterotaxy and Situs Inversus. *Am. J. Med. Genet. Part C Semin. Med. Genet.* **2009**, *151c*, 307–317. [CrossRef]
20. Schmutzer, K.J.; Linde, L.M. Situs Inversus Totalis Associated with Complex Cardiovascular Anomalies. *Am. Heart J.* **1958**, *56*, 761–768. [CrossRef]
21. Chen, W.; Guo, Z.; Qian, L.; Wang, L. Comorbidities in situs inversus totalis: A hospital-based study. *Birth Defects Res* **2020**, *112*, 418–426. [CrossRef]
22. Kennedy, M.P.; Omran, H.; Leigh, M.W.; Dell, S.; Morgan, L.; Molina, P.L.; Robinson, B.V.; Minnix, S.L.; Olbrich, H.; Severin, T.; et al. Congenital heart disease and other heterotaxic defects in a large cohort of patients with primary ciliary dyskinesia. *Circulation* **2007**, *115*, 2814–2821. [CrossRef]
23. Gabriel, G.C.; Lo, C.W. Left-right patterning in congenital heart disease beyond heterotaxy. *Am. J. Med. Genet. Part C Semin. Med. Genet.* **2020**, *184*, 90–96. [CrossRef]
24. Deng, H.; Xia, H.; Deng, S. Genetic basis of human left-right asymmetry disorders. *Expert Rev. Mol. Med.* **2015**, *16*, e19. [CrossRef]
25. Postema, M.C.; Carrion-Castillo, A.; Fisher, S.E.; Vingerhoets, G.; Francks, C. The genetics of situs inversus without primary ciliary dyskinesia. *Sci. Rep.* **2020**, *10*. [CrossRef]
26. Lucas, J.S.; Burgess, A.; Mitchison, H.M.; Moya, E.; Williamson, M.; Hogg, C.; Serv, N.P. Diagnosis and management of primary ciliary dyskinesia. *Arch. Dis. Child.* **2014**, *99*, 850–856. [CrossRef]
27. Afzelius, B.A. Human Syndrome Caused by Immotile Cilia. *Science* **1976**, *193*, 317–319. [CrossRef]
28. Bartoloni, L.; Blouin, J.L.; Pan, Y.Z.; Gehrig, C.; Maiti, A.K.; Scamuffa, N.; Rossier, C.; Jorissen, M.; Armengot, M.; Meeks, M.; et al. Mutations in the DNAH11 (axonemal heavy chain dynein type 11) gene cause one form of situs inversus totalis and most likely primary ciliary dyskinesia. *Proc. Natl. Acad. Sci. USA* **2002**, *99*, 10282–10286. [CrossRef]
29. Leigh, M.W.; Pittman, J.E.; Carson, J.L.; Ferkol, T.W.; Dell, S.D.; Davis, S.D.; Knowles, M.R.; Zariwala, M.A. Clinical and genetic aspects of primary ciliary dyskinesia/Kartagener syndrome. *Genet Med.* **2009**, *11*, 473–487. [CrossRef]
30. Zariwala, M.A.; Omran, H.; Ferkol, T.W. The emerging genetics of primary ciliary dyskinesia. *Proc. Am. Thorac. Soc.* **2011**, *8*, 430–433. [CrossRef]
31. Bisgrove, B.W.; Yost, H.J. The roles of cilia in developmental disorders and disease. *Development* **2006**, *133*, 4131–4143. [CrossRef]
32. Afzelius, B.A.; Stenram, U. Prevalence and genetics of immotile-cilia syndrome and left-handedness. *Int. J. Dev. Biol.* **2006**, *50*, 571–573. [CrossRef]
33. Bush, A.; Cole, P.; Hariri, M.; Mackay, I.; Phillips, G.; O'Callaghan, C.; Wilson, R.; Warner, J.O. Primary ciliary dyskinesia: Diagnosis and standards of care. *Eur. Respir. J.* **1998**, *12*, 982–988. [CrossRef]
34. Kartagener, M. Zur Pathogenese der Bronchiektasien bei Situs viscerum inversus. *Beiträge Klin. Tuberk.* **1933**, *83*, 489–501. [CrossRef]
35. Engesaeth, V.G.; Warner, J.O.; Bush, A. New Associations of Primary Ciliary Dyskinesia Syndrome. *Pediatric Pulmonol.* **1993**, *16*, 9–12. [CrossRef] [PubMed]
36. Noone, P.G.; Bali, D.; Carson, J.L.; Sannuti, A.; Gipson, C.L.; Ostrowski, L.E.; Bromberg, P.A.; Boucher, R.C.; Knowles, M.R. Discordant organ laterality in monozygotic twins with primary ciliary dyskinesia. *Am. J. Med. Genet.* **1999**, *82*, 155–160. [CrossRef]
37. Vandenberg, L.N.; Levin, M. A unified model for left-right asymmetry? Comparison and synthesis of molecular models of embryonic laterality. *Dev. Biol.* **2013**, *379*, 1–15. [CrossRef] [PubMed]
38. Komatsu, Y.; Mishina, Y. Establishment of left-right asymmetry in vertebrate development: The node in mouse embryos. *Cell. Mol. Life Sci.* **2013**, *70*, 4659–4666. [CrossRef] [PubMed]
39. Hirokawa, N.; Tanaka, Y.; Okada, Y. Left-Right Determination: Involvement of Molecular Motor KIF3, Cilia, and Nodal Flow. *Cold Spring Harb. Perspect. Biol.* **2009**, 1. [CrossRef]
40. Vandenberg, L.N.; Levin, M. Far From Solved: A Perspective on What We Know About Early Mechanisms of Left-Right Asymmetry. *Dev. Dyn.* **2010**, *239*, 3131–3146. [CrossRef] [PubMed]
41. Basu, B.; Bruedner, M. Cilia: Multifunctional Organelles at the Center of Vertebrate Left-Right Asymmetry. *Ciliary Funct. Mamm. Dev.* **2008**, *85*, 151–174. [CrossRef]

42. Nonaka, S.; Shiratori, H.; Saijoh, Y.; Hamada, H. Determination of left-right patterning of the mouse embryo by artificial nodal flow. *Nature* **2002**, *418*, 96–99. [CrossRef] [PubMed]
43. Watanabe, D.; Saijoh, Y.; Nonaka, S.; Sasaki, G.; Ikawa, Y.; Yokoyama, T.; Hamada, H. The left-right determinant Inversin is a component of node monocilia and other 9+0 cilia. *Development* **2003**, *130*, 1725–1734. [CrossRef]
44. Okada, Y.; Nonaka, S.; Tanaka, Y.; Saijoh, Y.; Hamada, H.; Hirokawa, N. Abnormal nodal flow precedes situs inversus in iv and inv mice. *Mol. Cell* **1999**, *4*, 459–468. [CrossRef]
45. Smith, D.J.; Montenegro-Johnson, T.D.; Lopes, S.S. Symmetry-Breaking Cilia-Driven Flow in Embryogenesis. *Annu. Rev. Fluid Mech.* **2019**, *51*, 105–128. [CrossRef]
46. Rogers, L.J.; Zucca, P.; Vallortigara, G. Advantages of having a lateralized brain. *Proc. R. Soc. B Biol. Sci.* **2004**, *271*, S420–S422. [CrossRef] [PubMed]
47. Vallortigara, G.; Rogers, L.J.; Bisazza, A. Possible evolutionary origins of cognitive brain lateralization. *Brain Res. Rev.* **1999**, *30*, 164–175. [CrossRef]
48. Cai, Q.; Van der Haegen, L.; Brysbaert, M. Complementary hemispheric specialization for language production and visuospatial attention. *Proc. Natl. Acad. Sci. USA* **2013**, *110*, E322–E330. [CrossRef]
49. Vingerhoets, G.; Alderweireldt, A.S.; Vandemaele, P.; Cai, Q.; Van der Haegen, L.; Brysbaert, M.; Achten, E. Praxis and language are linked: Evidence from co-lateralization in individuals, with atypical language dominance. *Cortex* **2013**, *49*, 172–183. [CrossRef]
50. Gerrits, R.; Van der Haegen, L.; Brysbaert, M.; Vingerhoets, G. Laterality for recognizing written words and faces in the fusiform gyrus covaries with language dominance. *Cortex J. Devoted Study Nerv. Syst. Behav.* **2019**, *117*, 196–204. [CrossRef]
51. Gerrits, R.; Verhelst, H.; Vingerhoets, G. Mirrored brain organization: Statistical anomaly or reversal of hemispheric functional segregation bias? *Proc. Natl. Acad. Sci. USA* **2020**, *117*, 14057–14065. [CrossRef]
52. Bryden, M.P.; Hecaen, H.; Deagostini, M. Patterns of Cerebral Organization. *Brain Lang.* **1983**, *20*, 249–262. [CrossRef]
53. McNeely, H.E.; Parlow, S.E. Complementarity of linguistic and prosodic processes in the intact brain. *Brain Lang.* **2001**, *79*, 473–481. [CrossRef]
54. Floel, A.; Buyx, A.; Breitenstein, C.; Lohmann, H.; Knecht, S. Hemispheric lateralization of spatial attention in right- and left-hemispheric language dominance. *Behav. Brain Res.* **2005**, *158*, 269–275. [CrossRef]
55. Whitehouse, A.J.O.; Bishop, D.V.M. Hemispheric division of function is the result of independent probabilistic biases. *Neuropsychologia* **2009**, *47*, 1938–1943. [CrossRef]
56. Badzakova-Trajkov, G.; Haberling, I.S.; Roberts, R.P.; Corballis, M.C. Cerebral Asymmetries: Complementary and independent processes. *PLoS ONE* **2010**, *5*, e9682. [CrossRef]
57. Rosch, R.E.; Bishop, D.V.M.; Badcock, N.A. Lateralised visual attention is unrelated to language lateralisation, and not influenced by task difficulty—A functional transcranial Doppler study. *Neuropsychologia* **2012**, *50*, 810–815. [CrossRef]
58. Groen, M.A.; Whitehouse, A.J.O.; Badcock, N.A.; Bishop, D.V.M. Does cerebral lateralization develop? A study using functional transcranial Doppler ultrasound assessing lateralization for language production and visuospatial memory. *Brain Behav.* **2012**, *2*, 256–269. [CrossRef]
59. Zago, L.; Petit, L.; Mellet, E.; Jobard, G.; Crivello, F.; Joliot, M.; Mazoyer, B.; Tzourio-Mazoyer, N. The association between hemispheric specialization for language production and for spatial attention depends on left-hand preference strength. *Neuropsychologia* **2016**, *93*, 394–406. [CrossRef]
60. Francks, C. Exploring human brain lateralization with molecular genetics and genomics. *Ann. N. Y. Acad. Sci.* **2015**, *1359*, 1–13. [CrossRef]
61. Vingerhoets, G.; Gerrits, R.; Bogaert, S. Atypical brain functional segregation is more frequent in situs inversus totalis. *Cortex* **2018**, *106*, 12–25. [CrossRef]
62. Mazoyer, B.; Zago, L.; Jobard, G.; Crivello, F.; Joliot, M.; Perchey, G.; Mellet, E.; Petit, L.; Tzourio-Mazoyer, N. Gaussian Mixture Modeling of Hemispheric Lateralization for Language in a Large Sample of Healthy Individuals Balanced for Handedness. *PLoS ONE* **2014**, *9*, e101165. [CrossRef]
63. Zago, L.; Herve, P.Y.; Genuer, R.; Laurent, A.; Mazoyer, B.; Tzourio-Mazoyer, N.; Joliot, M. Predicting Hemispheric Dominance for Language Production in Healthy Individuals Using Support Vector Machine. *Hum. Brain Mapp.* **2017**, *38*, 5871–5889. [CrossRef]
64. Bernal, B.; Ardila, A. Bilateral representation of language: A critical review and analysis of some unusual cases. *J. Neurolinguist.* **2014**, *28*, 63–80. [CrossRef]
65. Janecek, J.K.; Swanson, S.J.; Sabsevitz, D.S.; Hammeke, T.A.; Raghavan, M.; Rozman, M.E.; Binder, J.R. Language lateralization by fMRI and Wada testing in 229 patients with epilepsy: Rates and predictors of discordance. *Epilepsia* **2013**, *54*, 314–322. [CrossRef] [PubMed]
66. Moddel, G.; Lineweaver, T.; Schuele, S.U.; Reinholz, J.; Loddenkemper, T. Atypical language lateralization in epilepsy patients. *Epilepsia* **2009**, *50*, 1505–1516. [CrossRef]
67. Sun, T.; Walsh, C.A. Molecular approaches to brain asymmetry and handedness. *Nat. Rev. Neurosci.* **2006**, *7*, 655–662. [CrossRef] [PubMed]
68. Thomas, S.; Boutaud, L.; Reilly, M.L.; Benmerah, A. Cilia in hereditary cerebral anomalies. *Biol. Cell* **2019**, *111*, 217–231. [CrossRef]
69. Medina, L. Evolution and Embryological Development of Forebrain. In *Encyclopedia of Neuroscience*; Binder, M.D., Hirokawa, N., Windhorst, U., Eds.; Springer: Berlin/Heidelberg, Germany, 2009; pp. 1172–1192. [CrossRef]

70. Sun, T.; Patoine, C.; Abu-Khalil, A.; Visvader, J.; Sum, E.; Cherry, T.J.; Orkin, S.H.; Geschwind, D.H.; Walsh, C.A. Early asymmetry of gene transcription in embryonic human left and right cerebral cortex. *Science* **2005**, *308*, 1794–1798. [CrossRef]
71. Fukuchi-Shimogori, T.; Grove, E.A. Neocortex patterning by the secreted signaling molecule FGF8. *Science* **2001**, *294*, 1071–1074. [CrossRef] [PubMed]
72. Puelles, L.; Martínez, S.; Martinez-de-la-Torre, M.; Rubenstein, J.L.R. Gene maps and related histogenic domains in the forebrain and midbrain. In *The rat Nervous System*; Paxinos, G., Ed.; Elsevier: Amsterdam, The Netherlands, 2004; pp. 3–25.
73. De Kovel, C.G.F.; Lisgo, S.N.; Fisher, S.E.; Francks, C. Subtle left-right asymmetry of gene expression profiles in embryonic and foetal human brains. *Sci. Rep.* **2018**, *8*. [CrossRef]
74. De Kovel, C.G.F.; Lisgo, S.; Karlebach, G.; Ju, J.; Cheng, G.; Fisher, S.E.; Francks, C. Left-Right Asymmetry of Maturation Rates in Human Embryonic Neural Development. *Biol. Psychiatry* **2017**, *82*, 204–212. [CrossRef]
75. Schmitz, J.; Lor, S.; Klose, R.; Gunturkun, O.; Ocklenburg, S. The Functional Genetics of Handedness and Language Lateralization: Insights from Gene Ontology, Pathway and Disease Association Analyses. *Front. Psychol.* **2017**, *8*, 1144. [CrossRef] [PubMed]
76. Abu-Rustum, R.S.; Ziade, M.F.; Abu-Rustum, S.E. Reference Values for the Right and Left Fetal Choroid Plexus at 11 to 13 Weeks An Early Sign of "Developmental" Laterality? *J. Ultrasound Med.* **2013**, *32*, 1623–1629. [CrossRef] [PubMed]
77. Vasung, L.; Rollins, C.K.; Yun, H.J.; Velasco-Annis, C.; Zhang, J.; Wagstyl, K.; Evans, A.; Warfield, S.K.; Feldman, H.A.; Grant, P.E.; et al. Quantitative In vivo MRI Assessment of Structural Asymmetries and Sexual Dimorphism of Transient Fetal Compartments in the Human Brain. *Cereb. Cortex* **2020**, *30*, 1752–1767. [CrossRef]
78. Kasprian, G.; Langs, G.; Brugger, P.C.; Bittner, M.; Weber, M.; Arantes, M.; Prayer, D. The Prenatal Origin of Hemispheric Asymmetry: An In Utero Neuroimaging Study. *Cereb. Cortex* **2011**, *21*, 1076–1083. [CrossRef]
79. Habas, P.A.; Scott, J.A.; Roosta, A.; Rajagopalan, V.; Kim, K.; Rousseau, F.; Barkovich, A.J.; Glenn, O.A.; Studholme, C. Early Folding Patterns and Asymmetries of the Normal Human Brain Detected from in Utero MRI. *Cereb. Cortex* **2012**, *22*, 13–25. [CrossRef]
80. Yun, H.J.; Vasung, L.; Tarui, T.; Rollins, C.K.; Ortinau, C.M.; Grant, P.E.; Im, K. Temporal Patterns of Emergence and Spatial Distribution of Sulcal Pits During Fetal Life. *Cereb. Cortex* **2020**, *30*, 4257–4268. [CrossRef]
81. Dubois, J.; Benders, M.; Lazeyras, F.; Borradori-Tolsa, C.; Leuchter, R.H.V.; Mangin, J.F.; Huppi, P.S. Structural asymmetries of perisylvian regions in the preterm newborn. *Neuroimage* **2010**, *52*, 32–42. [CrossRef]
82. Karlebach, G.; Francks, C. Lateralization of gene expression in human language cortex. *Cortex* **2015**, *67*, 30–36. [CrossRef]
83. Brandler, W.M.; Morris, A.P.; Evans, D.M.; Scerri, T.S.; Kemp, J.P.; Timpson, N.J.; St Pourcain, B.; Smith, G.D.; Ring, S.M.; Stein, J.; et al. Common Variants in Left/Right Asymmetry Genes and Pathways Are Associated with Relative Hand Skill. *PLoS Genet.* **2013**, *9*, e1003751. [CrossRef]
84. Brandler, W.M.; Paracchini, S. The genetic relationship between handedness and neurodevelopmental disorders. *Trends Mol. Med.* **2014**, *20*, 83–90. [CrossRef]
85. Moher, D.L.A.; Tetzlaff, J.; Altman, D.G.; The PRISMA Group. Preferred reporting items for systematic reviews and meta-analyses: The PRISMA statement. *PLoS Med.* **2009**, *6*, e1000097. [CrossRef]
86. Kienast, P.; Schwartz, E.; Diogo, M.C.; Gruber, G.M.; Brugger, P.C.; Kiss, H.; Ulm, B.; Bartha-Doering, L.; Seidl, R.; Weber, M.; et al. The Prenatal Origins of Human Brain Asymmetry: Lessons Learned from a Cohort of Fetuses with Body Lateralization Defects. *Cereb. Cortex* **2021**. [CrossRef] [PubMed]
87. McManus, I.C. Reversed bodies, reversed brains, and (some) reversed behaviors: Of zebrafish and men. *Dev. Cell* **2005**, *8*, 796–797. [CrossRef]
88. Cockayne, E.A. The genetics of transposition of the viscera. *Q. J. Med. New Ser.* **1938**, *27*, 479–493.
89. Torgersen, J. Situs inversus, asymmetry and twinning. *Am. J. Hum. Genet.* **1950**, *2*, 361–370.
90. Papadatou-Pastou, M.; Ntolka, E.; Schmitz, J.; Martin, M.; Munafo, M.R.; Ocklenburg, S.; Paracchini, S. Human Handedness: A Meta-Analysis. *Psychol. Bull.* **2020**, *146*, 481–524. [CrossRef]
91. Everett, H.C. Situs Inversus Totalis—A Survey of Laterality and Some Observations on Frequency of Mental Disorder. *Am. J. Psychiatry* **1963**, *119*, 884. [CrossRef]
92. Matsumoto, T.; Kuriya, N.; Akagi, T.; Ohbu, K.; Toyoda, O.; Morita, J.; Ichikawa, K.; Matsuishi, T.; Hayashi, M.; Kato, H. Handedness and laterality of the viscera. *Neurology* **1997**, *49*, 1751. [CrossRef] [PubMed]
93. Mcmanus, I.C.; Martin, N.; Stubbings, G.F.; Chung, E.M.K.; Mitchison, H.M. Handedness and situs inversus in primary ciliary dyskinesia. *Proc. R. Soc. B Biol. Sci.* **2004**, *271*, 2579–2582. [CrossRef]
94. Woods, R.P. Brain Asymmetries in Situs-Inversus—A Case-Report and Review of the Literature. *Arch. Neurol.* **1986**, *43*, 1083–1084. [CrossRef]
95. Tanaka, S.; Kanzaki, R.; Yoshibayashi, M.; Kamiya, T.; Sugishita, M. Dichotic listening in patients with situs inversus: Brain asymmetry and situs asymmetry. *Neuropsychologia* **1999**, *37*, 869–874. [CrossRef]
96. Kennedy, D.N.; O'Craven, K.M.; Ticho, B.S.; Goldstein, A.M.; Makris, N.; Henson, J.W. Structural and functional brain asymmetries in human situs inversus totalis. *Neurology* **1999**, *53*, 1260–1265. [CrossRef]
97. Schuler, A.L.; Kasprian, G.; Schwartz, E.; Seidl, R.; Diogo, M.C.; Mitter, C.; Langs, G.; Prayer, D.; Bartha-Doering, L. Mens inversus in corpore inverso? Language lateralization in a boy with situs inversus totalis. *Brain Lang* **2017**, *174*, 9–15. [CrossRef]
98. Vingerhoets, G.; Li, X.; Hou, L.; Bogaert, S.; Verhelst, H.; Gerrits, R.; Siugzdaite, R.; Roberts, N. Brain structural and functional asymmetry in human situs inversus totalis. *Brain Struct. Funct.* **2018**, *223*, 1937–1952. [CrossRef]

99. Cohen, L.; Geny, C.; Hermine, O.; Gray, F.; Degos, J.D. Crossed Aphasia with Visceral Situs-Inversus. *Ann. Neurol.* **1993**, *33*, 215–218. [CrossRef]
100. Ihara, A.; Hirata, M.; Fujimaki, N.; Goto, T.; Umekawa, Y.; Fujita, N.; Terazono, Y.; Matani, A.; Wei, Q.; Yoshimine, T.; et al. Neuroimaging study on brain asymmetries in situs inversus totalis. *J. Neurol. Sci.* **2010**, *288*, 72–78. [CrossRef]
101. Xiang, L.; Crow, T.; Roberts, N. Cerebral torque is human specific and unrelated to brain size. *Brain Struct. Funct.* **2019**, *224*, 1141–1150. [CrossRef]
102. Balzeau, A.; Gilissen, E.; Grimaud-Herve, D. Shared Pattern of Endocranial Shape Asymmetries among Great Apes, Anatomically Modern Humans, and Fossil Hominins. *PLoS ONE* **2012**, *7*. [CrossRef]
103. Tubbs, R.S.; Wellons, J.C.; Salter, G.; Blount, J.P.; Oakes, W.J. Intracranial anatomic asymmetry in situs inversus totalis. *Anat. Embryol.* **2003**, *206*, 199–202. [CrossRef]
104. Leroy, F.; Cai, Q.; Bogart, S.L.; Dubois, J.; Coulon, O.; Monzalvo, K.; Fischer, C.; Glasel, H.; Van der Haegen, L.; Benezita, A.; et al. New human-specific brain landmark: The depth asymmetry of superior temporal sulcus. *Proc. Natl. Acad. Sci. USA* **2015**, *112*, 1208–1213. [CrossRef] [PubMed]
105. Mannaert, L.; Verhelst, H.; Gerrits, R.; Bogaert, S.; Vingerhoets, G. White matter asymmetries in human situs inversus totalis. *Brain Struct. Funct.* **2019**, *224*, 2559–2565. [CrossRef] [PubMed]

 MDPI

Review

How Asymmetries Evolved: Hearts, Brains, and Molecules

Michael C. Corballis

School of Psychology, University of Auckland, Auckland 1142, New Zealand; m.corballis@auckland.ac.nz; Tel.: +642-211365674

Abstract: Humans belong to the vast clade of species known as the bilateria, with a bilaterally symmetrical body plan. Over the course of evolution, exceptions to symmetry have arisen. Among chordates, the internal organs have been arranged asymmetrically in order to create more efficient functioning and packaging. The brain has also assumed asymmetries, although these generally trade off against the pressure toward symmetry, itself a reflection of the symmetry of limbs and sense organs. In humans, at least, brain asymmetries occur in independent networks, including those involved in language and manual manipulation biased to the left hemisphere, and emotion and face perception biased to the right. Similar asymmetries occur in other species, notably the great apes. A number of asymmetries are correlated with conditions such as dyslexia, autism, and schizophrenia, and have largely independent genetic associations. The origin of asymmetry itself, though, appears to be unitary, and in the case of the internal organs, at least, may depend ultimately on asymmetry at the molecular level.

Keywords: bilateria; cerebral asymmetry; handedness; language; molecular asymmetry; situs

Citation: Corballis, M.C. How Asymmetries Evolved: Hearts, Brains, and Molecules. *Symmetry* **2021**, *13*, 914. https://doi.org/10.3390/sym13060914

Academic Editors: Sebastian Ocklenburg, Onur Güntürkün and Chiara Spironelli

Received: 15 April 2021
Accepted: 17 May 2021
Published: 21 May 2021

Publisher's Note: MDPI stays neutral with regard to jurisdictional claims in published maps and institutional affiliations.

1. The Symmetrical Background

The evolution of asymmetry should be understood in relation to its opposite, the overwhelming bilateral symmetry which characterises the vast clade of organisms to which we belong. These are the bilateria. They go back at least to the Cambrian, beginning some 541 million years ago, and probably slightly earlier into the late Protozeroic [1]. Bilateral symmetry emerged in species that move in space, and depends on the prior establishment of two bodily axes. The antero–posterior axis may have arisen first in relation to feeding, involving openings at head and tail separated by a through-gut [2], as in worms that burrow. The demands of locomotion led further to sense organs, such as eyes and nose, oriented toward the direction of motion, and the limbs were shaped to facilitate linear motion in a consistent direction, further defining the antero–posterior axis. The dorsal–ventral axis evolved later through the influence of gravity and the demands of locomotion, creating consistent differences between top and bottom, such as eyes placed high for distance vision and feet touching the ground. The formation of these two axes, with their distinctive asymmetries, appears to be highly conserved genetically, at least across vertebrates and arthropods [3].

Only when these two axes are established can the left–right axis be defined, and the body remains highly symmetrical along this axis. The great British scientist Sir Isaac Newton remarked that this symmetry, with the exception only of the bowels, proved "the counsel and contrivance of an Author." There is no need, though, to appeal to a deity; bilateral symmetry can be understood in evolutionary terms. As an animal moves around, the environment it encounters is largely indifferent to whether things are on the left or right. Predators and prey and obstructions to movement can occur on either side. With respect to movement and orientation in space, there seem to be no contingencies favouring differences between the left and right sides of animals.

Bilateral symmetry, though, is not merely a matter of default; it also enhances biological fitness. In animals that move freely, locomotion is almost universally dependent on paired limbs, be they legs, flippers, or wings, and symmetry ensures linear movement, which provides the most efficient way to journey between two points. Having one leg longer than the other, or functionally more efficient, might leave an animal moving in circles, or at least making multiple corrections. Animals also need to be as sensitive to features on their left as on the right if they are to respond optimally to danger or to exploit what the environment has to offer. This means that sense organs, such as eyes, ears, and skin receptors, are symmetrically placed.

Much of behaviour is a matter of programming movement or processing information provided by the senses, creating evolutionary pressure for the brain itself to be symmetrical. Indeed, for much of human history the two sides of the brain were considered duplicates, albeit mirror images. Descartes [4], for example, observed "the brain to be double" (p. 275). In terms of gross anatomy, at least, the left and right sides did seem to be mirror images, causing the French physician Marie Francois Xavier Bichat (1771–1802) to formulate the "law of symmetry." Bichat died at the age of 30 and was not widely known at the time, but his law of symmetry gained wide currency in the 19th century, especially through the influence of Franz Joseph Gall (1758–1828) [5].

There is even pressure for the brain to preserve its bilateral symmetry in the face of asymmetrical experience. It is well established that animals have more difficulty learning to distinguish left–right mirror images than up-down mirror images, and tend to treat left–right mirror images as though they were the same [6]. In one classic experiment, people shown 2500 pictures were later able to recognise them with surprising accuracy, except that they were as likely to report a picture as familiar when it was the left–right reverse of the original as when it was the original itself [7]. So-called left–right equivalence is especially evident in young children learning to read; up until the age of six or so, they frequently write letters or words backwards, despite being shown them only in correct orientation [8]. Animals, too, have much more difficulty discriminating left–right mirror images than in discriminating up-down mirror images. Left–right equivalence is adaptive in the natural world, where objects or animals can occur in opposite profiles, and events on one side of the body might next time occur on the other side.

The equivalence of left–right mirror images can be attributed to a process of interhemispheric reversal in the formation of memories, so that memories are held both in the veridical format and the mirrored one [9]. Logothetis et al. [10] found that some single cells in the inferotemporal cortex of two adult rhesus monkeys responded equivalently to meaningless mirror-image shapes, and remarked that "Distinguishing mirror images has no apparent usefulness to any animal" (p. 360). It can be an impediment, though, in some human activities, notably in learning to read and write scripts written in a consistent left–right direction. With specific training, it can be overcome. Torres et al. [11] found that three weeks of training first-grade children to discriminate mirror-image letters, such as b and d, led to a doubling of reading speed: "a simple and cost-effective way to unleash the reading fluency potential of millions of children worldwide" (p. 742).

2. Emerging Asymmetries

2.1. Internal Organs

Bilateral symmetry, then, is a striking feature of nearly all animals, but there are also longstanding asymmetries. The most extreme example is situs solitus, the asymmetrical placement of thoracic and abdominal organs. In the vast majority of humans, for example, the heart is displaced to the left, along with the stomach, spleen, and aorta, while the liver, gall bladder and trilobed lung are displaced to the right. Approximately one in 10,000 individuals have situs inversus totalis, in which these asymmetries are reversed [12] and where it does occur, it seems to arise as a matter of chance when the normal directional influence is lacking [13,14]. The asymmetries are fundamentally the same in all vertebrates [15], and more generally in chordates, suggesting that they have a common origin and go far back in evolution [16].

In vertebrates, at least, imposing asymmetry on internal organs is adaptive. For example, a mass of muscle such as the heart, achieves much greater efficiency of pumping from a spirally coiled form than from a simple tube [17]. Beyond that, it is probably essentially a matter of efficient packaging in the human body. Just as it would be inefficient to pack a suitcase while maintaining perfect symmetry of the contents, so it is that the internal organs are arranged asymmetrically in the body. Similarly, design of an automobile abandons symmetry in its internal engine and controls, while largely maintaining symmetry of external body shape. Manufacturing has adopted a design long evident in biological evolution.

Deviations from bilateral symmetry can occur through random influences—no animal is perfectly symmetrical, even discounting the asymmetries of the internal organs. However, reliance on random or fluctuating asymmetry for internal organs would run the risk of error, so consistent asymmetry was stamped in early in evolution. Situs solitus, is clearly an ancestral condition, and is all but universal. Bilateral symmetry and asymmetry therefore coexist in a trade off, with pressure toward one vying with pressure toward the other.

2.2. Handedness

The clearest evidence of a trade off comes from use of the hands or forelimbs, which in some species is symmetrical while in others there seems a clear species-wide preference for one or the other in certain actions. For most animals, bilateral symmetry of the limbs is adaptive, especially in movement, but in bipedal animals the forelimbs are freed from locomotion and are potentially open to specialization. Symmetry of action can still be adaptive in reaching and grasping with the hands, but in more complex actions, biological fitness may benefit if the hands adopt complementary roles, such as one hand holding an object while the other operates on it. In some cases, one hand assumes a dominant role. For example, bipedal marsupials, such as kangaroos, show a 90 percent preference for the left hand when feeding, whereas quadrupedal marsupials, such as the sugar glider or grey short-tailed opossum, show no preference at the population level [18]. Cats and dogs, too, show no bias at the population level, but individual animals often show a consistent preference for one or other paw in activities such as reaching [19]. (For a more general review of limb preferences in non-human vertebrates, see [20].) Our closest non-human relatives, chimpanzees, are less consistently bipedal than are we humans, and correspondingly show lower right-hand preference, at approximately 65–70 percent, in intricate manual actions [21]. Gorillas are predominantly right-handed in bimanual actions, where the non-dominant hand holds a food-related object and the dominant hand performs actions on it, such as dipping, stripping, or extracting [22].

At least one study has shown a slight right-hand advantage for rhesus monkeys but no bias in capuchins [23]. It is not restricted to primates; for example, some 77% of walruses display a preference for the right flipper when feeding [24]. Some creatures, though, are clearly left-handed—or left-"limbed." In some species of parrot, approximately 90% of individuals show a preference for using the left foot when picking up pieces of food [25], and as we have seen bipedal kangaroos are predominantly left-handed. The preference for one or other limb being dominant is seldom if ever absolute, with the dominance ranging from approximately 65 to approximately 90 percent [26].

In humans, bipedalism is obligate and the hands are correspondingly less involved in locomotion and more available for asymmetrical activities such as tool manufacture, throwing, and writing, in all of which the right hand is dominant in some 90 percent of the population. Yet, the symmetry between the hands is largely preserved in their basic anatomy as well as in simple operations, such as reaching and grasping, and even catching. People can intercept a moving object equally well with either hand, but throw much more efficiently with just one hand, usually the right [27]. Most cricketers or baseball players, for example, can make one-handed catches with either hand, but few can throw even adequately with the non-preferred hand. The trade off between symmetry and asymmetry is therefore well illustrated in the way we use our hands.

2.3. Cerebral Asymmetry in Humans

Perhaps the first intimation of an exception to the law of symmetry as applied to the brain arose at a meeting in Montpellier in 1836, when an obscure French physician called Marc Dax produced evidence that speech was localised in the left hemisphere. This was largely disregarded, but some twenty-five years later, a more eminent physician called Paul Broca [28,29] showed that speech was disrupted following damage to the portion of the left prefrontal cortex since labelled as Broca's area, confirming the left-hemispheric dominance for speech. At that point, Dax's son recognised the significance of his father's work and arranged to have the early manuscript published, along with further evidence from 140 patients [30]. Evidence also emerged that comprehension of speech was impaired after damage in the left superior temporal gyrus, in the area since known as Wernicke's area. [31]. By the late 19th century, then, the brain was understood to exhibit some fundamental asymmetries, at least in function, in spite of its seeming anatomical symmetry. At this point, it was recognised that handedness itself was due to brain asymmetry, adding to the notion that the left hemisphere was the dominant or major hemisphere, with the right relegated to minor status. With some hesitation from the French medical establishment [32], the law of symmetry was overturned.

These developments also led to the view that the two sides of the brain were not simply uneven, but functioned in some ways as complementary opposites. In the most extreme versions, the left hemisphere was said to harness humanity, volition, masculinity, and reason, while animality, instinct, femininity, and madness were closeted in the right. This phase of speculation is well described by the historian Anne Harrington [33], who observed that it probably owed more to the social prejudices of the time than to the neurological facts. She wrote, "It is interesting that, once one has given the two hemispheres sexual identities, the idea of cerebral dominance becomes a rather apt metaphor for the social and economic domination of men over women in 19th-century Europe" (p. 624).

These extreme notions seemed to subside after the turn of the 20th century, but a second wave of speculation followed the split-brain research of the 1960s, when a series of patients underwent section of the forebrain commissures for the relief of intractable epilepsy. Again, the left hemisphere was shown to be dominant for language [34], and in 1981, Roger W. Sperry belatedly received the Nobel Prize in Physiology or Medicine "for his discoveries concerning the functional specialization of the cerebral hemispheres." There again followed a barrage of speculation about the duality of mind, with the left brain described as logical, rational and mechanistic, and the right brain intuitive, emotional and creative [35]. The social and political pressures of the time were different from those of the previous century, and the protests against the war in Vietnam, feminism, and anti-establishment movements seemed generally to anoint the right brain as favoured over the militaristic left. In his Nobel address, Sperry [36] himself noted, "The left-right dichotomy in cognitive mode is an idea with which it is very easy to run wild" (p. 1226). The dichotomy is still with us in popular culture—and indeed often runs wild.

Brain asymmetry, then, was a comparatively recent discovery in human history, and a revelation against the general assumption of bilateral symmetry. It was linked, moreover, to specifically human aspects of thought. This has led to a tendency to regard it as uniquely human (e.g., [37]), and perhaps even a species-defining feature [38]. This is also implicit in the view that language itself is unique to our species (e.g., [39]). The idea that brain asymmetry emerged only in *Homo sapiens* has no doubt dampened efforts to understand its evolutionary origins, although this has begun to change with the realisation that asymmetries are ubiquitous in biology.

It is also commonly assumed that brain asymmetry is unidimensional, to the extent that individuals are often described as being either left- or right-brained, implying that the dominance of one or other hemisphere operates as a whole. It has become clear, though, that there are several, perhaps many, dimensions of laterality. Handedness, too, is effectively a cerebral asymmetry, not a manual one, and is commonly associated with the left-hemispheric dominance for speech. The correlation is in fact much weaker than previously assumed [40]. Some 95 percent of right-handers are left-cerebrally dominant for language, but so are 70–80 percent of left-handers [41]. Situs inversus totalis does not seem to reverse normal handedness or functional brain asymmetry, with the exception of the Yakoklevian torque—an anatomical asymmetry normally characterised as a protrusion of the frontal lobe on the right and occipital lobe on the left. This is reversed in cases of situs inversus [42].

Overall, the brain shows multiple anatomical asymmetries. In a study of 171,141 brains scans derived from 99 data sets worldwide, Kong et al. [43] divided the brain into 34 distinct regions, with overall thickness of the cortex larger on the left and overall surface area larger on the right. On both measures, as many regions showed leftward as showed rightward asymmetry, with only a small minority showing no measurable asymmetry. The two measures, though, showed different associations. The frontal regions tended to be thicker on the left while the posterior one tended to be thicker on the eight, a pattern which the authors suggest may derive from the Yakoklevian torque. It was surface area, though, which showed greater association with functional asymmetries. The largest asymmetries in surface area were within language-related areas, including a leftward advantage in a posterior region of Broca's area and the transverse temporal gyrus (part of Wernicke's area), and a rightward advantage in an anterior region of Broca's area. The opposite asymmetries within Broca's area suggest two different circuits involved in language, with the leftward circuit connecting Broca's and Wernicke's areas involved in phonology and syntax. The role of the rightward circuit is not so clear.

Functionally, too, it is becoming increasingly evident that there are several, perhaps many, independent dimensions of laterality. Liu et al. [44] factor analysed laterality indices derived from intrinsic brain activity in the resting brain, revealing four independent factors. Two were left-lateralized, one corresponding to the language network and the other the default-mode network, and the other two were right-lateralized corresponding to a visual network and an attentional one. Badzakova-Trajkov et al. [45] similarly carried out a factor analysis of functional asymmetries while participants undertook language tasks, an attentional task, and a face-recognition task, which yielded three independent factors, a left-lateralized one corresponding to the language network and two right-lateralized networks corresponding to the face-processing network and the attentional network. The right-lateralized face-processing network was largely homologous with the left-lateralized language network, yet uncorrelated with it.

Häberling et al. [46] undertook a further factor analysis of laterality indices while participants performed various left-lateralized tasks, and found three independent factors, representing a language circuit, a gesture-related circuit associated with handedness, and another gesture-related circuit independent of handedness. These finding raised speculation as to how the mirror-neuron system might have lateralized and fissioned into separate subcircuits in the process of hominin evolution.

Orthogonal factor analysis provides a convenient way to identify lateralized networks that are independent of one another and, at least as a first approximation, provide a useful means of determining just how many dimensions of laterality there are.

2.4. Cerebral Asymmetry in Animals

Evidence for cerebral asymmetries in a wide variety of animals is now abundant (see [47] for review). One general finding is a right-hemisphere dominance for emotion, which seems to be present in all primates so far investigated, including humans [48]. It seems to be true of other animals as well, including dogs [49], horses [50], and birds [51], and probably goes far back in the evolution of vertebrates. Right-hemisphere biases also appear to be unrelated to handedness or motor asymmetries [51]. From an evolutionary perceptive, it may reflect a left-hemispheric disposition to approach and the right hemisphere to avoidance [52].

In humans, the planum temporale overlaps with Wernicke's area, one of the major language areas, and is larger on the left than on the right [53], but the same asymmetry is present in great apes [54–56], and in both adult [57] and infant baboons [58]. This asymmetry may therefore date back at least to the common ancestor of humans, great apes and Old World monkeys, 30–40 million years ago, and is not specifically connected to language.

The other major cortical language area, Broca's area, is more complex. Its anterior portion, area 44 (*pars opercularis*) is part of the language network in humans, and is larger on the left [59] (According to Kong et al. [43] the other portion, area 45 (*pars triangularis*) is larger on the right, while Keller et al. [59] find no asymmetry). Cantalupo and Hopkins [60] report that the homolog of Broca's area in chimpanzees is also larger on the left. Graïc et al. [61] report a structural asymmetry in area 44 of the chimpanzee characterised by smaller neurons, perhaps suggesting increased computational capacity. In this and other respects, the cyto-architectural structure of area 44 seems to resemble closely that in humans.

The emergence of language in humans, though, may be not so much a question of the size of Broca's or Wernicke's areas as of their connectivity. Berwick and Chomsky [39] suggest that two circuits connecting these areas, both present in the chimpanzee, are connected ("a slight rewiring") in the human brain to create a loop that gave us syntax. This occurred, they say, uniquely in humans within the last 100,000 years, "in barely a flick of an eye in evolutionary time" (p. 67). This seems to be more or less pure conjecture. Friederici [62] has suggested similarly but more cautiously that humans evolved a stronger left dorsal connection between these areas than in non-human primates, and that it was this left-sided circuit that enabled the hierarchical structure of language.

From a functional perspective, Friederici's analysis is based on studies showing that humans can detect the hierarchical embedding in sequences of the form $(A_3(A_2(A_1B_1)B_2)B_3)$ (double embedding of this type, when applied to sentences, can be very difficult even for humans to process—an example is *The cat that the dog that the man kicked chased miaowed*), whereas non-human primates cannot [63], and that human processing of such sequences activates area 44. A difficulty with this analysis is that processing sequences of this kind need not involve any understanding of embedding at all; one might simply note that three As are followed by three Bs [64,65]. It is not yet entirely clear how seemingly similar fronto-temporal circuits can give rise to language in humans but not in non-human primates, or whether there is indeed a critical difference between apes and humans in this circuitry.

2.5. Cerebellar Asymmetries

The cerebellum is often neglected in accounts of brain asymmetry, but it too shows functional and structural asymmetries, which tend to mirror asymmetries of the cerebrum; that is, leftward activity accompanies rightward activity in the cortex, and vice versa. In a follow up from the study by Liu et al. [44] of cortical asymmetries in the resting brain, activity on each side of the cerebellum correlated with activity in the association cortex on the opposite side [66]. This implied large-scale circuits combining cerebellum and cortex, with the cerebellum mapping in roughly homotopic fashion onto the association cortex. Cerebellar asymmetry also mirrored cortical asymmetry during a language task, but did not map onto asymmetries of the motor cortex itself. In a similar follow up from the study by Badzakova-Trajkov et al. [45], factor analysis of asymmetrical brain activity induced by language tasks and observations of manual gestures revealed two independent networks, one right lateralized in the cerebellum and left lateralized in the language areas of the brain, and the other associated with handedness and gesture but with no cerebellar involvement [67].

The role of the cerebellum in the hemispheric asymmetry for language gains further support from a recent study showing a correlation between left-hemispheric dominance for perception of dichotically presented syllables, and a rightward asymmetry in the number of voxels in lobule VI of the cerebellum [68]. The dichotic asymmetry also correlated with a leftward asymmetry of the number of voxels in the amygdala, and to a lesser extent with a leftward voxel asymmetry posterior superior temporal cortex. Although dichotic listening provides a less reliable index of functional asymmetry than does brain imaging itself, the results suggest that subcortical areas contribute more to brain asymmetries than is commonly realised. The authors also note that the human cerebellum has a surface area approximately four-fifths of the neocortex, whereas the proportion in the macaque is only about one-third [69]. This invites the speculation that the cerebellum, generally considered to have its primary role in motor coordination, may have expanded in the course of hominin evolution to play a part in the emergence of language.

In chimpanzees, the cerebellum generally follows the pattern of the Yakoklevian torque observed in the human brain [70]. In a sample of chimpanzees studied by Phillips and Hopkins [71] this pattern was reversed, and there was a rightward bias in the volume of the posterior cerebellum in chimpanzees. This was unrelated to handedness as measured in a coordinated manual task. (Curiously, using the same measures, the authors did find that a leftward bias of the posterior cerebellum was associated with right-handedness in capuchins. Unlike chimpanzees, though, capuchins do not appear to show species-wide handedness, nor do they show the Yakoklevian torque.) A subsequent analysis, though, showed an association of this asymmetry with handedness determined from a tool-using task designed to simulate termite fishing [72]. The authors speculate that the asymmetry associated with tool use may have served as the foundation for the emergence of language.

Aside from the question of asymmetry, a recent study reports epigenetic modifications of DNA in the human cerebellum that sets it apart from that in the chimpanzee or macaque, and may suggest a role in the development of language and cognition [73]. GPS methylation at genes known to be involved in neurodevelopment and synaptic plasticity was even more distinctively human in the cerebellum than in the prefrontal cortex. The author suggest that their results "highlight the value of tissue-specific species comparisons of methylation and are consistent with an important role for the cerebellum in human brain evolution.

3. The Genetics of Laterality

3.1. Handedness

Historically, attempts to discover the genetic basis of functional laterality have focused largely on handedness, presumably because it is easier to measure than brain asymmetry. Although left-handedness is associated with cultural influences, it is also highly polygenic, as indicated by genome-wide studies of the association between handedness and genetic loci, e.g., [74–76]. These studies clearly rule out single-gene models that have hitherto been popular, e.g., [77,78]. The largest study to date examined individuals from 1,766,671 individuals, combined from the UK Biobank [79] and the International Handedness Consortium, found 41 loci associated with left-handedness, and 7 different loci associated with ambidexterity [80]. A total of 11.9 percent of males were left-handed or ambidextrous, compared with only 9.3 percent of females, a difference comparable to that found in other large-scale studies. Left-handedness was also associated with genetic loci implicated in a number of phenotypical conditions, including schizophrenia, autism, bipolar disorder, neuroticism, mood swings, and educational attainment.

Using an additive model, the authors estimated that genetic effects accounted for 11.9 percent of the variance, shared environment accounted for 4.6 percent, but the largest portion, 83.6 percent, came from individual environmental effects. Dropping shared environment from the model raised the genetic component to 19.7 percent, closer to the 25 percent estimated from twin studies [75,81]. There appears to be still some uncertainty as to how to assess the genetic contribution.

Ambidexterity has often been lumped together with left-handedness, but the two were unrelated genetically. Ambidexterity also showed a different profile of associations with other traits, including a negative genetic correlation with educational attainment. Earlier studies had shown decrements in educational attainment among the ambidextrous relative to left- or right-handers [82,83].

In an overlapping analysis of 501,730 individuals from the UK Biobank, de Kovel et al. [84] revealed that left-handedness was higher in those with lower birthweight, among multiple births, those born in certain seasons of birth, children with lower incidence of breastfeeding, and males, with each of these effects being significant independently of all the others. Others have reported an association of left-handedness with schizophrenia [85], autism [86] and dyslexia [87]. De Kovel et al. refer to a similar analysis based on a large US cohort showing similar association, with the addition of increased left-handedness, among African Americans and those with an older mother [88]. As in the larger study described above, a genome-wide association analysis showed left-handedness to be significantly but only weakly heritable genetically. The bias toward right-handedness, then, may be universal, but subject to variation and possible reversal through extraneous influences, some cultural, some pathological, and some genetic.

This idea of a universal bias is not without precedent. Laland [89,90] suggested that all humans are born with a biological bias to be right-handed, but that deviations result from external pressures. The primary pressure comes from parents, consistent with evidence that the incidence of left-handedness is increased if one parent is left-handed, and more so if both are left-handed. This association has also been taken to support a genetic basis for left-handedness (or the absence of right-handedness), but may equally be due to parental influence. Given the evidence summarised above, though, there are probably additional influences. As a first approximation, then, there may be a universal bias toward right-handedness, but malleable enough to permit variations without undue disadvantage.

Although genetic studies show multiple genetic associations with handedness, these genes may represent different conditions that influence handedness, rather than being intrinsic to handedness itself. An example is the LRRTM1 gene, a maternally suppressed gene associated paternally with handedness and dyslexia; when inherited through the father a particular haplotype consisting of minor alleles at three locations significantly shifted handedness toward the left [91]—a finding partially confirmed elsewhere [92], This same haplotype was over-transmitted paternally in those with schizophrenia. These effects were discovered in dyslexic samples, and were not evident in a Chinese sample or in other samples from the general population, including the large-scale study described above [79].

3.2. Cerebral Asymmetry

Estimates of cerebral asymmetry based on brain imaging paint a similar picture. In a brain-wide genome-wide analysis in 32,256 individuals, Sha et al. [93] found 41 locations for cerebral asymmetry, parcellated into 34 cortical regions per hemisphere and 7 subcortical regions. Among these, they found 21 distinct, highly significant genomic loci for the different aspects of brain asymmetry. Ten of these were associated with cytoskeletal development, while the remaining 11 were mostly with brain development. These included significant genetic overlaps with autism, schizophrenia, and educational achievement. Earlier studies had shown direct associations of cerebral asymmetry with dyslexia [94], Alzheimer's disease (e.g., [95], ADHD [96], and depression [97]. In all cases, the negative aspects were associated with deviations away from normal asymmetries. Although some of these variables also correlated with handedness in Sha et al.'s study, there was no significant genetic overlap between handedness and structural brain asymmetries, although five individual markers (SNPs) were associated with both. Many of the asymmetries were strong, but their heritabilities were low. As mentioned earlier, situs inversus does not systematically reverse handedness or the normal cerebral asymmetries, with the exception of the Yakoklevian torque.

Again, these findings concur with those based on handedness in suggesting a fundamental but universal bias, with variations imposed by environmental and other conditions, some of possibly genetic origin. Sha et al. conclude from their findings that the development of brain asymmetry is "tightly constrained and largely genetically invariant in the population." The most parsimonious conclusion is that this universal bias also underlies the situs of the internal organs; Brandler and Paracchini [98] suggest that "the mechanisms for establishing LR asymmetry in the body are reused for brain midline development, which in turn influences traits such as handedness and reading ability" (p. 88).

This scenario need not contradict the evidence of relative independence among handedness, different dimensions of cerebral asymmetry, and situs of internal organs. The fundamental asymmetry is invoked where it proves adaptive, even though against the pressure toward symmetry in the bilateria. This is especially true of situs, but less so in handedness and or the various aspects of brain asymmetry where there may be some advantage to maintaining variation—a possibility explored by Ghirlanda and Vallortigara [26]. Evolution itself depends on variation, and within social species such as our own, variations in demeanour, cognition, skill, and personality provide for effective social living, allowing individuals to take multiple specialized roles. Száthmary [99] writes that language, itself strongly lateralized and subject to individual variation, was one of the seven major transitions in evolution, offering something unprecedented—the "negotiated division of labour" (p. 10,109). Whether it was indeed a major transition, or simply a result of progressive evolution is a moot point, and the evolution of complex societies depends not only on language but also on individual differences in other domains as well, including spatial abilities, creativity, athleticism, and computational abilities. We need, or have needed, butchers, bakers, candlestick makers, and software engineers. Genetically, such diversity need not be construed as group selection, but rather as a loosening of genetic determinism.

The universal bias toward asymmetry, then, appears to be most strongly expressed in the situs of internal organs, where deviations from asymmetry are maladaptive. It is also strongly expressed in cerebral asymmetry for language, where deviations may result in language disorders. The bias itself may be universal with deviations only due to extraneous conditions, some pathological, some cultural, and some themselves genetic. For example, a mutation of the FOXP2 gene results in a severe speech impediment, and brain imaging showed that members of an extended family affected by the mutation, unlike their unaffected relatives, showed no activation in Broca's area while covertly generating verbs [100]; the activation seemed to be scattered and to exhibit no consistent asymmetry. Handedness, though, seems to be largely unaffected, with one study showing 12 of the 15 members of the family to be right-handed [101].

The universal bias seems to be less strongly expressed in handedness, where deviations may be adaptive if maintained in a minority. It probably varies across species, but is absent in most animals, where there is no species-wide difference in dominance or preference between left and right forelimbs. That is, the ratio is approximately 0.5, with variations from around equality due only to chance. Laland [89] estimates a bias of 0.78 in humans, so that in the absence of extraneous influences 78 percent of the population would be right-handed, but parental or cultural influences increase it to approximately 90 percent overall. He suggests ratios of 0.8 to 0.9 in Neanderthals, 0.61 in Middle Pleistocene hominins, 0.57 in Lower Pleistocene hominids, and 0.56 in chimpanzees. The bias may be overestimated in Neanderthals, who may have been sufficiently human-like for a cultural influence increasing the overall incidence of right-handedness itself. The bias runs counter to the otherwise general bilateral symmetry of the limbs, and may be largely restricted to bipedal species.

If there is indeed a fundamental bias underling situs as well as handedness and cerebral asymmetries, what is its origin? Morgan and I [102] (readers tempted to consult this article should ignore the Abstract, which was inadvertently substituted from another article) once suggested that it was coded in the oocyte rather than in the genes themselves, and favoured development on the left. It may even depend on the chirality (left–right asymmetry) at the molecular level [103–105]. The asymmetries of the internal organs are governed at the earliest stages by an asymmetry of the cilia, hair-like organelles on the surface of cells, and this directs the asymmetry of a genetic sequence (the Nodal-Lefty-Pitx2 cascade) [106]. Cooke [107] outlines a scenario whereby the asymmetry of the cilia themselves is governed by the alignment of chiral molecules, creating a leftward flow of morphogenes across the embryo, which in turn guides the asymmetrical morphogenes of internal organs through a cascade of genetic influences. These ideas remain speculative, but imply that asymmetry—or symmetry breaking—is not restricted to humans, or even to vertebrates, but is a fundamental property of living matter.

Whether the asymmetry of the cilia can account for right-handedness, though, remains uncertain. Afzelius and Stenram [108] report on 239 cases of immotile-cilia syndrome, a rare condition in which the cilia are either absent or stationary. In these cases, one might expect random asymmetry, such that 50 percent would have situs inversus and be left-handed. In fact the figures were 44 percent and 14 percent, respectively. This suggest a bias other than that due to ciliary motility, especially in the case of handedness, where the bias was only slightly above the 10–12 percent found in the normal population. Cultural or familial influences may be strong even in the absence of a biological bias.

That said, asymmetries of the hands and brain are clearly more variable than that of situs, where departures from normal asymmetry are often maladaptive. Immotile cilia syndrome, with its high incidence of situs inversus, is accompanied by disorders of the respiratory tract, including sinusitis, rhinitis and bronchitis, and the combination of these with situs inversus is known as Kartagener syndrome, afflicting approximately one in 22,000 [108]. Departures from right-handedness and left-cerebral representation of language are far less drastic, and may even be adaptive in giving rise to special minority talents, as suggested earlier.

This raises the question as to whether disorders associated with lateralization are truly "disorders," or simply part of the fabric of human existence. Dyslexia is often associated with creativity, and even a number of well-known authors, such as Agatha Christie, Gustave Flaubert, and Evelyn Waugh, are said to have been dyslexic. Normal reading depends on an area known as the visual word form area usurping the left side of the occipito-temporal region brain concerned with visual shape analysis. This implies that visual processing can be diminished, or at least altered, when children learn to read [109]. This might explain the special talents of artists, such as Andy Warhol, Pablo Picasso o0 Robert Rauschenberg, who are also said to have been dyslexic. Leonardo da Vinci is often mentioned as another example, although his mirror writing might have been not so much a disability as a disguise. He was, however, left-handed, at least when writing.

Even mental illnesses may be adaptive, or once were so. Kauffman [110] points out that hallucinations were at one time considered normal, and played a part in the lives of visionaries, such as Jesus of Nazareth, St Paul of Tarsus, and even Socrates, and suggests that it was through the writing of Voltaire, Darwin, and Freud that they began to be associated with psychiatric illness. Creativity, too, has long been associated with schizophrenia and bipolar disorders, and research also suggests a genetic link [111]. Nature and culture may have combined to maintain a diversity and creativity of benefit to the species.

4. Conclusions

The emergence of animals that move created pressure toward bilateral symmetry, and the establishment of the vast clade of animals known as the bilateria. This pressure was due largely to the absence of asymmetrical influences from the natural environment—or what physicists call the conservation of parity. Departures from bilateral symmetry in movement or sensory input could be perilous; Martin Gardner [112] once put it like this:

> The slightest loss of bilateral symmetry, such as the loss of a right eye, would have immediate negative value for the survival of any animal. An enemy could sneak up unobserved on the right!
>
> (p. 70).

Nevertheless, bodily asymmetry is ubiquitous, especially in the placement of internal organs. It applies to all chordates and presumably far goes back in evolution. Its fundamental basis may even go back close to the origins of life itself, with the emergence of chiral molecules. At the molecular level, we are steeped in asymmetry.

The brain has largely retained its bilaterian symmetry. Over the course of evolution, though, it has also evolved computational functions not directly constrained by inputs from, or outputs on, the immediate environment. This may include emotion, which seems to be universally characterised by a bias toward the right hemisphere. Operations on the environment seem more likely to be asymmetrically programmed than are reactions to it, and generally favour the left hemisphere. Examples include throwing, the manufacture and use of tools, and language, whether in the form of speech, gesture, or writing. Again, there may be packaging constraints, with face recognition and perhaps music shifted to the right as compensation for the left-sided representation of language. In the large-scale brain-imaging study by Kong et al. [43], the great majority of the 34 regions examined were asymmetrical one way or the other, yet each region was identifiable on either side, and they were all packaged in such a way as to retain an overall symmetry. Indeed for most of the history of medicine the brain was thought to conform to the law of symmetry.

The genetic orchestration of the asymmetries remains elusive. The most parsimonious solution is that they are ultimately dependent on the same fundamental bias that underlies situs of the bodily organs, but are then expressed by the genetic cascades that create the various specializations, each of which may be expressed or perturbed independently. Even if the various cerebral asymmetries so far identified are not dependent on a single underlying event, they may still hark back to the chirality of biological molecules.

Funding: This research received no external funding.

Institutional Review Board Statement: Not applicable.

Informed Consent Statement: Not applicable.

Data Availability Statement: Not applicable.

Conflicts of Interest: The author declares no conflict of interest.

References

1. Carroll, S.B. Chance and necessity: The evolution of morphological complexity and diversity. *Nature* **2001**, *409*, 1102–1109. [CrossRef] [PubMed]
2. Evans, S.D.; Hughes, I.V.; Gehling, J.G.; Droser, M.L. Discovery of the oldest bilaterian from the Ediacaran of South Australia. *Proc. Natl. Acad. Sci. USA* **2020**, *117*, 7845–7850. [CrossRef] [PubMed]
3. Bridi, J.C.; Ludlow, Z.N.; Kottler, B.; Hartmann, B.; Broeck, L.V.; Dearlove, J.; Göker, M.; Strausfeld, N.J.; Callaerts, P.; Hirth, F. Ancestral regulatory mechanisms specify conserved midbrain circuitry in arthropods and vertebrates. *Proc. Natl. Acad. Sci. USA* **2020**, *117*, 19544–19555. [CrossRef]
4. Descartes, R. Treatise on the Passions of the Soul. In *Descartes: Philosophical Writings*; Trans. N. K. Smith; Modern Library: New York, NY, USA, 1649/1958.
5. Eling, P.; Finger, S. Franz Joseph Gall on hemispheric symmetries. *J. Hist. Neurosci.* **2020**, *29*, 325–338. [CrossRef]
6. Corballis, M.C.; Beale, I.L. Bilateral symmetry and behavior. *Psychol. Rev.* **1970**, *77*, 451–464. [CrossRef]
7. Standing, L.; Conezio, J.; Haber, R.N. Perception and memory for pictures: Single-trial learning of 2500 visual stimuli. *Psychon. Sci.* **1970**, *19*, 73–74. [CrossRef]
8. Fischer, J.-P.; Koch, A.-M. Mirror writing in typically developing children: A first longitudinal study. *Cogn. Dev.* **2016**, *38*, 114–124. [CrossRef]
9. Corballis, M.C. Mirror-image equivalence and interhemispheric mirror-image reversal. *Front. Hum. Neurosci.* **2018**, *12*, 140. [CrossRef] [PubMed]
10. Logothetis, N.K.; Pauls, J.; Poggio, T. Shape representation in the inferior temporal cortex of monkeys. *Curr. Biol.* **1995**, *5*, 552–563. [CrossRef]
11. Torres, A.R.; Mota, N.B.; Adamy, N.; Naschold, A.; Lima, T.; Copelli, M.; Weissheimer, J.; Pegado, F.; Ribeiro, S. Selective inhibition of mirror invariance for letters consolidated by sleep doubles reading fluency. *Curr. Biol.* **2021**, *31*, 742–752. [CrossRef]
12. Torgersen, J. Situs inversus, asymmetry, and twinning. *Am. J. Hum. Genet.* **1950**, *2*, 361–370. [PubMed]
13. Douard, R.; Feldman, A.; Bargy, F.; Loric, S.; Delmas, V. Anomalies of lateralization in man a case of total situs inversus. *Surg. Radiol. Anat.* **2000**, *22*, 293–297. [CrossRef] [PubMed]
14. Layton, W.M., Jr. Random determination of a developmental process. *J. Hered.* **1976**, *67*, 336–338. [CrossRef]
15. Patten, B.M. The formation of the cardiac loop in the chick. *Am. J. Anat.* **1922**, *30*, 373–379. [CrossRef]
16. Boorman, C.J.; Shimeld, S.M. Pitx homeobox genes in *Ciona* and amphioxus show left–right asymmetry is a conserved chordate character and define the ascidian adenohypophysis. *Evol. Dev.* **2002**, *4*, 354–365. [CrossRef]
17. Kilner, P.J.; Yang, G.-Z.; Wilkes, A.J.; Mohiaddin, R.H.; Firmin, D.N.; Yacoub, M.H. Asymmetric redirection of flow through the heart. *Nature* **2000**, *404*, 759–761. [CrossRef] [PubMed]
18. Giljov, A.; Karenina, K.; Ingram, J.; Malashichev, Y. Parallel emergence of true handedness in the evolution of marsupials and placentals. *Curr. Biol.* **2015**, *14*, 1778–1884. [CrossRef] [PubMed]

19. Ocklenburg, S.; Isparta, S.; Peterburs, J.; Papadatou-Pastou, M. Paw preferences in cats and dogs: Meta-analysis. *Laterality* **2019**, *24*, 647–677. [CrossRef] [PubMed]
20. Ströckens, F.; Güntürkün, O.; Ocklenburg, S. Limb preferences in non-human vertebrates. *Laterality* **2013**, *18*, 536–575. [CrossRef]
21. Hopkins, W.D.; Wesley, M.J.; Izard, M.K.; Hook, M.; Schapiro, S.J. Chimpanzees (*Pan troglodytes)* are predominantly right-handed: Replication in three populations of apes. *Behav. Neurosci.* **2004**, *118*, 659–663. [CrossRef]
22. Tabiowo, E.; Forrester, G.S. Structured bimanual actions and hand transfers reveal population-level right-handedness in captive gorillas. *Anim. Behav.* **2013**, *86*, 1049–1057. [CrossRef]
23. Westergaard, G.C.; Suomi, S.J. Hand preference for a bimanual task in tufted capuchins (*Cebus apella*) and rhesus macaques (*Macaca mulatta*). *J. Comp. Psychol.* **1996**, *110*, 406–411. [CrossRef] [PubMed]
24. Levermann, N.; Galatius, A.; Ehlme, G.; Rysgaard, S.; Born, E.W. Feeding behaviour of free-ranging walruses with notes on apparent dextrality of flipper use. *BMC Ecol.* **2003**, *3*, 9. [CrossRef] [PubMed]
25. Friedman, H.; Davis, M. "Left-handedness" in parrots. *Auk* **1938**, *35*, 478–480. [CrossRef]
26. Ghirlanda, S.; Vallortigara, G. The evolution of brain lateralization: A game-theoretical analysis of population structure. *Proc. Biol. Sci.* **2004**, *271*, 853–857. [CrossRef]
27. Watson, N.V.; Kimura, D. Right-hand superiority for throwing but not for intercepting. *Neuropsychologia* **1989**, *27*, 1399–1414. [CrossRef]
28. Broca, P. Remarques sur le siège de la faculté du langage articulé, suivies d'une observation d'aphémie (perte de la parole). *Bull. Société Anatomique* **1861**, *36*, 330–357.
29. Broca, P. Sur le siège de la faculté du langage articulé. *Bull. Société d'Anthropol.* **1865**, *63*, 77–93. [CrossRef]
30. Dax, M. Lésions de la moitié gauche de l'encéphale coïncident avec l'oubli des signes de la pensée (lu à Montpellier en 1836). *Bull. Hebdomadaire Médecine Chirurgie, 2me Série* **1865**, *2*, 259–262.
31. Wernicke, K. *Der Aphasische Symptomencomplex. Eine Psychologische Studie auf Anatomischer Basis*; M. Cohn U. Weigart: Breslau, Poland, 1874.
32. Leblanc, R. *Fearful asymmetry: Bouillaud, Dax, Broca, and the Localization of Language, Paris, 1825–1879*; McGill-Queen's University Press: Kingston, ON, Canada, 2017.
33. Harrington, A. Nineteenth-century ideas on hemisphere differences and duality of mind. *Behav. Brain Sci.* **1985**, *8*, 617–634. [CrossRef]
34. Gazzaniga, M.S.; Sperry, R.W. Language after section of the cerebral commissures. *Brain* **1967**, *90*, 131–148. [CrossRef] [PubMed]
35. Ornstein, R.E. *The Psychology of Consciousness*; Freeman: San Francisco, CA, USA, 1972.
36. Sperry, R.W. Some effects of disconnecting the cerebral hemisphere. *Science* **1982**, *217*, 1223–1226. [CrossRef] [PubMed]
37. Corballis, M.C. *The Lopsided Ape*; Oxford University Press: New York, NY, USA, 1991.
38. Crow, T.J. Why cerebral asymmetry is the key to the origin of *Homo sapiens*: How to find the gene or eliminate the theory. *Curr. Psychol. Cogn.* **1998**, *17*, 1237–1277.
39. Berwick, R.C.; Chomsky, N. *Why Only Us? Language and Evolution*; The MIT Press: Cambridge, MA, USA, 2016.
40. Packheiser, J.; Schmidt, J.; Arning, L.; Beste, C.; Güntürkün, O.; Ocklenburg, S. A large-scale estimate on the relationship between language and motor lateralization. *Sci. Rep.* **2020**, *10*, 13027. [CrossRef] [PubMed]
41. Badzakova-Trajkov, G.; Häberling, I.S.; Roberts, R.P.; Corballis, M.C. Cerebral asymmetries: Complementary and independent processes. *PLoS ONE* **2010**, *5*, e9682. [CrossRef] [PubMed]
42. Vingerhoets, G.; Li, X.; Hou, L.; Bogaert, S.; Verhelst, H.; Gerrits, R.; Siugzdaite, R.; Roberts, N. Brain structural and functional asymmetry in human situs inversus totalis. *Brain Struct. Funct.* **2018**, *223*, 1937–1952. [CrossRef]
43. Kong, X.Z.; Mathias, S.R.; Guadalupe, T.; Group, E.L.W.; Glahn, D.C.; Franke, B.; Crivello, F.; Tzourio-Mazoyer, N.; Fisher, S.E.; Thompson, P.M.; et al. Mapping cortical brain asymmetry in 17,141 healthy individuals worldwide via the ENIGMA Consortium. *Proc. Natl. Acad. Sci. USA* **2018**, *115*, E1154–E1163. [CrossRef]
44. Liu, H.; Stufflebeam, S.M.; Sepulcre, J.; Hedden, T.; Buckner, R.L. Evidence from intrinsic activity that asymmetry of the human brain is controlled by multiple factors. *Proc. Natl. Acad. Sci. USA* **2009**, *106*, 20499–20503. [CrossRef] [PubMed]
45. Badzakova-Trajkov, G.; Corballis, M.C.; Häberling, I.S. Complementarity or independence of hemispheric specializations? A brief review. *Neuropsychologia* **2016**, *93*, 386–393. [CrossRef]
46. Häberling, I.S.; Corballis, P.M.; Corballis, M.C. Language, gesture, and handedness: Evidence for independent lateralized networks. *Cortex* **2016**, *82*, 72–85. [CrossRef]
47. Rogers, L.J.; Vallortigara, G.; Andrew, R.J. *Divided Brains: The Biology and Behaviour of Brain Asymmetries*; Cambridge University Press: New York, NY, USA, 2013.
48. Lindell, A.K. Continuities in emotion lateralization in human and nonhuman primates. *Front. Hum. Neurosci.* **2013**, *7*, 464. [CrossRef] [PubMed]
49. Siniscalchi, M.; Quaranta, A.; Rogers, L.J. Hemispheric specialization in dogs for processing different acoustic stimuli. *PLoS ONE* **2008**, *3*, e3349. [CrossRef] [PubMed]
50. Smith, A.V.; Proops, L.; Grounds, K.; Wathan, J.; Scott, S.K.; McComb, K. Domestic horses (*Equus caballus*) discriminate between negative and positive human nonverbal vocalisations. *Sci. Rep.* **2018**, *8*, 13052. [CrossRef] [PubMed]
51. Rogers, L.J. Asymmetry of motor behavior and sensory perception: Which comes first? *Symmetry* **2020**, *12*, 690. [CrossRef]

52. Rutherford, H.J.V.; Lindell, A.K. Thriving and surviving: Approach and avoidance motivation and lateralization. *Emotion Rev.* **2011**, *3*, 333–343. [CrossRef]
53. Geschwind, N.; Levitsky, W. Human brain: Left-right asymmetries in temporal speech region. *Science* **1968**, *161*, 186–187. [CrossRef]
54. Gannon, P.J.; Holloway, R.L.; Broadfield, D.C.; Braun, A.R. Asymmetry of chimpanzee planum temporale: Humanlike pattern of Wernicke's brain language area homolog. *Science* **1998**, *279*, 220–222. [CrossRef]
55. Hopkins, W.D.; Nir, T.M. Planum temporale surface area and grey matter asymmetries in chimpanzees (Pan troglodytes): The effect of handedness and comparison with findings in humans. *Behav. Brain Res.* **2010**, *208*, 436–443. [CrossRef]
56. Spocter, M.A.; Sherwood, C.C.; Schapiro, S.J.; Hopkins, W.D. Reproducibility of leftward planum temporale asymmetries in two genetically isolated populations of chimpanzees (pan troglodytes). *Proc. Royal Soc. B* **2020**, *287*, 20201320. [CrossRef]
57. Marie, D.; Roth, M.; Lacoste, R.; Nazarian, B.; Bertello, A.; Anton, J.-L.; Meguerditchian, A. left brain asymmetry of the planum temporale in a nonhominid primate: Redefining the origin of brain specialization for language. *Cereb. Cort.* **2018**, *28*, 1808–1815. [CrossRef]
58. Becker, Y.; Sein, J.; Velly, L.; Giacomino, L.; Renaud, L.; Lacoste, R.; Anton, J.-L.; Nazarian, B.; Berne, C.; Meguerditchian, A. Early left-planum temporale asymmetry in newborn monkeys (*Papio anubis*): A longitudinal structural MRI study at two stages of development. *NeuroImage* **2021**, *227*, 117575. [CrossRef] [PubMed]
59. Keller, S.S.; Highley, J.R.; Garcia-Finana, M.; Sluming, V.; Rezaie, R.; Roberts, N. Sulcal variability, stereological measurement and asymmetry of Broca's area on MR images. *J. Anat.* **2007**, *211*, 534–555. [CrossRef] [PubMed]
60. Cantalupo, C.; Hopkins, W.D. Asymmetric Broca's area in great apes. *Nature* **2001**, *414*, 505. [CrossRef] [PubMed]
61. Graïc, J.-M.; Peruffo, A.; Corain, L.; Centelleghe, C.; Granato, A.; Zanellato, E.; Cozzi, B. Asymmetry in the cytoarchitecture of the area 44 homolog of the brain of the chimpanzee *pan troglodytes*. *Front. Neuroanat.* **2020**, *14*, 56. [CrossRef]
62. Friederici, A.D. Hierarchy processing in human neurobiology: How specific is it? *Phil. Trans. R. Soc. B* **2019**, *375*, 20180391. [CrossRef]
63. Fitch, W.T.; Hauser, M.D. Computational constraints on syntactic processing in a nonhuman primate. *Science* **2004**, *303*, 377–380. [CrossRef]
64. Corballis, M.C. On phrase-structure and brain responses: A comment on Bahlmann, Gunter, and Friederici (2006). *J. Cogn. Neurosci.* **2007**, *19*, 1581–1583. [CrossRef]
65. Corballis, M.C. Recursion, language, and starlings. *Cogn. Sci.* **2007**, *31*, 697–704. [CrossRef]
66. Wang, D.; Buckner, R.L.; Liu, H. Cerebellar asymmetry and its relation to cerebral asymmetry estimated by intrinsic functional connectivity. *J. Neurophysiol.* **2013**, *109*, 46–57. [CrossRef]
67. Häberling, I.S.; Corballis, M.C. Cerebellar asymmetry, cortical asymmetry and handedness: Two independent networks. *Laterality Asymmetries Body Brain Cogn.* **2016**, *21*, 397–414. [CrossRef]
68. Guadalupe, T.; Kong, T.; Kong, X.-Z.; Akkermans, S.E.A.; Fisher, S.E.; Francks, C. Relations between hemispheric asymmetries of grey matter and auditory processing of spoken syllables in 281 healthy adults. *Brain Struct. Funct.* **2021**. [CrossRef]
69. Sereno, M.I.; Diedrichsen, J.; Tachrount, M.; Testa-Silva, G.; d'Arceuil, H.; De Zeeuw, C. The human cerebellum has almost 80% of the surface area of the neocortex. *Proc. Natl. Acad. Sci. USA* **2020**, *117*, 19538–19543. [CrossRef] [PubMed]
70. Pilcher, D.; Hammock, L.; Hopkins, W.D. Cerebral volume asymmetries in non-human primates as revealed by magnetic resonance imaging. *Laterality* **2001**, *6*, 165–180. [CrossRef] [PubMed]
71. Phillips, K.A.; Hopkins, W.D. Exploring the relationship between cerebellar asymmetry and handedness in chimpanzees (*Pan troglodytes*) and capuchins (*Cebus paella*). *Neuropsychologia* **2007**, *45*, 2333–2339. [CrossRef] [PubMed]
72. Cantalupo, C.; Freeman, H.; Rodes, W.; Hopkins, W.D. Handedness for Tool Use Correlates with Cerebellar Asymmetries in Chimpanzees (*Pan troglodytes*). *Behav. Neurosci.* **2008**, *122*, 191–198. [CrossRef] [PubMed]
73. Guevara, E.E.; Hopkins, W.D.; Hof, P.R.; Ely, J.J.; Bradley, B.J.; Sherwood, C.C. Comparative analysis reveals distinctive epigenetic features of the human cerebellum. *PLoS Genet.* **2021**, *17*, e1009506. [CrossRef] [PubMed]
74. McManus, I.C.; Davison, A.; Armour, J.A.L. Multilocus genetic models of handedness closely resemble single-locus models in explaining family data and are compatible with genome-wide association studies. *Ann. N. Y. Acad. Sci.* **2013**, *1288*, 48–58. [CrossRef] [PubMed]
75. Medland, S.E.; Duffy, D.L.; Wright, M.J.; Geffen, G.M.; Hay, D.A.; Levy, F.; Van-Beijsterveldt, C.E.; Willemsen, G.; Townsend, G.C.; White, V.; et al. Genetic influences on handedness: Data from 25,732 Australian and Dutch twin families. *Neuropsychologia* **2009**, *47*, 330–337. [CrossRef] [PubMed]
76. De Kovel, C.G.F.; Francks, C. The molecular genetics of hand preference revisited. *Sci. Rep.* **2019**, *9*, 5986. [CrossRef]
77. Annett, M. *Handedness and Brain Asymmetry: The Right Shift Theory*; Psychology Press: Hove, UK, 2002.
78. McManus, C. *Right Hand, Left Hand*; Weidenfeld & Nicolson: London, UK, 2002.
79. Sudlow, C.; Gallacher, J.; Allen, N.; Beral, V.; Burton, P.; Danesh, J.; Downey, P.; Elliott, P.; Green, J.; Landray, M.; et al. UK Biobank: An open access resource for identifying the causes of a wide range of complex diseases of middle and old age. *PLoS Med.* **2015**, *12*, e1001779. [CrossRef]
80. Cuellar-Partida, G.; Tung, J.Y.; Eriksson, N.; Albrecht, E.; Aliev, F.; Andreassen, O.A.; Barroso, I.; Beckmann, J.S.; Boks, M.P.; Boomsma, D.I.; et al. Genome-wide association study identifies 48 common genetic variants associated with handedness. *Nature Hum. Behav.* **2021**, *5*, 59–70. [CrossRef] [PubMed]

81. Medland, S.E.; Duffy, D.L.; Wright, M.J.; Geffen, G.M.; Martin, N.G. Handedness in twins: Joint analysis of data from 35 samples. *Twin Res. Human Genet.* **2006**, *9*, 46–53. [CrossRef]
82. Corballis, M.C.; Hattie, J.; Fletcher, R. Handedness and intellectual achievement: An even-handed look. *Neuropsychologia* **2008**, *46*, 374–378. [CrossRef]
83. Crow, T.J.; Crow, L.R.; Done, D.J.; Leask, S. Relative hand skill predicts academic ability: Global deficits at the point of hemispheric indecision. *Neuropsychologia* **1998**, *36*, 1275–1282. [CrossRef]
84. De Kovel, C.G.F.; Carrión-Castillo, A.; Francks, C. A large-scale population study of early life factors influencing left-handedness. *Sci. Rep.* **2019**, *9*, 584. [CrossRef]
85. Deep-Soboslay, T.M.; Calicott, J.P.; Lener, M.S.; Verchinski, B.A.; Apud, J.A.; Weinberger, D.R.; Apud, J.A.; Weinberger, D.R.; Elvevåg, B. Handedness, heritability, neurocognition and brain asymmetry in schizophrenia. *Brain* **2010**, *133*, 3113–3122. [CrossRef]
86. Markou, P.; Ahtam, B.; Papadatou-Pastou, M. Elevated levels of atypical handedness in autism: Meta-analyses. *Neuropsychol. Rev.* **2017**, *27*, 258–283. [CrossRef]
87. Scerri, T.S.; Brandler, W.M.; Paraccini, S.; Morris, A.P.; Ring, S.M.; Richardson, A.J.; Talcott, J.B.; Stein, J.; Monaco, A.P. PCSK6 is associated with handedness in individuals with dyslexia. *Hum. Mol. Genet.* **2011**, *20*, 608–614. [CrossRef]
88. Johnston, D.W.; Nicholls, M.E.R.; Shah, M.A.; Shields, M.A. *Handedness, Health and Cognitive Development: Evidence from Children in the NLSY*; Forschungsinstitut zur Zukunft der Arbeit: Bonn, Germany, 2010.
89. Laland, K.N. Exploring gene–culture interactions: Insights from handedness, sexual selection and niche-construction case studies. *Phil. Trans. R. Soc. B* **2008**, *363*, 3577–3589. [CrossRef]
90. Laland, K.N.; Kumm, J.; Van Horn, J.D.; Feldman, M.W. A gene-culture model of handedness. *Behav. Genet.* **1995**, *25*, 433–445. [CrossRef]
91. Francks, C.; Maegawa, S.; Lauren, J.; Abrahams, B.S.; Velayos-Baeza, A.; Medland, S.E.; Colella, S.; Groszer, M.; McAuley, E.Z.; Caffrey, T.M.; et al. LRRTM1 on chromosome 2p12 is a maternally suppressed gene that is associated paternally with handedness and schizophrenia. *Mol. Psychiat.* **2007**, *12*, 1129–1139. [CrossRef] [PubMed]
92. Ludwig, K.U.; Mattheisen, M.; Muhleisen, T.W.; Roeske, D.; Schmal, C.; Breuer, R.; Schulte-Körne, G.; Müller-Myhsok, B.; Nöthen, N.N.; Hoffmann, P.; et al. Supporting evidence for LRRMT1 imprinting in schizophrenia. *Mol. Psychiat.* **2009**, *14*, 743–745. [CrossRef] [PubMed]
93. Sha, Z.; Schijven, D.; Carrion-Castillo, A.; Joliot, M.; Mazoyer, B.; Fisher, S.E.; Crivello, F.; Francks, C. The genetic architecture of structural left-right asymmetry of the human brain. *Nat. Hum. Behav.* **2021**. [CrossRef]
94. Altarelli, I.; Leroy, F.; Monzalvo, K.; Fluss, J.; Billard, C.; Dehaene-Lambertz, G.; Galaburda, A.M.; Ramus, F. Planum temporale asymmetry in developmental dyslexia: Revisiting an old question. *Hum. Brain Mapp.* **2014**, *35*, 5717–5735. [CrossRef] [PubMed]
95. Thompson, P.M.; Moussai, J.; Zohoori, S.; Goldkorn, A.; Khan, A.A.; Mega, M.S.; Small, G.W.; Cummings, J.L.; Toga, A.W. Cortical variability and asymmetry in normal aging and Alzheimer's disease. *Cereb. Cort.* **1998**, *8*, 492–509. [CrossRef]
96. Shaw, P.; Lalonde, F.; LePage, C.; Rabin, C.; Eckstrand, K.; Sharp, W.; Greenstein, D.; Evans, A.; Giedd, J.N.; Rapoport, J. Development of cortical asymmetry in typically developing children and its disruption in attention-deficit/hyperactivity disorder. *Arch. Gen. Psychiat.* **2009**, *66*, 888–896. [CrossRef]
97. Tomarkenand, A.J.; Keener, A.D. Frontal brain asymmetry and depression: A self-regulatory perspective. *Cognit. Emot.* **1998**, *12*, 387–420. [CrossRef]
98. Brandler, W.M.; Paracchini, S. The genetic relationship between handedness and neurodevelopmental disorders. *Trends Mol. Genet.* **2014**, *20*, 83–90. [CrossRef]
99. Szathmáry, E. Toward major evolutionary transitions theory 2.0. *Proc. Natl. Acad. Sci. USA* **2005**, *102*, 10104–10111. [CrossRef] [PubMed]
100. Liègeois, F.; Baldeweg, T.; Connelly, A.; Gadian, D.G.; Mishkin, M.; Vargha-Khadem, V. fMRI abnormalities associated with FOXP2 gene mutation. *Nat. Neurosci.* **2003**, *6*, 1230–1237. [CrossRef]
101. Alcock, K.J.; Passingham, R.E.; Watkins, K.E.; Vargha-Khadem, F. Oral dyspraxia in inherited speech and language impairment. *Brain Lang.* **2000**, *75*, 17–33. [CrossRef]
102. Morgan, M.J.; Corballis, M.C. On the biological basis of human laterality: II. The mechanisms of inheritance. *Behav. Brain Sci.* **1978**, *2*, 269–277. [CrossRef]
103. Afzelius, B.A. A human syndrome caused by immotile cilia. *Science* **1976**, *193*, 317–319. [CrossRef] [PubMed]
104. Almirantis, V. Left-right asymmetry in vertebrates. *Bioessays* **1995**, *17*, 79–83. [CrossRef] [PubMed]
105. Brown, N.A.; Wolpert, L. The development of handedness in left/right asymmetry. *Development* **1990**, *109*, 1–9. [CrossRef] [PubMed]
106. Dasgupta, A.; Amack, J.D. Cilia in vertebrate left-right patterning. *Philosoph. Transact. Royal Soc. B-Biol. Sci.* **2016**, *371*, 20150410. [CrossRef]
107. Cooke, J. The evolutionary origins and significance of vertebrate left–right organisation. *BioEssays* **2004**, *26*, 413–421. [CrossRef]
108. Afzelius, B.A.; Stenram, U. Prevalence and genetics of immotile-cilia syndrome and left-handedness. *Int. J. Dev. Biol.* **2006**, *50*, 571–573. [CrossRef] [PubMed]
109. Dehaene, S.; Pegado, F.; Braga, L.W.; Ventura, P.; Nunes Filho, G.; Jobert, A.; Dehaene-Lambertz, G.; Kolinsky, R.; Morais, J.; Cohen, L. How learning to read changes the cortical networks for vision and language. *Science* **2010**, *330*, 1359–1364. [CrossRef]

110. Kauffman, P.R. Might hallucinations have social utility? A proposal for scientific study. *J. Nerv. Ment. Dis.* **2016**, *204*, 702–712. [CrossRef]
111. Power, R.A.; Steinberg, S.; Bjornsdottir, G.; Rietveld, C.A.; Abdellaoui, A.; Nivard, M.M.; Johannesson, M.; Galesloot, T.E.; Hottenga, J.J.; Willemsen, G.; et al. Polygenic risk scores for schizophrenia and bipolar disorder predict creativity. *Nat. Neurosci.* **2015**, *18*, 953–955. [CrossRef] [PubMed]
112. Gardner, M. *The Ambidextrous Universe*; Allen Lane, The Penguin Press: London, UK, 1967.

Review

It Is Not Just in the Genes

Martina Manns 

Research Division Experimental and Molecular Psychiatry, Department of Psychiatry, Psychotherapy and Preventive Medicine, LWL University Hospital, Ruhr-University Bochum, 44780 Bochum, Germany; Martina.manns@rub.de; Tel.: +49-234-32-21628; Fax: +49-234-32-14377

Abstract: Asymmetries in the functional and structural organization of the nervous system are widespread in the animal kingdom and especially characterize the human brain. Although there is little doubt that asymmetries arise through genetic and nongenetic factors, an overarching model to explain the development of functional lateralization patterns is still lacking. Current genetic psychology collects data on genes relevant to brain lateralizations, while animal research provides information on the cellular mechanisms mediating the effects of not only genetic but also environmental factors. This review combines data from human and animal research (especially on birds) and outlines a multi-level model for asymmetry formation. The relative impact of genetic and nongenetic factors varies between different developmental phases and neuronal structures. The basic lateralized organization of a brain is already established through genetically controlled embryonic events. During ongoing development, hemispheric specialization increases for specific functions and subsystems interact to shape the final functional organization of a brain. In particular, these developmental steps are influenced by environmental experiences, which regulate the fine-tuning of neural networks via processes that are referred to as ontogenetic plasticity. The plastic potential of the nervous system could be decisive for the evolutionary success of lateralized brains.

Keywords: avian brain; brain asymmetries; hemispheric lateralization; ontogeny; epigenetic; neuronal plasticity; visual system

Citation: Manns, M. It Is Not Just in the Genes. *Symmetry* **2021**, *13*, 1815. https://doi.org/10.3390/sym13101815

Academic Editor: Sebastian Ocklenburg

Received: 4 September 2021
Accepted: 23 September 2021
Published: 29 September 2021

1. The Functional Organization of Brain Asymmetries and Its Development

"A number of embryonic events make up an integrated overture to the post-hatching expression of lateralization" Lesley Rogers [1]

1.1. Lateralization Patterns of Neuronal Systems across the Animal Kingdom

A fundamental organizational principle of our brain is its asymmetries, which encompass both structural and functional differences between the two hemispheres. This characteristic has led to numerous hypotheses and research projects, which have attempted to elucidate the evolutionary and developmental origins of this specific trait [2,3]. However, lateralization of the brain is not specific to humans, but is present in many species across the animal kingdom. Not only vertebrates, but also many invertebrates, such as flies, bees, octopuses or nematodes, show left–right differences in neural organization and behavior [3–9], which suggests that lateralization is a common feature of metazoan nervous systems [10]. Neuronal asymmetries can be observed in all areas of information processing, including perception, cognition, emotion, homeostatic regulation or motor control and are based on neuroanatomical as well as physiological left–right differences [7,11,12]. Lateralization can be present at the individual level, with left-sided dominance for a certain function in half of a population and right-sided dominance in the other half. In other cases, the direction of a lateralized function within a population is aligned, so that lateralization is present at population level [2,11,13]. Comparative studies indicate that some aspects of functional brain lateralizations share a common evolutionary history [3,7,8,14]. It has been suggested that the vertebrate brain is characterized by specific functional dichotomy, with the left hemisphere more strongly involved in routine and approach behavior, while the right hemisphere dominates detection and response to unexpected, novel and potentially pivotal stimuli [15–17]. For example, several species of fish, amphibians and birds react faster when a predator approaches from the left, indicating that right-hemispheric networks are specialized for the detection of potential dangers, while foraging is controlled by the left hemisphere [3,7,8,18]. The processing of social stimuli, such as faces, is also dominated by the right hemisphere [19] in humans [20], sheep [21,22] and chicks [23]. On the other hand, at least in mammals, communicative signals, such as spoken language in humans [24] or other forms of conspecific vocalizations [25,26], are typically processed within the left hemisphere. A widespread behavioral indicator of hemispheric lateralization is the preferred use of one extremity, which has been documented in a variety of vertebrate and invertebrate species at individual and population levels [5,27–31]. In humans, handedness is the most obvious asymmetry with about 90% of individual preferring to use their right hand for complex manual tasks like fine-tuned object manipulation or writing [32]. Handedness is related to other behavioral asymmetries, like preferential foot use [33] or cradling bias [34], while deviations from the typical pattern are associated with psychiatric or developmental disorders [2,35–37]. Therefore, handedness is used as the favorite measure for correlating functional lateralization with structural left–right differences and genetic variations ([e.g., [12,32,36,38]).

1.2. Understanding Ontogeny of Neuronal Asymmetries—An Unfinished Business

Despite increasing knowledge about the relationships between different functional lateralizations and their structural foundations, our understanding of the underlying ontogenetic mechanisms is still limited. The presence of population-level lateralizations and cross-species similarities makes it likely that neuronal asymmetries have developed under phylogenetic pressure and, therefore, have a genetic basis [7,8,39,40]. However, human and animal research currently differ in the approaches and methods used in investigating the mechanisms guiding the development of a functional lateralized brain and, therefore, there has only been limited integration of knowledge between research approaches [41]. In some animal models, the genetically controlled events that drive the development of neuronal asymmetries have been studied in detail. In the nervous system of the nematode *Caenorhabditis elegans*, for example, there are pairs of chemoperceptive neurons, which are characterized by molecular left–right differences and different connectivity patterns that are related to their differential functional embedding [42,43]. Molecular genetic studies identified a complex regulation network comprising transcription factors, microRNAs, chromatin regulators and intercellular signals, which determine the asymmetric features of these specific neurons [44]. A second well-studied example is the epithalamus of vertebrate brains, which connects limbic regions of the forebrain with hindbrain motor circuits and which is characterized by evolutionarily conserved asymmetries within the pineal complex and the adjacent habenular nuclei. The molecular pathways that control these asymmetries have mainly been elucidated in studies with larvae of the zebrafish. Here, too, it is a chain of gene expression cascades that underlie the development of lateralization in this area [42,45,46]. Other genes are persistently expressed asymmetrically within the adult forebrain of zebrafish [47].

In human research, however, the first popular models, such as the right shift [48] or dextral chance [49] theory, suggested a single gene origin for human brain lateralization and attributed the left-hemispheric dominance for language processing and hand use to a common genetic factor. Their predictions fit data on the prevalence of handedness and language lateralization, but they did not explain the nature, as well as the action, of such a factor. However, recent meta-analysis studies have shown that the associations between language lateralization and motor asymmetries are much weaker than previously assumed [50]. Currently, research concentrates on the identification of genes that regulate functional and structural lateralization using large-scale heritability and genome-wide association (GWAS), or single nucleotide (SNP) variation studies to find associations between gene variants and phenotypic lateralizations. These studies have reported an increasing number of genes and their variants related to lateralization pattern. One recent study even identified multifaceted gene networks associated with different aspects of anatomical brain asymmetries [51]. It has also been suggested that the impact of single genes is small and functional lateralizations are polygenic traits [38,52–56]. A recent study, for instance, detected 41 gene loci associated with left-handedness and seven associated with ambidexterity [52]. This also suggests that different manifestations of a trait can be controlled by different types of genes, which are either relevant during different phases of development or which influence discrete differentiation processes of the underlying neural networks.

In general, additive genetic effects account for less than one quarter of the variance in human handedness data, while nonshared environmental factors explain the remaining variance [52,57]. This is not surprising since neuronal systems always differentiate in close interactions with environmental experiences and genes alone do not explain the functional organization of neuronal systems [58]. This implies that the emergence of a functional lateralization pattern can only be understood by elucidating how genes and the environment interact to shape the functional organization of a lateralized brain [2,3,14,59–63]. It must also be considered how noncoding microRNA [64], or epigenetic mechanisms, which affect gene activity and expression by modifying DNA accessibility or chromatin structure, mediate long-term effects of gene–environment interactions [37,63].

Research has reported a potpourri of environmental factors influencing lateralization patterns in humans, including sex hormones [65], stress experience [66], sensory input, learning, birthweight, location and season of birth, breast feeding and cultural constraints [32,53]. These influences underline, on the one hand, the general role of environmental factors, while on the other hand, the lack of specificity of some factors suggests that deviations from genetically controlled patterns simply reflect random stochastic asymmetry [67].

1.3. Structural Foundations of Functional Asymmetries

Since neuronal functions are based on the organization of specific neuronal networks or cells, it is necessary to clarify how exactly structural and functional asymmetries are related. An increasing number studies have reported structure–function associations, but have also provided an inconsistent pattern. However, it is important to differentiate that left–right differences in the structure of neural circuits can be realized on different organizational levels, from the cellular to the macroscopic level.

An obvious global shape asymmetry of the human brain is the so called "cerebral torque", which refers to a counter-clockwise twist of the whole brain along the anterior–posterior axis [68,69]. At macroscopic level, the left hemisphere has a thicker cortex but a smaller surface area relative to the right hemisphere [70]. Region-specific left–right differences are reported in size and shape [70–72] and connectivity [73,74], as well the cellular and molecular organization [75,76]. Similar cortical asymmetries are also present in chimpanzees [77–79]. The left-hemispheric dominance of language processing is related to left–right differences in the microcircuitry of cortical columns in the posterior part of the superior temporal gyrus [80]. Moreover, there are function-related asymmetries in the hippocampus and subcortical structures in humans [71] and other mammals [81]. Handedness for instance is related to asymmetries within the nigrostriatal dopaminergic system in humans [82] and rodents [83–85].

Cortical left–right differences emerge early during development in humans [86,87], but also in nonhuman primates [88]. The cortical torque can be detected by the second trimester of gestation [68,89], while asymmetry of perisylvian language-related cortical regions appears during the third trimester [90–92]. Motor asymmetries can be observed even earlier. Human fetuses tend to make more movements with their right arms and preferentially suck the right thumb from the 12th gestational week onwards [93]. These motor asymmetries are related to postnatal handedness [94]. In relation to this behavioral lateralization, the fetal spinal cord segments innervating hands and arms display asymmetries in gene expression and DNA methylation at the end of the first trimester [95].

In sum, average left–right differences of global brain anatomy, which emerge early during development, suggest a developmental program that is genetically determined [96]. However, when analyzing specific cognitive functions, gene–structure interrelations are less detectable. Twin studies, for instance, indicate that pre- and postnatal events can affect asymmetry during development of the planum temporal [97,98]. Accordingly, a recent large-scale study did not find significant associations between cortical asymmetries and language lateralization [99]. There is also no significant relation between cortical asymmetries and handedness [70]. The lack of correlations may not come as a surprise since the macroscopic cortical features do not necessarily represent the internal microscopic organization. It is conceivable that functional asymmetries only emerge on the cellular, synaptic or neurophysiological level. This means that it is necessary to understand how neuronal asymmetries arise at precisely this cellular level. To this end, findings from developmental neurosciences have to be integrated into models of asymmetry formation. Experiments with animal models have shown that activity-dependent processes triggered by internal or external signals are decisive for the functional maturation of neural networks [100–103]. In the following, I will, therefore, first summarize what is known about the role of genetic factors for asymmetry formation during different developmental phases. I then illustrate the possible effects of environmental factors as suggested by the light-dependent development of visual asymmetries in birds.

2. Potential Roles of Genetic Factors for Asymmetry Formation

The relative importance of genes and the environment depends on the species examined, the specific neuronal function and their developmental trajectories, as well as the level of analysis [61]. This means that we have to differentiate the action of gene–environment interactions depending on the development phase. The development of the nervous system can be roughly divided into three phases, during which the degree of hemispheric specialization increases (Figure 1). The first phase comprises the earliest embryological steps, in which the axes of the body plan are determined. The second phase includes the differentiation of neural systems and networks, while processes mediating the refinement of neural connections dominate the third phase. During these phases, different cellular processes dominate development and genes can influence the action of epigenetic factors in different ways, which affect the developing organism (Figure 1):

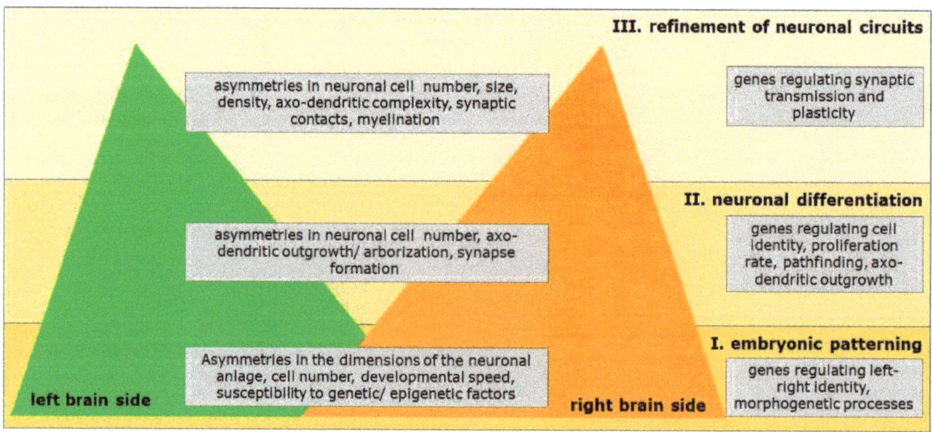

Figure 1. Model of the hierarchical development of brain lateralization—the three main phases of neuronal development are dominated by different cellular processes, which lead to an increasingly lateralized functional organization of the two hemispheres (indicated by the green and orange triangles). In each phase, certain types of genes regulate differentiation and, thus, asymmetry formation.

I. Specific genes can account for left–right differences in the amount of neuronal substrate. An asymmetrical number of neuronal and/or glial precursor cells can result in gross morphological asymmetries and can be related to differences in cellular identity or cell-type-specific proteins.
II. Specific genes can control left–right differences in developmental dynamics. Differences in the maturation of left- or right-hemispheric neurons or networks could lead to asymmetries in the susceptibility to epigenetic factors like hormones, sensory input or motor activity.
III. Specific genes can regulate asymmetrical morphogenetic events leading to asymmetric body positions or craniofacial asymmetries, which bias sensory experiences.
IV. Specific genes can control asymmetrical differentiation of neuronal elements like growth or arborization of axons and dendrites or development of synapses.
V. Specific genes involved in synaptogenesis or signal transmission can lead to left–right differences in the degree of neuronal plasticity.

In the meantime, a number of genes have been discovered that mediate at least one of these actions during asymmetry formation:

2.1. Embryonic Patterning

Asymmetry formation within neuronal systems starts with breaking the symmetry of the body plan during early embryogenesis in all bilaterian animals, when the primary axes and tissue layers form. Complex cascades of genetic and epigenetic interactions lead to an asymmetrical placement of internal organs, but also induce asymmetries of paired organs like the lungs or the nervous system [104–109]. Determination of the left–right body axis is coordinated by a midline structure called the node. In several species, including humans, symmetry is broken by the rotation of motile cilia, which generate a directed flow that acts as a signal for the asymmetrical expression of a gene cascade, the Nodal signaling pathway. This pathway is remarkably conserved within bilaterian evolution [10,106,108].

This implies that asymmetry formation of body and brain starts with the action of cilia and, therefore, genes controlling generation and motility of cilia could play an early role in the development of neuronal asymmetries [38,110]. Some studies have actually provided evidence for the involvement of cilial genes for handedness—however, only in specific humans populations [38,54,111].

A second critical mechanism during this early phase is the lateralized action of the Nodal pathway. One key player in this signaling cascade is PCSK6, which cleaves the Nodal protoprotein into its biologically active form [10,110,112]. *PCSK6* polymorphism has been associated with human handedness [38,113], but also with structural asymmetries in temporal cortical areas, indicating a potential role of *PCSK6* not only for motoric but also language networks [114].

However, when symmetry breaking processes of visceral and neuronal structures share the same developmental route, one should assume that individuals with reversed visceral organization also display reversed brain asymmetries. A test case involves individuals with situs inversus, where the visceral organs are organized as a mirror image of the default organ position. Situs inversus can occur in, but does not depend on, ciliary dyskinesia [115]. While the typical gross morphological asymmetry of the human brain–cerebral torque is actually reversed in situs inversus, functional and cortical lateralizations are not [115–119], although atypical functional segregation can be more frequent in participants with visceral reversal [115,120]. Similarly, in less complex animals, such as the nematode *C. elegans*, motor lateralization is independent from left–right body asymmetry [121] and zebrafish with situs inversus develop reversed lateralization of some but not all structural and behavioral lateralizations [122]. This suggests that early embryonic patterning processes regulate, to some degree, the establishment of basic brain asymmetries, but lateralization of specific functional modules are presumably shaped by specific cellular mechanisms later during development [119,123].

2.2. Regionalization of Neuronal Substrate

When the neuronal anlage starts to differentiate region-specific differences, genes playing a role in symmetry breaking of the embryo are also involved in the generation of specific brain asymmetries. The best known example is the Nodal pathway, whereby asymmetrical left-sided Nodal signaling within the developing dorsal diencephalon is required for determining the direction of epithalamic asymmetries [42,45,112,124]. It is conceivable that laterality signals result in asymmetrical expression of neuron-type-specific gene batteries, which are responsible for cell-type-specific structural and functional properties [125].

2.3. Differential Developmental Dynamics

One consequence of the early left–right patterning is that the left and right hemispheres develop at different speeds. In human embryos, the right hemisphere tends to develop a little earlier than the left one [86] and the lateralized gradient of brain development might contribute to the development of the cerebral torque [69]. Differences in developmental speed of cortical subareas are indicated by specific lateralized gene expression profiles from the fifth week postconception onwards [126]. The early appearance of asymmetrical arm movements in human fetuses can be explained by left–right differences in the differentiation of spinal neurons, since the cortex and spinal cord are not connected at this age [127,128]. As a result of the asymmetrical developmental gradients of the two hemispheres, it is possible that a nongenetic factor, which acts on the developing organism at a certain point in time, differentially influences left- and right-hemispherical neuronal structures. There is, for instance, some evidence that the right hemisphere of human fetuses is generally less subject to external influences than the left one [86].

2.4. Differentiation of Hemisphere-Specific Neuronal Elements

When the nervous tissue starts to differentiate region-specific neurons and connections, specific genes regulate proliferation, migration and growth of axonal and/or dendritic fibers. Therefore, asymmetrical expression of these genes can account for the asymmetrical differentiation of specific brain regions.

Sun et al. [129,130] identified a couple of genes in perisylvian regions of the human cortex, which are asymmetrically expressed at the end of the first trimester and, therefore, before a neuroanatomical asymmetry of this area can be detected [90,91]. Intriguingly, most of these asymmetrically expressed genes function in signal transduction and gene expression regulation [129,130].

One of these genes is the transcription factor *LMO4*, which is consistently more highly expressed in the right perisylvian cortex of 12–16-week human fetuses and, hence, during a period of high proliferation and migration rate [129,130]. *LMO4* displays higher expression level also in the right forebrain of zebrafish [47], while in the mouse cortex, *LMO4* expression is not constantly lateralized to one side [129,130]. Expression of *LMO4* is confined to postmitotic neurons [131] and regulates key aspects of neuronal differentiation, radial migration of newborn nerve cells and acquisition of neuronal identities [132,133].

Another example is the transcription factor forkhead box P2 gene *FOXP2*, which is involved in neural development and, in particular, in regulating neurogenesis of the embryonal cortex. It is expressed in distinct brain areas from gestational week six onwards and is related to speech development [134]. Intriguingly, *FOXP2* polymorphism is associated with the interindividual variability in hemispheric asymmetries for speech perception [135].

2.5. Ontogenetic Plasticity

After the establishment of the basic brain organization, neuronal networks typically sharpen their functional efficiency. Growth, stabilization or reduction of synaptic contacts or cell death occur in an activity-dependent manner and are triggered by sensory experience [100,136]. This critical period is likely to amplify expression of genes and proteins that mediate synaptic plasticity. Accordingly, genes that are involved in regulating ontogenetic plasticity can affect the asymmetrical development during specific sensitive phases. Asymmetrical expression of these genes can result in a differential sensitivity of left- and right-hemispheric circuits towards stimulation. Karlebach and Francks [137], for instance, identified several asymmetrically expressed genes in the human cortex that are likely to fine-tune electrophysiological and neurotransmission properties of cortical circuits during different phases of development. Additionally, in the rat hippocampus, a dynamic pattern of asymmetrically expressed genes has been identified during the first postnatal weeks, with a large percentage of genes being associated with synaptic function [138]. One example could be the transmembrane molecule LRRTM1 (leucine-rich repeat transmembrane neuronal 1). It interacts at synapses with the extracellular matrix as a regulator of neuronal plasticity [139]. Gene variations have been associated with handedness [53,140,141].

Crucial mediators of ontogenetic plasticity are neurotrophic factors like BDNF (brain-derived neurotrophic factor), which mediates activity-dependent synaptic stabilization, axo-dendritic growth, arborization and cell survival [142,143]. It is, therefore, intriguing that BDNF is asymmetrically expressed in the hippocampus of rats, specifically during the first two weeks after birth when neurogenesis rate is high [144]. BDNF might also mediate stress effects in the brain and could, therefore, regulate the well-known action of stress hormones onto brain lateralization [145].

To sum it up, neuronal development is controlled at very different levels of differentiation by genes that are either asymmetrically expressed or whose variants are associated with specific phenotypes. The same function (e.g., handedness) can, therefore, be regulated during different developmental phases by different types of genes. Asymmetrical expression of single genes can be confined to specific developmental phases, while other genes are lateralized up until adulthood. At all levels, nongenetic factors can modulate genetic effect and thereby change the direction and/or degree of lateralization. However, little is yet understood about the neuronal processes through which environmental factors can influence the differentiation of the complex functional organization of lateralized brains. One of the few models in which the influence of a specific environmental factor has been examined in more detail is the visual system of birds. Research on chicks and pigeons has delineated a chain of events that begins with asymmetrical photic stimulation of the embryo in the egg and ends in a lateralized organization of visual processing and cognition [1–3,14,40,59–61,146,147]. This model suggests critical steps for the formation of asymmetries that can serve as a blueprint for a better understanding of the ontogenesis of brain asymmetries in general. These developmental steps are summarized below (Figure 2) and are complemented by findings in other species, especially in humans.

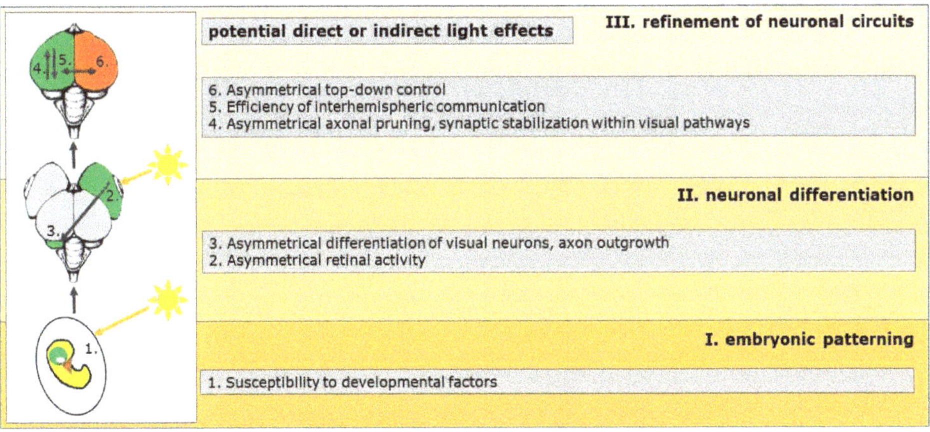

Figure 2. The developing visual systems of chicks and pigeons exemplifies how one environmental factor—in this case light—affects the development of brain asymmetries during the three main phases of neuronal development (see text for details).

3. The Avian Visual System as a Model for Ontogenetic Plasticity

The visual system of chicks and pigeons is lateralized with a pattern that is similar to the lateralization of the human brain. The left hemisphere dominates the discrimination of small optic details, rule learning, categorization and visuomotor control [59–61,147–149]. The right hemisphere on the contrary, is in charge of spatial attention [150] and aspects of social cognition [23]. These hemispherical specializations can be identified very easily by temporarily occluding one eye with an opaque cap. Since the optic nerves cross virtually completely in birds, information from the left eye is primarily directed to the right hemisphere and vice versa. A comparison of monocular and binocular testing, therefore, enables the investigation of hemispherical differences in performances or analysis strategies. Behavioral asymmetries are accompanied by anatomical left–right differences within the ascending visual pathways. In both pigeons and chicks, for example, differences in the projection strength between the two hemispheres can be observed. Major aspects of these asymmetries develop in response to asymmetrical visual stimulation during development. Therefore, light deprivation before and after hatching prevents or modifies visual lateralizations. The comparison of structural and behavioral lateralizations of light-exposed or light-deprived birds makes it possible to unravel critical neuronal processes that mediate light-dependent development (Figure 2) [1–3,14,40,59–61,146,147,151,152].

3.1. *Mechanisms during Embryonic Patterning (Phase I)*

As in all vertebrates, asymmetry formation in birds starts during embryonic body patterning [153,154], whereby symmetry breaking is independent from motile cilia [105,106]. At this point of development, light cannot directly affect visual lateralization patterns but there are at least three routes serving as starting points for the induction of asymmetries in the visual system:

1. Differences in left–right identity presumably determine asymmetries in the developing nervous system, which result in functional lateralizations when no other factors modify these predetermined ones. Early asymmetrical differentiation is indicated by a rightward torsion of the neuronal tube. Some evidence suggests that bending is caused by differential growth of the left and right neuronal tube side but physical mechanisms in relation to asymmetrical heart bending might also play a role [154]. This bending also occurs in mammalian embryos [155] and could contribute to the emergence of the cortical torque in the human brain. Presumably dependent on these early asymmetrical developmental processes, some visual lateralization develops independent from asymmetrical light stimulation. In chicks, visual choice to approach a social partner [156,157], uni-hemispheric sleep [158] or structural asymmetries of forebrain areas [159–162] are present in birds that are not exposed to biased visual stimulation. Interocular transfer of conditioned information [163] or lateralized visuospatial attention [164] emerge without embryonic light stimulation in pigeons but, interestingly, not in chicks [165,166]. It is currently unknown which genetic factors and which neuronal processes control the emergence of these visual asymmetries. However, endogenous asymmetries can be modulated by later visual experience [159,163,164,167].
2. As a consequence of the primarily Nodal-dependent left–right determination of the body, morphogenetic processes lead to bending of the head region, which turns to the left so that the beak points to the right and the right side of the head is facing the egg [168]. Due to the fact that the size of the embryo during the last embryonic phase does not allow free head rotations anymore, the left eye arrests on the body while the right eye is close to the egg shell and can be stimulated by light shining through the egg shell (Figure 2). This biased environmental stimulation triggers the second step in asymmetry formation, inducing asymmetrical differentiation processes, which involve neuronal mechanisms well known for ontogenetic plasticity [59,60,152] (see phase II). Such a rightward torsion occurs in all amniotes [168] including human embryos, which display a right-turn of their head during the last gestational phase

from week 38 onwards [169]. During this time, human fetuses are already responsive to sensory stimulation. They are able to memorize auditory stimuli from the external world by the last trimester of pregnancy, with a particular sensitivity to melody contour in both music and language [170,171]. Differential auditory input to the left and right ear because of postural asymmetries, therefore, might affect the development of language lateralizations [172–174].

3. Although visual systems are not developed, there is some evidence that during this phase, light stimulation already affects the establishment of some aspects of lateralization in both chickens and zebrafish [166,175–177] (Figure 2). Transduction mechanisms mediating these light effects are unknown but might include epigenetic mechanisms [166,177]. It is also possible that some genes unfold their actions only after photostimulation [178].

3.2. Mechanisms during Neuronal Differentiation (Phase II)

It is well known that the differentiation of visual networks is critically influenced by visual stimulation (e.g., [100]], and it is therefore not surprising that unbalanced light stimulation differentially affects left- and right-hemispheric developmental processes during species-specific sensitive phases [61,179]. Some behaviors and anatomical asymmetries only develop after embryonic light stimulation [180–182] and can be reversed by altered visual experience before (chicks, [183]) or after (pigeons [184,185]) hatching. In chicks, the outgrowth of visual fibers is influenced by light stimulation, resulting in a transiently stronger innervation of the right visual forebrain. Thereby, the action of light is modified by corticosterone, testosterone and estradiol [1,14,146,147,149,186]. The modulatory action of steroid hormones is in line with the often described sex- and stress-effects on human and nonhuman lateralization patterns [66,187,188]. In pigeons, left–right differences in cell size and projection strength differentiate in response to asymmetric photic stimulation [180,182,184,185,189]. Posthatch experimental manipulations have shown that starting with asymmetrical retinal activity [190], asymmetrical differentiation within the ascending visual system is mediated partly by BDNF-dependent processes [191,192].

The avian models exemplify how an environmental factor shapes the generation of neuronal asymmetries by modifying specific bottom-up systems. In a similar way, left–right differences in spectrotemporal selectivity of neurons in the auditory cortex of mice develop depending on hearing experience, which is related to the left-hemispheric dominance for the analysis of vocalization features [193]. In humans, visual experience can affect handedness [173], head turning preference [194] or lateralized face-processing competence [195,196].

However, an asymmetrical sensory trigger, such as light, not only enhances differentiation of the stronger stimulated hemisphere but also modifies the balance of left- and right-hemispheric development. A detailed analysis of light- and dark-incubated pigeons, for instance, revealed that light induces a left-hemispheric increase in visuoperceptual skills but simultaneously decreases visuomotor speed within the right hemisphere [182]. At the neuroanatomical level, embryonic light stimulation does not increase the bilateral innervation of the more strongly stimulated left brain side, but rather decreases input to the right side [180].

Presumably, interdependent left- and right-hemispheric developmental processes also play a role in the experience-dependent specialization of the human cortex, as indicated by the distribution of hemispheric language and face recognition processing. While the visual word form area in the left hemisphere becomes specialized while learning to read, the right hemisphere develops face recognition dominance. This suggests that the hemispheric organization of face recognition and of word recognition does not develop independently, and that word lateralization may precede and drive later face lateralization [196,197].

3.3. Consolidation of Functional Asymmetries (Phase III)

The ontogeny of visual asymmetries in birds is profoundly triggered within the developing ascending visual pathways but cognitive asymmetries emerge only at a higher (forebrain) processing level [60,148]. This means that asymmetries, which are induced within bottom-up systems, have to be transferred onto higher brain structures. At this level, they might interact with inherent or light-independent asymmetries (see above) and thereby sculpt and stabilize the final functional organization of the visual brain. In the pigeon, these processes mainly take place after hatching, when light input is normally symmetrical. During this phase, lateralization can still be modified by manipulating the visual experience [59,60,152,184,190]. It is likely that top-down as well as commissural mechanisms play a critical role in these stabilization processes [59,60,185,189,198,199]. As a consequence, relevant top-down and/or commissural systems develop their own asymmetrical properties for controlling asymmetrical decision-making and behaviors, but also for determining the degree of interhemispheric crosstalk. For example, left-hemispheric dominance for conflict choices is related to the asymmetrical action of top-down projections from the forebrain [198]. Light-dependent efficiency of interhemispheric integration has been shown in chicks, where only light-stimulated individuals can efficiently allocate food searching to the left and predator vigilance to the right hemisphere [200]. Also, only light-exposed chicks can use object (left-hemisphere)- as well as position (right-hemispheric)-dependent cues in food searching tasks [201,202]. A study with pigeons showed that only light-stimulated birds integrate hemispheric-specific knowledge for solving a task that cannot be correctly answered with information of one hemisphere alone [199]. Relevance of interhemispheric mechanisms for the generation and modulation of hemispheric-specific functions is in line with studies exploring the role of the corpus callosum for brain lateralizations [203,204]. The avian model suggests that top-down and commissural systems unfold their effects mainly at the end of asymmetry formation and modulate the interaction of more or less strongly lateralized neuronal networks in the left and right hemispheres [148]. To this regard, these processes shape the final functional organization of lateralized cognitive modules.

4. Conclusions

Studies on the genetic basis and/or environmental influences on the formation of asymmetries in humans and other animals have shown that the development of a lateralized functional architecture of the brain is to be understood as an example of ontogenetic plasticity. Genes and environmental factors play different but intertwined and complementary roles that can be specific to certain processing modules. The final functional lateralization pattern is then the result of hierarchical processes that build on one another. Genetically controlled early embryonic developmental steps set the framework for hemispherical differences and can be indicated by gross morphological asymmetries in volume and/ or shape of gray and white matter. Epigenetic processes lead to increasing hemispherical specialization and control dynamics of interhemispheric communication. This means that no factor alone can explain the variance of lateralization patterns in a population; it is the sum of individual experiences, which shape individual brain lateralization. It is possible to identify general roles of single genes or environmental factors, but only their interplay within a specific environment determines the functional outcome. Consequently, single factors can only explain limited variance in the lateralization pattern within a population.

This flexibility enables fluctuating lateralization patterns within a population depending on the ecological requirements. Recent field studies showed, for instance, that factors such as predator pressure, environmental pollutants or seasonal conditions can modify brain asymmetries [205–207]. Humans have cultural constraints affecting, for example, the prevalence of left-handedness [208]. It is conceivable that the specific ecological or social conditions account for population-level lateralization in humans, which is absent in other animals species [208]. Ontogenetic plasticity, however, can be a general mechanism that enhances the evolutionary benefit of brain asymmetries [61,208].

Funding: This work was funded by the DFG (grant number MA4485/2-2).

Institutional Review Board Statement: Not applicable.

Informed Consent Statement: Not applicable.

Data Availability Statement: Not applicable.

Conflicts of Interest: The author declares no conflict of interest.

References

1. Rogers, L.J. Factors Influencing Development of Lateralization. *Cortex* **2006**, *42*, 107–109. [CrossRef]
2. Güntürkün, O.; Ocklenburg, S. Ontogenesis of Lateralization. *Neuron* **2017**, *94*, 249–263. [CrossRef] [PubMed]
3. Güntürkün, O.; Ströckens, F.; Ocklenburg, S. Brain Lateralization: A Comparative Perspective. *Physiol. Rev.* **2020**, *100*, 1019–1063. [CrossRef]
4. Frasnelli, E.; Vallortigara, G.; Rogers, L.J. Left–right asymmetries of behaviour and nervous system in invertebrates. *Neurosci. Biobehav. Rev.* **2012**, *36*, 1273–1291. [CrossRef]
5. Niven, J.E.; Frasnelli, E. Insights into the evolution of lateralization from the insects. *Prog. Brain Res.* **2018**, *238*, 3–31. [CrossRef]
6. Rogers, L.J.; Vallortigara, G.; Andrew, R.J. *Divided Brains: The Biology and Behaviour of Brain Asymmetries*; University Press: Cambridge, UK, 2013.
7. Vallortigara, G.; Rogers, L.J. Survival with an asymmetrical brain: Advantages and disadvantages of cerebral lateralization. *Behav. Brain Sci.* **2005**, *28*, 575–589. [CrossRef]
8. Vallortigara, G.; Rogers, L.J. A function for the bicameral mind. *Cortex* **2020**, *124*, 274–285. [CrossRef]
9. Vallortigara, G.; Versace, E. Laterality at the neural, cognitive, and behavioral levels. In *APA Handbook of Comparative Psychology: Basic Concepts, Methods, Neural Substrate, and Behavior*; American Psychological Association: Washington, DC, USA, 2017; Volume 1, pp. 557–577. [CrossRef]
10. Heger, P.; Zheng, W.; Rottmann, A.; Panfilio, K.A.; Wiehe, T. The genetic factors of bilaterian evolution. *eLife* **2020**, *9*. [CrossRef]
11. Manns, M. Hemispheric Specialization. In *Encyclopedia of Animal Cognition and Behavior*; Vonk, J., Shackelford, T., Eds.; Springer International Publishing: Cham, Germany, 2019; pp. 1–10.
12. Manns, M.; El Basbasse, Y.; Freund, N.; Ocklenburg, S. Paw preferences in mice and rats: Meta-analysis. *Neurosci. Biobehav. Rev.* **2021**, *127*, 593–606. [CrossRef]
13. Frasnelli, E.; Vallortigara, G. Individual-Level and Population-Level Lateralization: Two Sides of the Same Coin. *Symmetry* **2018**, *10*, 739. [CrossRef]
14. Rogers, L.J. Asymmetry of brain and behavior in animals: Its development, function, and human relevance. *Genesis* **2014**, *52*, 555–571. [CrossRef] [PubMed]
15. Lippolis, G.; Joss, J.; Rogers, L. Australian Lungfish (*Neoceratodus forsteri*): A Missing Link in the Evolution of Complementary Side Biases for Predator Avoidance and Prey Capture. *Brain Behav. Evol.* **2009**, *73*, 295–303. [CrossRef] [PubMed]
16. MacNeilage, P.F.; Rogers, L.J.; Vallortigara, G. Origins of the Left & Right Brain. *Sci. Am.* **2009**, *301*, 60–67. [CrossRef] [PubMed]
17. Vallortigara, G. Comparative Neuropsychology of the Dual Brain: A Stroll through Animals' Left and Right Perceptual Worlds. *Brain Lang.* **2000**, *73*, 189–219. [CrossRef]
18. Rogers, L.J.; Andrew, R. *Comparative Vertebrate Lateralization*; Cambridge University Press: Cambridge, UK, 2002.
19. Salva, O.R.; Regolin, L.; Mascalzoni, E.; Vallortigara, G. Cerebral and Behavioural Asymmetries in Animal Social Recognition. *Comp. Cogn. Behav. Rev.* **2012**, *7*, 110–138. [CrossRef]
20. Gainotti, G. Laterality effects in normal subjects' recognition of familiar faces, voices and names. Perceptual and representational components. *Neuropsychologia* **2013**, *51*, 1151–1160. [CrossRef]
21. Kendrick, K.M. Brain asymmetries for face recognition and emotion control in sheep. *Cortex* **2006**, *42*, 96–98. [CrossRef]
22. Versace, E.; Morgante, M.; Pulina, G.; Vallortigara, G. Behavioural lateralization in sheep (*Ovis aries*). *Behav. Brain Res.* **2007**, *184*, 72–80. [CrossRef]
23. Daisley, J.N.; Mascalzoni, E.; Salva, O.R.; Rugani, R.; Regolin, L. Lateralization of social cognition in the domestic chicken (*Gallus gallus*). *Philos. Trans. R. Soc. B Biol. Sci.* **2008**, *364*, 965–981. [CrossRef]
24. Ocklenburg, S.; Beste, C.; Arning, L.; Peterburs, J.; Güntürkün, O. The ontogenesis of language lateralization and its relation to handedness. *Neurosci. Biobehav. Rev.* **2014**, *43*, 191–198. [CrossRef]
25. Levy, R.B.; Marquarding, T.; Reid, A.P.; Pun, C.M.; Renier, N.; Oviedo, H.V. Circuit asymmetries underlie functional lateralization in the mouse auditory cortex. *Nat. Commun.* **2019**, *10*, 2783. [CrossRef]
26. Ocklenburg, S.; Ströckens, F.; Güntürkün, O. Lateralisation of conspecific vocalisation in non-human vertebrates. *Laterality* **2013**, *18*, 1–31. [CrossRef] [PubMed]
27. Bell, A.T.A.; Niven, J.E. Individual-level, context-dependent handedness in the desert locust. *Curr. Biol.* **2014**, *24*, R382–R383. [CrossRef]
28. Bell, A.T.A.; Niven, J.E. Strength of forelimb lateralization predicts motor errors in an insect. *Biol. Lett.* **2016**, *12*, 20160547. [CrossRef] [PubMed]

29. Byrne, R.A.; Kuba, M.J.; Meisel, D.V.; Griebel, U.; Mather, J.A. Does Octopus vulgaris have preferred arms? *J. Comp. Psychol.* **2006**, *120*, 198–204. [CrossRef] [PubMed]
30. Ströckens, F.; Güntürkün, O.; Ocklenburg, S. Limb preferences in non-human vertebrates. *Laterality* **2013**, *18*, 536–575. [CrossRef] [PubMed]
31. Versace, E.; Vallortigara, G. Forelimb preferences in human beings and other species: Multiple models for testing hypotheses on lateralization. *Front. Psychol.* **2015**, *6*, 233. [CrossRef]
32. Papadatou-Pastou, M.; Ntolka, E.; Schmitz, J.; Martin, M.; Munafò, M.R.; Ocklenburg, S.; Paracchini, S. Human handedness: A meta-analysis. *Psychol. Bull.* **2020**, *146*, 481–524. [CrossRef]
33. Packheiser, J.; Schmitz, J.; Berretz, G.; Carey, D.P.; Paracchini, S.; Papadatou-Pastou, M.; Ocklenburg, S. Four meta-analyses across 164 studies on atypical footedness prevalence and its relation to handedness. *Sci. Rep.* **2020**, *10*, 1–21. [CrossRef]
34. Packheiser, J.; Schmitz, J.; Berretz, G.; Papadatou-Pastou, M.; Ocklenburg, S. Handedness and sex effects on lateral biases in human cradling: Three meta-analyses. *Neurosci. Biobehav. Rev.* **2019**, *104*, 30–42. [CrossRef]
35. Grimshaw, G.M.; Ecarmel, D. An asymmetric inhibition model of hemispheric differences in emotional processing. *Front. Psychol.* **2014**, *5*, 489. [CrossRef]
36. Ocklenburg, S.; Beste, C.; Güntürkün, O. Handedness: A neurogenetic shift of perspective. *Neurosci. Biobehav. Rev.* **2013**, *37*, 2788–2793. [CrossRef]
37. Schmitz, J.; Metz, G.A.; Güntürkün, O.; Ocklenburg, S. Beyond the genome—Towards an epigenetic understanding of handedness ontogenesis. *Prog. Neurobiol.* **2017**, *159*, 69–89. [CrossRef] [PubMed]
38. Brandler, W.M.; Morris, A.P.; Evans, D.M.; Scerri, T.S.; Kemp, J.P.; Timpson, N.J.; Pourcain, B.S.; Smith, G.D.; Ring, S.M.; Stein, J.; et al. Common Variants in Left/Right Asymmetry Genes and Pathways Are Associated with Relative Hand Skill. *PLoS Genet.* **2013**, *9*, e1003751. [CrossRef] [PubMed]
39. Laland, K.N. Exploring gene–culture interactions: Insights from handedness, sexual selection and niche-construction case studies. *Philos. Trans. R. Soc. B Biol. Sci.* **2008**, *363*, 3577–3589. [CrossRef] [PubMed]
40. Rogers, L.J.; Vallortigara, G. When and Why Did Brains Break Symmetry? *Symmetry* **2015**, *7*, 2181–2194. [CrossRef]
41. Ocklenburg, S.; Berretz, G.; Packheiser, J.; Friedrich, P. Laterality 2020: Entering the next decade. *Laterality* **2021**, *26*, 265–297. [CrossRef]
42. Concha, M.L.; Bianco, I.; Wilson, S. Encoding asymmetry within neural circuits. *Nat. Rev. Neurosci.* **2012**, *13*, 832–843. [CrossRef]
43. Hobert, O.; Johnston, R.J.; Chang, S. Left–right asymmetry in the nervous system: The Caenorhabditis elegans model. *Nat. Rev. Neurosci.* **2002**, *3*, 629–640. [CrossRef]
44. Hobert, O. Development of left/right asymmetry in the Caenorhabditis elegans nervous system: From zygote to postmitotic neuron. *Genesis* **2014**, *52*, 528–543. [CrossRef]
45. Concha, M.L.; Signore, I.A.; Colombo, A. Mechanisms of directional asymmetry in the zebrafish epithalamus. *Semin. Cell Dev. Biol.* **2009**, *20*, 498–509. [CrossRef] [PubMed]
46. Roberson, S.; Halpern, M.E. Development and connectivity of the habenular nuclei. *Semin. Cell Dev. Biol.* **2018**, *78*, 107–115. [CrossRef] [PubMed]
47. Messina, A.; Boiti, A.; Vallortigara, G. Asymmetric distribution of pallial-expressed genes in zebrafish (*Danio rerio*). *Eur. J. Neurosci.* **2021**, *53*, 362–375. [CrossRef] [PubMed]
48. Annett, M. Handedness and Cerebral Dominance. *J. Neuropsychiatry Clin. Neurosci.* **1998**, *10*, 459–469. [CrossRef]
49. McManus, I.C. Handedness, language dominance and aphasia: A genetic model. *Psychol. Med. Monogr. Suppl.* **1985**, *8*, 1–40. [CrossRef]
50. Packheiser, J.; Schmitz, J.; Arning, L.; Beste, C.; Güntürkün, O.; Ocklenburg, S. A large-scale estimate on the relationship between language and motor lateralization. *Sci. Rep.* **2020**, *10*, 1–10. [CrossRef]
51. Sha, Z.; Schijven, D.; Carrion-Castillo, A.; Joliot, M.; Mazoyer, B.; Fisher, S.E.; Crivello, F.; Francks, C. The genetic architecture of structural left–right asymmetry of the human brain. *Nat. Hum. Behav.* **2021**, 1–14. [CrossRef]
52. Cuellar-Partida, G.; Tung, J.Y.; Eriksson, N.; Albrecht, E.; Aliev, F.; Andreassen, O.A.; Barroso, I.; Beckmann, J.S.; Boks, M.P.; Boomsma, D.I.; et al. Genome-wide association study identifies 48 common genetic variants associated with handedness. *Nat. Hum. Behav.* **2021**, *5*, 59–70. [CrossRef]
53. De Kovel, C.G.F.; Carrión-Castillo, A.; Francks, C. A large-scale population study of early life factors influencing left-handedness. *Sci. Rep.* **2019**, *9*, 1–11. [CrossRef]
54. De Kovel, C.G.F.; Francks, C. The molecular genetics of hand preference revisited. *Sci. Rep.* **2019**, *9*, 1–9. [CrossRef]
55. McManus, C. Half a century of handedness research: Myths, truths; fictions, facts; backwards, but mostly forwards. *Brain Neurosci. Adv.* **2019**, *3*, 2398212818820513. [CrossRef]
56. McManus, I.C.; Davison, A.; Armour, J. Multilocus genetic models of handedness closely resemble single-locus models in explaining family data and are compatible with genome-wide association studies. *Ann. N. Y. Acad. Sci.* **2013**, *1288*, 48–58. [CrossRef] [PubMed]
57. Medland, S.E.; Duffy, D.L.; Wright, M.; Geffen, G.M.; Hay, D.A.; Levy, F.; Van-Beijsterveldt, C.E.; Willemsen, G.; Townsend, G.C.; White, V.; et al. Genetic influences on handedness: Data from 25,732 Australian and Dutch twin families. *Neuropsychology* **2009**, *47*, 330–337. [CrossRef] [PubMed]

58. Park, W.J.; Fine, I. New insights into cortical development and plasticity: From molecules to behavior. *Curr. Opin. Physiol.* **2020**, *16*, 50–60. [CrossRef] [PubMed]
59. Güntürkün, O.; Manns, M. The Embryonic Development of Visual Asymmetry in the Pigeon. In *The Two Halves of the Brain*; Hugdahl, K., Westerhausen, R., Eds.; The MIT Press: Cambridge, MA, USA, 2010; pp. 121–142.
60. Manns, M.; Güntürkün, O. Dual coding of visual asymmetries in the pigeon brain: The interaction of bottom-up and top-down systems. *Exp. Brain Res.* **2009**, *199*, 323–332. [CrossRef] [PubMed]
61. Manns, M.; Ströckens, F. Functional and structural comparison of visual lateralization in birds-similar but still different. *Front. Psychol.* **2014**, *5*, 206. [CrossRef]
62. Ocklenburg, S.; Güntürkün, O. Hemispheric Asymmetries: The Comparative View. *Front. Psychol.* **2012**, *3*, 5. [CrossRef]
63. Schmitz, J.; Güntürkün, O.; Ocklenburg, S. Building an Asymmetrical Brain: The Molecular Perspective. *Front. Psychol.* **2019**, *10*, 982. [CrossRef]
64. Miao, N.; Lai, X.; Zeng, Z.; Cai, W.; Chen, W.; Sun, T. Differential expression of microRNAs in the human fetal left and right cerebral cortex. *Mol. Biol. Rep.* **2020**, *47*, 6573–6586. [CrossRef]
65. Richards, G.; Beking, T.; Kreukels, B.P.; Geuze, R.H.; Beaton, A.A.; Groothuis, T. An examination of the influence of prenatal sex hormones on handedness: Literature review and amniotic fluid data. *Horm. Behav.* **2021**, *129*, 104929. [CrossRef]
66. Berretz, G.; Wolf, O.T.; Güntürkün, O.; Ocklenburg, S. Atypical lateralization in neurodevelopmental and psychiatric disorders: What is the role of stress? *Cortex* **2020**, *125*, 215–232. [CrossRef]
67. McManus, C. Is any but a tiny fraction of handedness variance likely to be due to the external environment? *Laterality* **2021**, *26*, 310–314. [CrossRef]
68. Kong, X.-Z.; Postema, M.; Schijven, D.; Castillo, A.C.; Pepe, A.; Crivello, F.; Joliot, M.; Mazoyer, B.; Fisher, S.E.; Francks, C. Large-Scale Phenomic and Genomic Analysis of Brain Asymmetrical Skew. *Cereb. Cortex* **2021**, *31*, 4151–4168. [CrossRef] [PubMed]
69. Xiang, L.; Crow, T.; Roberts, N. Cerebral torque is human specific and unrelated to brain size. *Brain Struct. Funct.* **2019**, *224*, 1141–1150. [CrossRef] [PubMed]
70. Kong, X.-Z.; Mathias, S.R.; Guadalupe, T.; Glahn, D.C.; Franke, B.; Crivello, F.; Tzourio-Mazoyer, N.; Fisher, S.E.; Thompson, P.M.; Francks, C.; et al. Mapping cortical brain asymmetry in 17,141 healthy individuals worldwide via the ENIGMA Consortium. *Proc. Natl. Acad. Sci. USA* **2018**, *115*, E5154–E5163. [CrossRef] [PubMed]
71. Guadalupe, T.; Zwiers, M.P.; Teumer, A.; Wittfeld, K.; Vasquez, A.A.; Hoogman, M.; Hagoort, P.; Fernandez, G.; Buitelaar, J.; Hegenscheid, K.; et al. Measurement and genetics of human subcortical and hippocampal asymmetries in large datasets. *Hum. Brain Mapp.* **2014**, *35*, 3277–3289. [CrossRef]
72. Maingault, S.; Tzourio-Mazoyer, N.; Mazoyer, B.; Crivello, F. Regional correlations between cortical thickness and surface area asymmetries: A surface-based morphometry study of 250 adults. *Neuropsychology* **2016**, *93*, 350–364. [CrossRef]
73. Ocklenburg, S.; Friedrich, P.; Güntürkün, O.; Genç, E. Intrahemispheric white matter asymmetries: The missing link between brain structure and functional lateralization? *Rev. Neurosci.* **2016**, *27*, 465–480. [CrossRef]
74. De Schotten, M.T.; Dell'Acqua, F.; Forkel, S.; Simmons, A.; Vergani, F.; Murphy, D.; Catani, M. A lateralized brain network for visuospatial attention. *Nat. Neurosci.* **2011**, *14*, 1245–1246. [CrossRef]
75. Chance, S.A. The cortical microstructural basis of lateralized cognition: A review. *Front. Psychol.* **2014**, *5*, 820. [CrossRef]
76. Schmitz, J.; Fraenz, C.; Schlüter, C.; Friedrich, P.; Jung, R.E.; Güntürkün, O.; Genç, E.; Ocklenburg, S. Hemispheric asymmetries in cortical gray matter microstructure identified by neurite orientation dispersion and density imaging. *NeuroImage* **2019**, *189*, 667–675. [CrossRef]
77. Cheng, L.; Zhang, Y.; Li, G.; Wang, J.; Sherwood, C.; Gong, G.; Fan, L.; Jiang, T. Connectional asymmetry of the inferior parietal lobule shapes hemispheric specialization in humans, chimpanzees, and rhesus macaques. *eLife* **2021**, *10*, e67600. [CrossRef] [PubMed]
78. Graïc, J.-M.; Peruffo, A.; Corain, L.; Centelleghe, C.; Granato, A.; Zanellato, E.; Cozzi, B. Asymmetry in the Cytoarchitecture of the Area 44 Homolog of the Brain of the Chimpanzee Pan troglodytes. *Front. Neuroanat.* **2020**, *14*, 55. [CrossRef] [PubMed]
79. Spocter, M.A.; Sherwood, C.C.; Schapiro, S.J.; Hopkins, W.D. Reproducibility of leftward planum temporale asymmetries in two genetically isolated populations of chimpanzees Pan troglodytes. *Proc. R. Soc. B Boil. Sci.* **2020**, *287*, 20201320. [CrossRef] [PubMed]
80. Galuske, R.A.W.; Schlote, W.; Bratzke, H.; Singer, W. Interhemispheric Asymmetries of the Modular Structure in Human Temporal Cortex. *Science* **2000**, *289*, 1946–1949. [CrossRef] [PubMed]
81. Hou, G.; Yang, X.; Yuan, T.-F. Hippocampal Asymmetry: Differences in Structures and Functions. *Neurochem. Res.* **2013**, *38*, 453–460. [CrossRef]
82. De la Fuente-Fernández, R.; Kishore, A.; Calne, D.B.; Ruth, T.J.; Stoessl, A. Nigrostriatal dopamine system and motor lateralization. *Behav. Brain Res.* **2000**, *112*, 63–68. [CrossRef]
83. Budilin, S.Y.; Midzyanovskaya, I.S.; Shchegolevskii, N.V.; Ioffe, M.E.; Bazyan, A.S. Asymmetry in dopamine levels in the nucleus accumbens and motor preference in rats. *Neurosci. Behav. Physiol.* **2008**, *38*, 991–994. [CrossRef]
84. Nielsen, D.M.; Visker, K.E.; Cunningham, M.J.; Keller, R.W.; Glick, S.D.; Carlson, J.N. Paw Preference, Rotation, and Dopamine Function in Collins HI and LO Mouse Strains. *Physiol. Behav.* **1997**, *61*, 525–535. [CrossRef]

85. Schwarting, R.; Nagel, J.A.; Huston, J.P. Asymmetries of brain dopamine metabolism related to conditioned paw usage in the rat. *Brain Res.* **1987**, *417*, 75–84. [CrossRef]
86. Bisiacchi, P.; Cainelli, E. Structural and functional brain asymmetries in the early phases of life: A scoping review. *Brain Struct. Funct.* **2021**, 1–18, Online ahead of print. [CrossRef]
87. Vasung, L.; Rollins, C.K.; Velasco-Annis, C.; Yun, H.J.; Zhang, J.; Warfield, S.K.; Feldman, H.A.; Gholipour, A.; Grant, P.E. Spatiotemporal Differences in the Regional Cortical Plate and Subplate Volume Growth during Fetal Development. *Cereb. Cortex* **2020**, *30*, 4438–4453. [CrossRef] [PubMed]
88. Xia, J.; Wang, F.; Wu, Z.; Wang, L.; Zhang, C.; Shen, D.; Li, G. Mapping hemispheric asymmetries of the macaque cerebral cortex during early brain development. *Hum. Brain Mapp.* **2019**, *41*, 95–106. [CrossRef] [PubMed]
89. Weinberger, D.R.; Luchins, D.J.; Morihisa, J.; Wyatt, R.J. Asymmetrical volumes of the right and left frontal and occipital regions of the human brain. *Ann. Neurol.* **1982**, *11*, 97–100. [CrossRef] [PubMed]
90. Chi, J.G.; Dooling, E.C.; Gilles, F.H. Left-Right Asymmetries of the Temporal Speech Areas of the Human Fetus. *Arch. Neurol.* **1977**, *34*, 346–348. [CrossRef]
91. Hering-Hanit, R.; Achiron, R.; Lipitz, S.; Achiron, A. Asymmetry of fetal cerebral hemispheres: In utero ultrasound study. *Arch. Dis. Child.-Fetal Neonatal Ed.* **2001**, *85*, 194–196. [CrossRef]
92. Kasprian, G.; Langs, G.; Brugger, P.C.; Bittner, M.; Weber, M.; Arantes, M.; Prayer, D. The Prenatal Origin of Hemispheric Asymmetry: An In Utero Neuroimaging Study. *Cereb. Cortex* **2011**, *21*, 1076–1083. [CrossRef]
93. Hepper, P.G. The developmental origins of laterality: Fetal handedness. *Dev. Psychobiol.* **2013**, *55*, 588–595. [CrossRef]
94. Hepper, P.G.; Wells, D.L.; Lynch, C. Prenatal thumb sucking is related to postnatal handedness. *Neuropsychologia* **2005**, *43*, 313–315. [CrossRef]
95. Ocklenburg, S.; Schmitz, J.; Moinfar, Z.; Moser, D.; Klose, R.; Lor, S.; Kunz, G.; Tegenthoff, M.; Faustmann, P.; Francks, C.; et al. Epigenetic regulation of lateralized fetal spinal gene expression underlies hemispheric asymmetries. *eLife* **2017**, *6*, e22784. [CrossRef]
96. Francks, C. Exploring human brain lateralization with molecular genetics and genomics. *Ann. N. Y. Acad. Sci.* **2015**, *1359*, 1–13. [CrossRef]
97. Eckert, M.A.; Leonard, C.M.; Molloy, E.A.; Blumenthal, J.; Zijdenbos, A.; Giedd, J.N. The Epigenesis of Planum Temporale Asymmetry in Twins. *Cereb. Cortex* **2002**, *12*, 749–755. [CrossRef]
98. Steinmetz, H.; Herzog, A.; Schlaug, G.; Huang, Y.; Jäncke, L. Brain (A)Symmetry in Monozygotic Twins. *Cereb. Cortex* **1995**, *5*, 296–300. [CrossRef]
99. Tzourio-Mazoyer, N.; Crivello, F.; Mazoyer, B. Is the planum temporale surface area a marker of hemispheric or regional language lateralization? *Brain Struct. Funct.* **2017**, *223*, 1–12. [CrossRef]
100. Cohen-Cory, S. The Developing Synapse: Construction and Modulation of Synaptic Structures and Circuits. *Science* **2002**, *298*, 770–776. [CrossRef]
101. Katz, L.C.; Shatz, C.J. Synaptic Activity and the Construction of Cortical Circuits. *Science* **1996**, *274*, 1133–1138. [CrossRef]
102. Pratt, K.G.; Hiramoto, M.; Cline, H.T. An Evolutionarily Conserved Mechanism for Activity-Dependent Visual Circuit Development. *Front. Neural Circuits* **2016**, *10*, 79. [CrossRef]
103. Sur, M.; Rubenstein, J.L.R. Patterning and Plasticity of the Cerebral Cortex. *Science* **2005**, *310*, 805–810. [CrossRef] [PubMed]
104. Gilmour, D.; Rembold, M.; Leptin, M. From morphogen to morphogenesis and back. *Nat. Cell Biol.* **2017**, *541*, 311–320. [CrossRef] [PubMed]
105. Grimes, D.T. Making and breaking symmetry in development, growth and disease. *Development* **2019**, *146*, dev170985. [CrossRef] [PubMed]
106. Grimes, D.T.; Burdine, R.D. Left–Right Patterning: Breaking Symmetry to Asymmetric Morphogenesis. *Trends Genet.* **2017**, *33*, 616–628. [CrossRef]
107. Levin, M. Left–right asymmetry in embryonic development: A comprehensive review. *Mech. Dev.* **2005**, *122*, 3–25. [CrossRef]
108. Nakamura, T.; Hamada, H. Left-right patterning: Conserved and divergent mechanisms. *Development* **2012**, *139*, 3257–3262. [CrossRef]
109. Zinski, J.; Tajer, B.; Mullins, M.C. TGF-β Family Signaling in Early Vertebrate Development. *Cold Spring Harb. Perspect. Biol.* **2018**, *10*, a033274. [CrossRef] [PubMed]
110. Brandler, W.M.; Paracchini, S. The genetic relationship between handedness and neurodevelopmental disorders. *Trends Mol. Med.* **2014**, *20*, 83–90. [CrossRef] [PubMed]
111. Schmitz, J.; Fraenz, C.; Schlüter, C.; Friedrich, P.; Kumsta, R.; Moser, D.; Güntürkün, O.; Genç, E.; Ocklenburg, S. Schizotypy and altered hemispheric asymmetries: The role of cilia genes. *Psychiatry Res. Neuroimaging* **2019**, *294*, 110991. [CrossRef] [PubMed]
112. Signore, I.A.; Palma, K.; Concha, M.L. Nodal signalling and asymmetry of the nervous system. *Philos. Trans. R. Soc. B: Biol. Sci.* **2016**, *371*, 20150401. [CrossRef]
113. Arning, L.; Ocklenburg, S.; Schulz, S.; Ness, V.; Gerding, W.M.; Hengstler, J.G.; Falkenstein, M.; Epplen, J.T.; Güntürkün, O.; Beste, C. PCSK6 VNTR Polymorphism Is Associated with Degree of Handedness but Not Direction of Handedness. *PLoS ONE* **2013**, *8*, e67251. [CrossRef]

114. Berretz, G.; Arning, L.; Gerding, W.M.; Friedrich, P.; Fraenz, C.; Schlüter, C.; Epplen, J.T.; Güntürkün, O.; Beste, C.; Genç, E.; et al. Structural Asymmetry in the Frontal and Temporal Lobes Is Associated with PCSK6 VNTR Polymorphism. *Mol. Neurobiol.* **2019**, *56*, 7765–7773. [CrossRef] [PubMed]
115. Postema, M.C.; Carrion-Castillo, A.; Fisher, S.E.; Vingerhoets, G.; Francks, C. The genetics of situs inversus without primary ciliary dyskinesia. *Sci. Rep.* **2020**, *10*, 1–11. [CrossRef]
116. Ihara, A.; Hirata, M.; Fujimaki, N.; Goto, T.; Umekawa, Y.; Fujita, N.; Terazono, Y.; Matani, A.; Wei, Q.; Yoshimine, T.; et al. Neuroimaging study on brain asymmetries in situs inversus totalis. *J. Neurol. Sci.* **2010**, *288*, 72–78. [CrossRef] [PubMed]
117. Kennedy, D.N.; O'Craven, K.M.; Ticho, B.S.; Goldstein, A.M.; Makris, N.; Henson, J.W. Structural and functional brain asymmetries in human situs inversus totalis. *Neurology* **1999**, *53*, 1260. [CrossRef] [PubMed]
118. Schuler, A.-L.; Kasprian, G.; Schwartz, E.; Seidl, R.; Diogo, M.C.; Mitter, C.; Langs, G.; Prayer, D.; Bartha-Doering, L. Mens inversus in corpore inverso? Language lateralization in a boy with situs inversus totalis. *Brain Lang.* **2017**, *174*, 9–15. [CrossRef]
119. Vingerhoets, G.; Li, X.; Hou, L.; Bogaert, S.; Verhelst, H.; Gerrits, R.; Siugzdaite, R.; Roberts, N. Brain structural and functional asymmetry in human situs inversus totalis. *Brain Struct. Funct.* **2018**, *223*, 1–16. [CrossRef]
120. Vingerhoets, G.; Gerrits, R.; Bogaert, S. Atypical brain functional segregation is more frequent in situs inversus totalis. *Cortex* **2018**, *106*, 12–25. [CrossRef]
121. Downes, J.C.; Birsoy, B.; Chipman, K.C.; Rothman, J.H. Handedness of a Motor Program in C. elegans Is Independent of Left-Right Body Asymmetry. *PLoS ONE* **2012**, *7*, e52138. [CrossRef]
122. Barth, K.A.; Miklosi, A.; Watkins, J.; Bianco, I.H.; Wilson, S.W.; Andrew, R.J. fsi Zebrafish Show Concordant Reversal of Laterality of Viscera, Neuroanatomy, and a Subset of Behavioral Responses. *Curr. Biol.* **2005**, *15*, 844–850. [CrossRef]
123. McManus, C. Reversed Bodies, Reversed Brains, and (Some) Reversed Behaviors: Of Zebrafish and Men. *Dev. Cell* **2005**, *8*, 796–797. [CrossRef] [PubMed]
124. Halpern, M. Leaning to the left: Laterality in the zebrafish forebrain. *Trends Neurosci.* **2003**, *26*, 308–313. [CrossRef]
125. Hobert, O. Homeobox genes and the specification of neuronal identity. *Nat. Rev. Neurosci.* **2021**, *22*, 627. [CrossRef] [PubMed]
126. De Kovel, C.G.F.; Lisgo, S.N.; Fisher, S.; Francks, C. Subtle left-right asymmetry of gene expression profiles in embryonic and foetal human brains. *Sci. Rep.* **2018**, *8*, 12606. [CrossRef]
127. Hepper, P.G.; Mccartney, G.R.; Shannon, E. Lateralised behaviour in first trimester human foetuses. *Neuropsychologia* **1998**, *36*, 531–534. [CrossRef]
128. McCartney, G.; Hepper, P. Development of lateralized behaviour in the human fetus from 12 to 27 weeks' gestation. *Dev. Med. Child Neurol.* **1999**, *41*, 83–86. [CrossRef]
129. Sun, T.; Collura, R.V.; Ruvolo, M.; Walsh, C.A. Genomic and Evolutionary Analyses of Asymmetrically Expressed Genes in Human Fetal Left and Right Cerebral Cortex. *Cereb. Cortex* **2006**, *16*, i18–i25. [CrossRef]
130. Sun, T.; Patoine, C.; Abu-Khalil, A.; Visvader, J.; Sum, E.; Cherry, T.J.; Orkin, S.H.; Geschwind, D.H.; Walsh, C.A. Early Asymmetry of Gene Transcription in Embryonic Human Left and Right Cerebral Cortex. *Science* **2005**, *308*, 1794–1798. [CrossRef]
131. Cederquist, G.Y.; Azim, E.; Shnider, S.J.; Padmanabhan, H.; Macklis, J.D. Lmo4 Establishes Rostral Motor Cortex Projection Neuron Subtype Diversity. *J. Neurosci.* **2013**, *33*, 6321–6332. [CrossRef]
132. Asprer, J.S.T.; Lee, B.; Wu, C.S.; Vadakkan, T.; Dickinson, M.E.; Lu, H.-C.; Lee, S.-K. LMO4 functions as a co-activator of neurogenin 2 in the developing cortex. *Development* **2011**, *138*, 2823–2832. [CrossRef]
133. Kashani, A.H.; Qiu, Z.; Jurata, L.; Lee, S.-K.; Pfaff, S.; Goebbels, S.; Nave, K.-A.; Ghosh, A. Calcium Activation of the LMO4 Transcription Complex and Its Role in the Patterning of Thalamocortical Connections. *J. Neurosci.* **2006**, *26*, 8398–8408. [CrossRef] [PubMed]
134. Nudel, R.; Newbury, D.F. Foxp2. *Wiley Interdiscip. Rev. Cogn. Sci.* **2013**, *4*, 547–560. [CrossRef] [PubMed]
135. Ocklenburg, S.; Arning, L.; Gerding, W.M.; Epplen, J.T.; Güntürkün, O.; Beste, C. FOXP2 variation modulates functional hemispheric asymmetries for speech perception. *Brain Lang.* **2013**, *126*, 279–284. [CrossRef] [PubMed]
136. Wong, R.O.L.; Ghosh, A. Activity-dependent regulation of dendritic growth and patterning. *Nat. Rev. Neurosci.* **2002**, *3*, 803–812. [CrossRef] [PubMed]
137. Karlebach, G.; Francks, C. Lateralization of gene expression in human language cortex. *Cortex* **2015**, *67*, 30–36. [CrossRef] [PubMed]
138. Moskal, J.R.; Kroes, R.A.; Otto, N.J.; Rahimi, O.; Claiborne, B.J. Distinct patterns of gene expression in the left and right hippocampal formation of developing rats. *Hippocampus* **2006**, *16*, 629–634. [CrossRef] [PubMed]
139. Ribic, A.; Biederer, T. Emerging Roles of Synapse Organizers in the Regulation of Critical Periods. *Neural Plast.* **2019**, *2019*, 1–9. [CrossRef]
140. Beste, C.; Arning, L.; Gerding, W.M.; Epplen, J.T.; Mertins, A.; Röder, M.C.; Bless, J.J.; Hugdahl, K.; Westerhausen, R.; Güntürkün, O.; et al. Cognitive Control Processes and Functional Cerebral Asymmetries: Association with Variation in the Handedness-Associated Gene LRRTM. *Mol. Neurobiol.* **2018**, *55*, 2268–2274. [CrossRef] [PubMed]
141. Francks, C.; Maegawa, S.; Laurén, J.; Abrahams, B.S.; Velayos-Baeza, A.; Medland, S.; Colella, S.; Groszer, M.; McAuley, E.Z.; Caffrey, T.M.; et al. LRRTM1 on chromosome 2p12 is a maternally suppressed gene that is associated paternally with handedness and schizophrenia. *Mol. Psychiatry* **2007**, *12*, 1129–1139. [CrossRef]
142. Berardi, N.; Maffei, L. From visual experience to visual function: Roles of neurotrophins. *J. Neurobiol.* **1999**, *41*, 119–126. [CrossRef]

143. Vicario-Abejón, C.; Owens, D.; McKay, R.; Segal, M. Role of neurotrophins in central synapse formation and stabilization. *Nat. Rev. Neurosci.* **2002**, *3*, 965–974. [CrossRef] [PubMed]
144. Sardar, R.; Zandieh, Z.; Namjoo, Z.; Soleimani, M.; Shirazi, R.; Hami, J. Laterality and sex differences in the expression of brain-derived neurotrophic factor in developing rat hippocampus. *Metab. Brain Dis.* **2021**, *36*, 133–144. [CrossRef]
145. Marrocco, J.; McEwen, B.S. Sex in the brain: Hormones and sex differences. *Dialog-Clin. Neurosci.* **2016**, *18*, 373–383. [CrossRef]
146. Deng, C.; Rogers, L.J. Factors affecting the development of lateralization in chicks. In *Comparative Vertebrate Lateralization*; Rogers, L.J., Andrew, R., Eds.; Cambridge University Press: Cambridge, UK, 2002; pp. 206–246.
147. Rogers, L.J. Development and function of lateralization in the avian brain. *Brain Res. Bull.* **2008**, *76*, 235–244. [CrossRef] [PubMed]
148. Manns, M.; Otto, T.; Salm, L. Pigeons show how meta-control enables decision-making in an ambiguous world. *Sci. Rep.* **2021**, *11*, 1–10. [CrossRef] [PubMed]
149. Rogers, L.J. A Matter of Degree: Strength of Brain Asymmetry and Behaviour. *Symmetry* **2017**, *9*, 57. [CrossRef]
150. Diekamp, B.; Regolin, L.; Güntürkün, O.; Vallortigara, G. A left-sided visuospatial bias in birds. *Curr. Biol.* **2005**, *15*, R372–R373. [CrossRef]
151. Manns, M. The riddle of nature and nurture-lateralization has an epigenetic trait. *Behav. Brain Sci.* **2005**, *28*, 602–603. [CrossRef]
152. Manns, M. The epigenetic control of asymmetry formation: Lessons from the avian visual system. In *Behavioral and Morphological Asymmetries in Vertebrates*; Malashichev, Y., Deckel, W., Eds.; Landes Bioscience: Georgetown, TX, USA, 2006; pp. 13–23.
153. Monsoro-Burq, A.H.; Levin, M. Avian models and the study of invariant asymmetry: How the chicken and the egg taught us to tell right from left. *Int. J. Dev. Biol.* **2018**, *62*, 63–77. [CrossRef]
154. Chen, Z.; Guo, Q.; Dai, E.; Forsch, N.; Taber, L.A. How the embryonic chick brain twists. *J. R. Soc. Interface* **2016**, *13*, 20160395. [CrossRef]
155. Fujinaga, M.; Hoffman, B.B.; Baden, J.M. Axial rotation in rat embryos: Morphological analysis and microsurgical study on the role of the allantois. *Teratology* **1995**, *51*, 94–106. [CrossRef]
156. Andrew, R.J.; Johnston, A.N.B.; Robins, A.; Rogers, L.J. Light experience and the development of behavioural lateralisation in chicks. II. Choice of familiar versus unfamiliar model social partner. *Behav. Brain Res.* **2004**, *155*, 67–76. [CrossRef]
157. Deng, C.; Rogers, L.J. Social recognition and approach in the chick: Lateralization and effect of visual experience. *Anim. Behav.* **2002**, *63*, 697–706. [CrossRef]
158. Mascetti, G.G.; Vallortigara, G. Why do birds sleep with one eye open? Light exposure of the chick embryo as a determinant of monocular sleep. *Curr. Biol.* **2001**, *11*, 971–974. [CrossRef]
159. Costalunga, G.; Kobylkov, D.; Rosa-Salva, O.; Vallortigara, G.; Mayer, U. Light-incubation effects on lateralisation of single unit responses in the visual Wulst of domestic chicks. *Brain Struct. Funct.* **2021**, 1–17. [CrossRef]
160. Johnston, A.; Rogers, L.; Dodd, P. [3H]MK-801 binding asymmetry in the IMHV region of dark-reared chicks is reversed by imprinting. *Brain Res. Bull.* **1995**, *37*, 5–8. [CrossRef]
161. Lorenzi, E.; Mayer, U.; Rosa-Salva, O.; Morandi-Raikova, A.; Vallortigara, G. Spontaneous and light-induced lateralization of immediate early genes expression in domestic chicks. *Behav. Brain Res.* **2019**, *368*, 111905. [CrossRef] [PubMed]
162. Morandi-Raikova, A.; Danieli, K.; Lorenzi, E.; Rosa-Salva, O.; Mayer, U. Anatomical asymmetries in the tectofugal pathway of dark-incubated domestic chicks: Rightwards lateralization of parvalbumin neurons in the entopallium. *Laterality* **2021**, *26*, 163–185. [CrossRef] [PubMed]
163. Letzner, S.; Patzke, N.; Verhaal, J.; Manns, M. Shaping a lateralized brain: Asymmetrical light experience modulates access to visual interhemispheric information in pigeons. *Sci. Rep.* **2014**, *4*, 4253. [CrossRef] [PubMed]
164. Letzner, S.; Güntürkün, O.; Lor, S.; Pawlik, R.J.; Manns, M. Visuospatial attention in the lateralised brain of pigeons—A matter of ontogenetic light experiences. *Sci. Rep.* **2017**, *7*, 15547. [CrossRef]
165. Chiandetti, C. Pseudoneglect and embryonic light stimulation in the avian brain. *Behav. Neurosci.* **2011**, *125*, 775–782. [CrossRef]
166. Chiandetti, C.; Galliussi, J.; Andrew, R.J.; Vallortigara, G. Early-light embryonic stimulation suggests a second route, via gene activation, to cerebral lateralization in vertebrates. *Sci. Rep.* **2013**, *3*, 2701. [CrossRef]
167. Bobbo, D.; Galvani, F.; Mascetti, G.G.; Vallortigara, G. Light exposure of the chick embryo influences monocular sleep. *Behav. Brain Res.* **2002**, *134*, 447–466. [CrossRef]
168. Zhu, L.; Marvin, M.J.; Gardiner, A.; Lassar, A.B.; Mercola, M.; Stern, C.D.; Levin, M. Cerberus regulates left–right asymmetry of the embryonic head and heart. *Curr. Biol.* **1999**, *9*, 931–938. [CrossRef]
169. Ververs, I.A.; De Vries, J.I.; Van Geijn, H.P.; Hopkins, B. Prenatal head position from 12–38 weeks. I. Developmental aspects. *Early Hum. Dev.* **1994**, *39*, 83–91. [CrossRef]
170. Mampe, B.; Friederici, A.D.; Christophe, A.; Wermke, K. Newborns' Cry Melody Is Shaped by Their Native Language. *Curr. Biol.* **2009**, *19*, 1994–1997. [CrossRef] [PubMed]
171. Partanen, E.; Kujala, T.; Näätänen, R.; Liitola, A.; Sambeth, A.; Huotilainen, M. Learning-induced neural plasticity of speech processing before birth. *Proc. Natl. Acad. Sci. USA* **2013**, *110*, 15145–15150. [CrossRef] [PubMed]
172. Fagard, J. The nature and nurture of human infant hand preference. *Ann. N. Y. Acad. Sci.* **2013**, *1288*, 114–123. [CrossRef]
173. Ocklenburg, S.; Bürger, C.; Westermann, C.; Schneider, D.; Biedermann, H.; Güntürkün, O. Visual experience affects handedness. *Behav. Brain Res.* **2010**, *207*, 447–451. [CrossRef] [PubMed]
174. Previc, F.H. A general theory concerning the prenatal origins of cerebral lateralization in humans. *Psychol. Rev.* **1991**, *98*, 299–334. [CrossRef] [PubMed]

175. Andrew, R.; Osorio, D.; Budaev, S. Light during embryonic development modulates patterns of lateralization strongly and similarly in both zebrafish and chick. *Philos. Trans. R. Soc. B Biol. Sci.* **2008**, *364*, 983–989. [CrossRef] [PubMed]
176. Budaev, S.; Andrew, R.J. Patterns of early embryonic light exposure determine behavioural asymmetries in zebrafish: A habenular hypothesis. *Behav. Brain Res.* **2009**, *200*, 91–94. [CrossRef] [PubMed]
177. Chiandetti, C.; Vallortigara, G. Distinct effect of early and late embryonic light-stimulation on chicks' lateralization. *Neuroscience* **2019**, *414*, 1–7. [CrossRef] [PubMed]
178. Caputto, B.L.; Guido, M.E. Immediate early gene expression within the visual system: Light and circadian regulation in the retina and the suprachiasmatic nucleus. *Neurochem. Res.* **2000**, *25*, 153–162. [CrossRef] [PubMed]
179. Ströckens, F.; Freund, N.; Manns, M.; Ocklenburg, S.; Güntürkün, O. Visual asymmetries and the ascending thalamofugal pathway in pigeons. *Brain Struct. Funct.* **2012**, *218*, 1197–1209. [CrossRef]
180. Letzner, S.; Manns, M.; Güntürkün, O. Light-dependent development of the tectorotundal projection in pigeons. *Eur. J. Neurosci.* **2020**, *52*, 3561–3571. [CrossRef] [PubMed]
181. Rogers, L.J. Light experience and asymmetry of brain function in chickens. *Nat. Cell Biol.* **1982**, *297*, 223–225. [CrossRef]
182. Skiba, M.; Diekamp, B.; Güntürkün, O. Embryonic light stimulation induces different asymmetries in visuoperceptual and visuomotor pathways of pigeons. *Behav. Brain Res.* **2002**, *134*, 149–156. [CrossRef]
183. Rogers, L.J. Light input and the reversal of functional lateralization in the chicken brain. *Behav. Brain Res.* **1990**, *38*, 211–221. [CrossRef]
184. Manns, M.; Güntürkün, O. Monocular deprivation alters the direction of functional and morphological asymmetries in the pigeon's (*Columba livia*) visual system. *Behav. Neurosci.* **1999**, *113*, 1257–1266. [CrossRef]
185. Manns, M.; Güntürkün, O. 'Natural' and artificial monocular deprivation effects on thalamic soma sizes in pigeons. *NeuroReport* **1999**, *10*, 3223–3228. [CrossRef]
186. Rogers, L.; Deng, C. Corticosterone treatment of the chick embryo affects light-stimulated development of the thalamofugal visual pathway. *Behav. Brain Res.* **2005**, *159*, 63–71. [CrossRef]
187. Hausmann, M. Why sex hormones matter for neuroscience: A very short review on sex, sex hormones, and functional brain asymmetries. *J. Neurosci. Res.* **2017**, *95*, 40–49. [CrossRef]
188. Hirnstein, M.; Hugdahl, K.; Hausmann, M. Cognitive sex differences and hemispheric asymmetry: A critical review of 40 years of research. *Laterality* **2019**, *24*, 204–252. [CrossRef]
189. Manns, M.; Güntürkün, O. Light experience induces differential asymmetry pattern of GABA- and parvalbumin-positive cells in the pigeon's visual midbrain. *J. Chem. Neuroanat.* **2003**, *25*, 249–259. [CrossRef]
190. Prior, H.; Diekamp, B.; Güntürkün, O.; Manns, M. Post-hatch activity-dependent modulation of visual asymmetry formation in pigeons. *NeuroReport* **2004**, *15*, 1311–1314. [CrossRef] [PubMed]
191. Manns, M.; Freund, N.; Leske, O.; Güntürkün, O. Breaking the balance: Ocular BDNF-injections induce visual asymmetry in pigeons. *Dev. Neurobiol.* **2008**, *68*, 1123–1134. [CrossRef] [PubMed]
192. Manns, M.; Güntürkün, O.; Heumann, R.; Blöchl, A. Photic inhibition of TrkB/Ras activity in the pigeon's tectum during development: Impact on brain asymmetry formation. *Eur. J. Neurosci.* **2005**, *22*, 2180–2186. [CrossRef] [PubMed]
193. Oviedo, H.V.; Bureau, I.; Svoboda, K.; Zador, A.M. The functional asymmetry of auditory cortex is reflected in the organization of local cortical circuits. *Nat. Neurosci.* **2010**, *13*, 1413–1420. [CrossRef]
194. Nava, E.; Güntürkün, O.; Röder, B. Experience-dependent emergence of functional asymmetries. *Laterality* **2013**, *18*, 407–415. [CrossRef]
195. Le Grand, R.; Mondloch, C.J.; Maurer, D.; Brent, H.P. Early visual experience and face processing. *Nat. Cell Biol.* **2001**, *410*, 890. [CrossRef]
196. Röder, B.; Ley, P.; Shenoy, B.H.; Kekunnaya, R.; Bottari, D. Sensitive periods for the functional specialization of the neural system for human face processing. *Proc. Natl. Acad. Sci. USA* **2013**, *110*, 16760–16765. [CrossRef]
197. Dehaene, S.; Pegado, F.; Braga, L.W.; Ventura, P.; Filho, G.N.; Jobert, A.; Dehaene-Lambertz, G.; Kolinsky, R.; Morais, J.; Cohen, L. How Learning to Read Changes the Cortical Networks for Vision and Language. *Science* **2010**, *330*, 1359–1364. [CrossRef]
198. Dundas, E.M.; Plaut, D.C.; Behrmann, M. An ERP investigation of the co-development of hemispheric lateralization of face and word recognition. *Neuropsychology* **2014**, *61*, 315–323. [CrossRef]
199. Freund, N.; Valencia-Alfonso, C.E.; Kirsch, J.; Brodmann, K.; Manns, M.; Güntürkün, O. Asymmetric top-down modulation of ascending visual pathways in pigeons. *Neuropsychology* **2016**, *83*, 37–47. [CrossRef]
200. Manns, M.; Römling, J. The impact of asymmetrical light input on cerebral hemispheric specialization and interhemispheric cooperation. *Nat. Commun.* **2012**, *3*, 696. [CrossRef] [PubMed]
201. Dharmaretnam, M.; Rogers, L. Hemispheric specialization and dual processing in strongly versus weakly lateralized chicks. *Behav. Brain Res.* **2005**, *162*, 62–70. [CrossRef] [PubMed]
202. Chiandetti, C.; Regolin, L.; Rogers, L.J.; Vallortigara, G. Effects of light stimulation of embryos on the use of position-specific and object-specific cues in binocular and monocular domestic chicks (*Gallus gallus*). *Behav. Brain Res.* **2005**, *163*, 10–17. [CrossRef] [PubMed]
203. Gazzaniga, M.S. Forty-five years of split-brain research and still going strong. *Nat. Rev. Neurosci.* **2005**, *6*, 653–659. [CrossRef]
204. Witelson, S.; Nowakowski, R. Left out axoms make men right: A hypothesis for the origin of handedness and functional asymmetry. *Neuropsychology* **1991**, *29*, 327–333. [CrossRef]

205. Besson, M.; Gache, C.; Bertucci, F.; Brooker, R.; Roux, N.; Jacob, H.; Berthe, C.; Sovrano, V.A.; Dixson, D.L.; Lecchini, D. Exposure to agricultural pesticide impairs visual lateralization in a larval coral reef fish. *Sci. Rep.* **2017**, *7*, 1–9. [CrossRef]
206. Broder, E.D.; Angeloni, L.M. Predator-induced phenotypic plasticity of laterality. *Anim. Behav.* **2014**, *98*, 125–130. [CrossRef]
207. Jozet-Alves, C.; Hébert, M. Embryonic exposure to predator odour modulates visual lateralization in cuttlefish. *Proc. R. Soc. B Boil. Sci.* **2013**, *280*, 20122575. [CrossRef]
208. Manns, M. Laterality for the next decade: Costs and benefits of neuronal asymmetries–putting lateralization in an evolutionary context. *Laterality* **2021**, *26*, 315–318. [CrossRef] [PubMed]

 symmetry

MDPI

Review

Structural Brain Asymmetries for Language: A Comparative Approach across Primates

Yannick Becker [1] and **Adrien Meguerditchian** [1,2,*]

[1] Laboratoire de Psychologie Cognitive, CNRS/Université Aix-Marseille UMR7290, 13003 Marseille, France; yannick.becker@univ-amu.fr
[2] Station de Primatologie CNRS-CELPHEDIA UAR846, 13790 Rousset, France
* Correspondence: adrien.meguerditchian@univ-amu.fr

Abstract: Humans are the only species that can speak. Nonhuman primates, however, share some 'domain-general' cognitive properties that are essential to language processes. Whether these shared cognitive properties between humans and nonhuman primates are the results of a continuous evolution [homologies] or of a convergent evolution [analogies] remain difficult to demonstrate. However, comparing their respective underlying structure—the brain—to determinate their similarity or their divergence across species is critical to help increase the probability of either of the two hypotheses, respectively. Key areas associated with language processes are the Planum Temporale, Broca's Area, the Arcuate Fasciculus, Cingulate Sulcus, The Insula, Superior Temporal Sulcus, the Inferior Parietal lobe, and the Central Sulcus. These structures share a fundamental feature: They are functionally and structurally specialised to one hemisphere. Interestingly, several nonhuman primate species, such as chimpanzees and baboons, show human-like structural brain asymmetries for areas homologous to key language regions. The question then arises: for what function did these asymmetries arise in non-linguistic primates, if not for language per se? In an attempt to provide some answers, we review the literature on the lateralisation of the gestural communication system, which may represent the missing behavioural link to brain asymmetries for language area's homologues in our common ancestor.

Keywords: hemispheric specialization; language evolution; lateralization; MRI; baboon; development; language areas

Citation: Becker, Y.; Meguerditchian, A. Structural Brain Asymmetries for Language: A Comparative Approach across Primates. *Symmetry* **2022**, *14*, 876. https://doi.org/10.3390/sym14050876

Academic Editors: Sebastian Ocklenburg and Onur Güntürkün

Received: 26 November 2021
Accepted: 10 April 2022
Published: 25 April 2022

1. Introduction

"Human being: n. a man, woman, or child of the species Homo sapiens, distinguished from other animals by superior mental development and language ... " [The Oxford Pocket Dictionary of Current English, Oxford University Press 2009].

This definition states the obvious: humans are the only species able to speak, and thus, this fascinating cognitive faculty is considered as a key feature which divides us from other animals [1]. However, in recent years, the conception of one holistic language faculty has been contrasted with an alternative view that language may be the result of an assembly of cognitive properties that are domain-general and not specific to language [2,3]. Therefore, since nonhuman animals, especially primates, have been shown to share some of these domain-general cognitive properties, the research on nonhuman primates was reconsidered as a critical model to investigate language evolution [3–5]. Whether these shared cognitive properties between humans and nonhuman primates are the result of a continuous evolution (homologies) or of a convergent evolution (analogies) remains difficult to demonstrate. However, comparing their respective underlying structure—the brain—to determinate their similarity or divergence across species is critical to help increase the probability of either of the two hypotheses, respectively. In humans, language models congruently describe a few key hubs of language processing, namely Broca's and

Wernicke's area, and their interconnection, the Arcuate Fasciculus [6–9]. However, in recent years, data have expanded to more distributed models, taking into account several different fibre tracts and regions crucial for language processing, as demonstrated by clinical cases [10] and questioned a localisationist viewpoint of language specialisation [11]. For example, the importance of regions in the interface between social cognition and communication like the Superior Temporal Sulcus and Geschwind's territory in the Inferior Parietal Lobe are highlighted [12], as well as the anterior Insula cortex [13] and the anterior and mid-Cingulate Sulci [14]. Most of these structures share one fundamental feature: they are functionally and structurally specialised towards one hemisphere, mostly the left [7,13,14]. Specifically, 'functional lateralisation' refers to a more pronounced activation in one hemisphere than the other. 'Structural lateralisation' refers to a more pronounced volume, size, surface measurements, or neuron density of a given region in one hemisphere than the other. Although the link between structural and functional asymmetry remains in question [15,16], most authors usually consider that structural asymmetry (the size of the roads) of these regions might reflect the functional asymmetry (the traffic) for language tasks [13,17,18].

Thus, the fundamental question of 'how language lateralisation has evolved' is more relevant than ever in the light of the aforementioned findings about language organisation in the brain. This is where comparative studies on our primate cousins are of importance. In this view of language evolution, the different cognitive components of language could have evolved gradually, in opposition to a saltatory emergence [19]. In this evolutionary process, it is not excluded that 'exaptation' phenomena, where an opportunistic selective adaptation is piggybacked onto preexisting structures, could have played a crucial role [3]. Because brain tissue hardly fossilises, one fruitful way to look for intermediate steps is the comparison of brains between primate species, including humans, in order to infer potential features inherited from their shared common ancestors [20]. In other words, one can determine which brain architectures and behaviours are shared between us and other primates, suggesting its pre-existence before the emergence of our species. Comparative brain approaches for primate species have particularly focussed on neuroanatomy, given the well-known limitation for functional studies in apes and monkeys (e.g., techniques, ethics, sample size, reproductivity). The advent of non-invasive techniques such as Magnetic Resonance Imaging (MRI) favours in-vivo acquisitions on anesthetised subjects, allowing no limitation in terms of sample sizes and species diversity [21]. As a result, primate brain research has benefited in recent years from considerable increase in available MRI databases on large cohorts in many primate species, including macaques: Prime-De [22] and chimpanzees: www.chimpanzeebrain.org (accessed on 12 April 2022) [23]. Just as in human brain research, research on primate brain anatomy allows morphological quantification of regions in each hemisphere and determination of possible interhemispheric asymmetries. In this quest, an increased body of evidence highlights that different nonhuman primate species, such as chimpanzees or baboons, show human-like structural brain asymmetries for areas homologous to key language regions [20,24].

Thus, the question arises: for which function have these asymmetries in non-linguistic primates developed, if not for language per se? In humans, handedness for manual actions was for a long time considered a behavioural reflection of language-related brain lateralisation [25]. However, more recent studies indicate that the direction of handedness for manual actions poorly predicts language lateralisation, especially in left-handed humans. Indeed, 96% of right-handers and 70% of left-handers have their left hemisphere functionally specialised for most language functions [25]. In fact, handedness for manual actions was shown to be associated with the contralateral lateralisation of the motor hand area within the Central Sulcus, rather than key language areas [26]. Altogether, it is now acknowledged that direction of handedness might be independent from direction of language lateralisation [27–29]. As a result, comparative research on handedness for manual actions across primates might not be suitable for investigating phylogenetical origins of hemispheric language specialisation. To do so, it has been suggested that studying the manual lateralisation of gestural communication in nonhuman primates—and not handedness for manipulative actions—might constitute a more fruitful approach [30]. In fact, following the evolutionary framework on the gestural origins of language [31], the gestural communication system in nonhuman primates was found to share key features of domain-general processes, important for language, such as intentionality, referentiality, and learning flexibility [32–34]. Interestingly, production of communicative manual gestures has been found highly lateralised in favour of the right hand in both baboons and chimpanzees. In contrast, the handedness patterns for non-communicative manual actions in chimpanzees and baboons were found to be different from those found for communicative gestures at both the populational and individual levels [30]. These findings supported the idea that gestural communication in nonhuman primates may be related to a specific lateralised system for communication, which might be different from handedness for manipulative actions [33].

In addition, the sequential and hierarchical motor actions that underlie tool-use behaviours have been used as the basis for the emergence of communicative systems [35–38], and share neural substrates with the human language network [39]. A recent functional MRI study with human subjects showed that the use of learning tools improved a language syntax task and there is neural overlap of both behaviours in the basal ganglia [40].

Therefore, there is the necessity to take stock of what the work on neuroanatomical correlates of gestural communication and tool-use in nonhuman primates has provided, in order to test its supposed continuity with language lateralisation.

In this paper, we aimed to review the literature comparing structural brain asymmetries across primates for areas related to language in humans. We focus on the classical perisylvian language regions, namely the Planum Temporale, Broca's Area, and the white matter tract that interconnects these two regions—The Arcuate Fasciculus. In this review, it is of importance to take also into account other key regions of the large, distributed language network beyond the perisylvian regions, such the Insula, the Cingulate Sulcus, and the Superior Temporal Sulcus (STS) as novel grey matter areas of interest. Finally, we consider the Central Sulcus (CS), which delimitates the primary motor from the primary somato-sensory cortex and include thus the mouth and lips motor areas as well as the motor hand area related to handedness (see Figure 1). After briefly describing their functions, we first review their structural and functional lateralisation in humans, including infants, to discuss whether structural markers can predict the functional lateralisation of language. In the next step, we compare these findings with nonhuman primates and discuss whether this asymmetric organisation is shared between species. Finally, we address the following question: If such brain lateralisation is shared with our non-linguistic primate cousins, for what behavioural functions did it evolve, if not for language? In an attempt to provide some elements of responses, we propose to review the literature about the lateralisation of the gestural communicative system, which could potentially constitute the ideal missing behavioural link with brain asymmetries for language in our common ancestor.

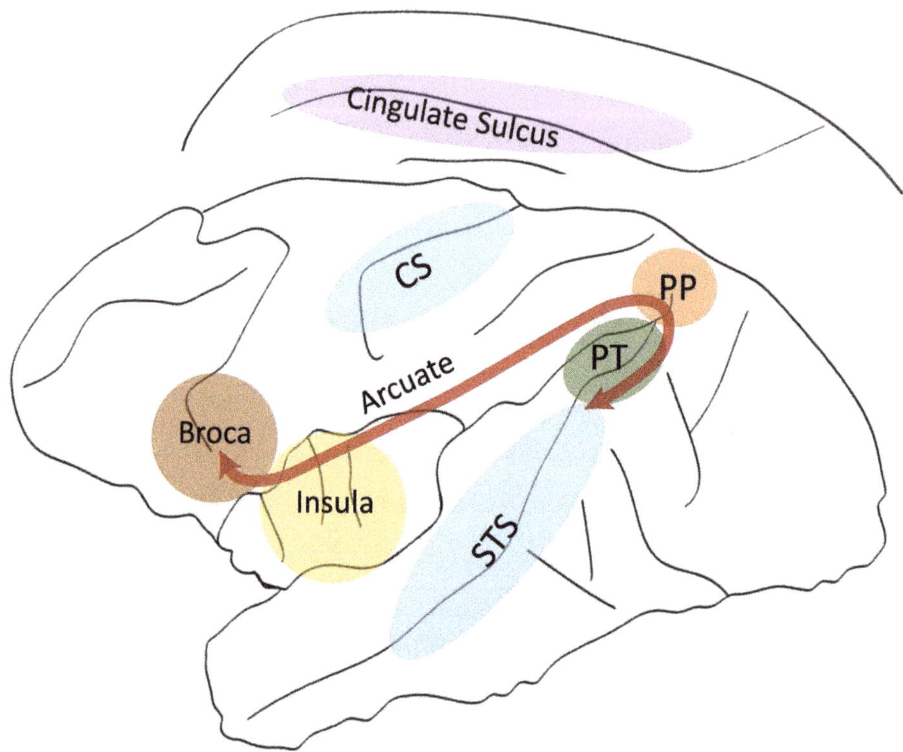

Figure 1. Illustration of the reviewed key structures in a monkey brain. In brown: Broca's area, in yellow: the Insula, in red: the Arcuate Fasciculus, in blue: the Central Sulcus, in green: the Planum Temporale, in light red: the Planum Parietale, in grey: the Superior Temporal Sulcus, in purple: the Cingulate Sulcus. Arcuate: Arcuate Fasciculus, CS: Central Sulcus, PT: Planum Temporale, PP: Planum Parietale, STS: Superior Temporal Sulcus.

2. Planum Temporale

The most emblematic marker of the lateralised language organisation is the Planum Temporale (PT), which is located within the Sylvian fissure and is part of the auditory association cortex [41] and Wernicke's area [42]. In adults, left hemispheric lesions in this region result in severe language comprehension and production deficits [43–45]. Therefore, many studies have shown the particular functional significance of the PT in the left hemisphere for a variety of auditory language processing [46], including the main perception component of the audio–motor loop for phonological processing [47]. Interestingly, in preverbal newborns, the functional implication of the left PT was highlighted from birth [17,48,49].

In a pioneering work by Geschwind and Levitsky [50], the PT was shown to be anatomically asymmetric: in 100 post-mortem brains, 65% of the left PT was larger than the right, which was confirmed by in-vivo MRI studies [46] and also highlighted in early development (Post-mortem Infants: [51,52]; in-vivo MRI infants: [53–56]; foetuses: [57,58]). These asymmetries later increase during development and are associated with language development [17,48,49,51]. PT asymmetry, is, therefore seen as a marker for human unique innate readiness to acquire language [17].

Indeed, in adults, a direct relationship was shown between the left PT's size and functional asymmetry of language tasks [59]. In addition, an absence or reversed PT asymmetry has been linked to several language-related pathologies like dyslexia [60,61]. This function-structure relationship is however debated [13,15,16,62–68]. For example,

Greve et al. [65] showed that regardless of the functional hemispheric dominance for language, the structural PT volume asymmetry is left biased. In contrast, Ocklenburg et al. [68] found that a higher density at a microstructural level of the left PT was associated with faster processing of auditory speed in the same area, as shown in EEG. More recently, Tzourio-Mazoyer et al. [16] demonstrated that, although the structural PT asymmetry is not predictive of its functional counterpart in a language task, an adjacent auditory area at the end of the Sylvian fissure is.

Several studies in nonhuman primates also showed striking human-like PT asymmetries in their homologous regions. Manual delineation of post-mortem brains [69], in-vivo MRI scans [70,71] as well as voxel-based morphology on MRI scans [23], showed larger left PTs in chimpanzees and in apes in general [72]. Recently, in-vivo MRI studies on adult and newborn baboons with manual PT delineation extended this finding to a shared feature between Old World monkeys and humans [24,73,74]. Interestingly, the asymmetry strength increased with age in this longitudinal study [74]. These results are questioning the PT asymmetry: (1) to be unique in humans and (2) to be marker for language development in newborns. Rather, the PT asymmetry might have evolved for a cognitive function shared between Old World monkeys, apes, and humans, which is at the core of language processing in humans. A potential candidate related to such a function may be communicative gesture. Indeed, Meguerditchian et al. [75] highlighted a relation between Planum Temporale grey matter volume asymmetry and hand preference for communicative gesture in chimpanzee. Moreover, the left PT asymmetry was also found related to handedness for tool-use but not for handedness for manipulative actions in chimpanzees, which required no structured sequence of motor actions [76].

3. Broca's Area

Broca area and its left hemispheric specialisation was historically considered as the center of speech production [77]. This modular view of the neuronal basis of language was progressively questioned by the view that language involves a plastic and large distributed network [11,78] and even implicates the two hemispheres. However, it is still well acknowledged that Broca's area in the left hemisphere remains a key component for language specialisation within its distributed neural network [79]. Interestingly, complementary work thereby highlighted Broca's area as a lateralised interface between speech and multimodal motor integration, including gesture and mouth movements [80]. Broca's area is also known for its involvement in motor planning, sequential and hierarchical organisation of behaviours, including syntax [81], tool-use [39,82], and sign language production, thus including manual and oro–facial gestures [83,84]. In infants, speech perception activates Broca's area from very early development on as highlighted in MEG or functional MRI studies [48,85,86]. This activation before the babbling stage suggested that activity of this area is not due to motor learning but might drive the learning of complex sequences [86].

In contrast to the PT [see section above], a clear structural leftward asymmetry has not been reproducibly demonstrated [87,88], which may be due to natural variability between subjects in sulcal contours defining this area [87,89]. Cytoarchitectonic analyses, however, reported a leftward asymmetry of some parts of area 44 and/or area 45, which together form Broca's area [90–92]. In contrast, other accounts state a rightward asymmetry for area 45 grey matter, which gets reduced during aging, especially due to a loss in the right hemisphere [93]. In development, an early structural primacy of right-sided dendrite systems shortly after birth and a progressive shift to left-sided primacy during years three to six was highlighted and related to critical periods for language acquisition [94]. A second study found leftward asymmetries on the cellular level very early from 1-year old infants on [92], which increased into an adult-like leftward asymmetry at 5 years for area 45 and 11 years for area 44. This maturational effect was suggested to be influenced by language practice and thus, the interhemispheric asymmetry of this area would continue to change throughout life [92]. Interestingly, because area 45 supports semantic processes and area 44 subserves syntactic processes in adults, some see in this maturational difference a neural underpinning of the earlier onset of semantics than syntax in children [79].

In great apes, a homologous region was documented in several studies, which described the precentral inferior sulcus, the inferior frontal sulcus, and the fronto-orbital sulcus as common borders of Broca's homologue [95–97]. Like in humans, inquiries about Broca's area's structural asymmetry on a population-level remain inconsistent in apes due to interindividual variation in location and cytoarchitecture [98]. Leftward lateralisation was found at a macrostructural level [95] but not at a cytoarchitectonic level in a relative smaller sample size ([98]; but see also [99]). In monkeys, no data of structural asymmetry for this region has been reported so far. In fact, determining Broca's homologue is challenging because the common borders of Broca's homologue in apes are absent in monkeys. Nevertheless, we know from few detailed cytoarchitectonic studies in macaques that the two parts of Broca's area 44 and 45 are respectively located in the fundus and lower caudal/posterior bank, and on the rostral/anterior side, of the most ventral part of the inferior arcuate sulcus IAS [100–102]. Electric stimulation of this region elicited oro-facial and finger movements. Therefore, together with cytoarchitectonic similarities, the region anterior to the ventral part of the IAS was proposed as an equivalent area 44 in macaques [101]. Recent studies even reported that Broca's homologue's activation preceded voluntary trained production of a vocalisation after intensive operant conditioning in juvenile rhesus monkeys [103]. Moreover, the use of positron emission tomography (PET) in three captive chimpanzees has revealed that begging food from a human by using either gestures, novel species atypical attention-getting sounds, or both simultaneously, activated a homologous region of Broca's area (IFG) predominantly in the left hemisphere [104]. Some functions associated with Broca's homologue's lateralisation in nonhuman primates have been proposed in relation to tool-use processing and communication gesture production. Regarding tool-use, chimpanzees that performed better a tool-use task with their right hand showed a greater left-lateralisation of Broca's homologue [105]. Such a link might be attributed to the typical Broca's function described in humans, namely hierarchical organisation of behaviours involved in tool-use [81,82]. Regarding gestural communication, Meguerditchian et al. [75] highlighted a relation between Broca's homologue grey matter volume asymmetry and hand preference for communicative gesture in chimpanzees. Similarly, in a recent study in baboons, we showed that variation in hand preference for communicative gesture—but not for non-communicative manipulative actions—is related to the anatomical variation of Broca's homologue. Indeed, the right Broca's portion of the IAS is deeper than the left for baboons communicating with their left hand and vice versa [106].

4. Arcuate Fasciculus

The Arcuate Fasciculus (AF) is a bundle of white matter, which arches dorsally around the Sylvian fissure, interconnecting Broca's area in the frontal lobe with the Planum Temporale in the temporal lobe [107]. It was highlighted that the connectivity between language areas, due to the AF, is crucial. For example, the integrity of the AF might be more important for lesion recovery (e.g., strokes) than the integrity of grey matter regions that it is connecting [11]. In addition, as neurological cases have shown, the AF plays a key role in language processing in the left hemisphere, with lesion of the direct pathway also causing conduction aphasia [10] in deaf signers [108]. The AF is already present at birth [109] but matures slowly until late childhood [109,110]. In contemporary language models, the AF (or also called Dorsal Pathway) is often opposed to the Ventral Pathway, which interconnects roughly the same regions, but travels ventrally around the Sylvian fissure [78]. In contrast to the AF, the Ventral Pathway matures more rapidly in development and was also described to be phylogenetically more ancient [79,110]. It is assumed that the late maturation of the AF is due to the frontal portion, which is connected to Broca's area and not fully myelinated until the age of seven [110,111]. In fact, controversy persists whether this portion is also already present at birth [109,112–115].

Regarding lateralisation, the human AF was found larger in the left hemisphere for a number of macroscopic and microscopic measurements like the number of streamlines, volume of the tract, fibre density, and mean fractional anisotropy in 60% of normal adult humans. The remaining 40% of the adult population shows either a reduced lateralisation to the left (20%) or no lateralisation at all (20%) [18,116–120]. In early development, the leftward AF is the most asymmetrical region of the developing white matter ([113,121], but see also [122]). Interestingly, the early leftward asymmetry in newborns was correlated with later language capacities in children [123–125]. Catani et al. [18] argued that the AF's asymmetry represents a better structural marker for functional language specialisation than PT asymmetry [18]. This structure–function relationship is, however, debated [15,126–129]. For example, Verhelst et al. [129] demonstrated in a fixel based analysis that structural AF asymmetry did not differ between subjects with either right or left functional language hemispheric dominance.

Axon tracing in monkey brains and diffusion MRI in chimpanzee and monkey brains have highlighted the existence of the Arcuate Fasciculus across primates that interconnects frontal and temporal areas [130–139]. It is debated to what extent the AF's strength, lateralization, and frontal and temporal termination sites differ between primate species. In fact, recent findings speak for a rather conserved organisation across primates (for a review, see [9]). Therefore, latest functional results suggest that language abilities allowing humans to name, conceptualise, and thus better remember sound would be shared across primates [139]. In order to highlight anatomical differences across primates, which could explain the human uniqueness for language, several authors conclude that the left AF lateralisation is the crux of the human-specific distinction [136,138,139]. In fact, Rilling et al. [133] did not find any AF asymmetry in three macaque and four chimpanzee subjects. However, by adding more chimpanzee subjects, the authors were able to report a left lateralised AF, which was still weaker than in humans [134]. This result remained unique regarding AF lateralisation in nonhuman primates [136,138,139]. In fact, this inconsistency across the literature about the presence or not of population-level leftward AF bias might be explained by the small sample size (i.e., only few subjects) usually included in those AF studies in apes, which makes it difficult to infer any bias at the population-level. Only studies including an increased sample size would help elucidate this debate.

5. Insula

The Insula cortex lies in the depth of the Sylvian fissure, which separates the temporal lobe from the parietal and frontal lobes. The anterior part of the Insula is implicated for different language processing functions (General: [43,140,141]; auditory processing: [142]; motor aspects: [143]; syntax: [144,145]; sign language: [108]). Interestingly, the Insula seems to be particularly involved in motor planning of speech as seen in pathologies [140,143]. Further, Lœvenbrueck et al. [146,147] highlighted in adults that prosodic pointing gesture activates Broca's area as well the left anterior Insula [146,147].

Moreover, at the structural level, the volume asymmetry of this region may be associated with hemispheric dominance for language. In fact, Keller et al. [13] found that the size of the Insula could predict functional lateralisation for language in the same hemisphere in the majority of individuals. Therefore, the Insula was proposed as a more reliable marker for functional language specialisation than the Planum Temporale [13,148]. Although little is known about the functional implication of the Insula during language development in newborns and infants, several studies highlighted an early lateralisation of the Insula towards the left hemisphere [56]. Thanks to all the aforementioned data, the (anterior) Insula was established as a region of interest for studying linguistic (motor) processing. Additionally, the anterior Insula comprises Von Economo (VEN) and Fork neurons [149] that were for a long time thought to be uniquely human and implicated in social awareness. Therefore, the insula VEN and Fork neurons are often used for theories about the social origins of language [150].

While larger in humans, the insular cortex also exists in apes and monkeys, where the anterior portion especially, expanded during primate brain evolution [151]. Von Economo neurons, which were thought to be exclusive to apes [152], are also present in the anterior insular cortex of monkeys [150]. The authors argue that two distinct insular regions could be implicated in monkey communication [150]. First, a specific sensory-motor organisation for body parts in one part of the anterior Insula (the Idfa region) was found. Electric stimulation of this region elicited vocal cord movements in macaques. This region happens to be juxtaposed to a dorsal region, which receives inputs from area 44 (Part of Broca's Area). Together, the two regions could be homologous to the human anterior insula, implicated in several language processing functions (see above). Second, Von Economo and Fork neurons in another part of the anterior insula [the Ial region] project into a region of the thalamus (PAG) that is involved in vocalisations [150]. However, the structural lateralisation of this region was poorly investigated in nonhuman primates. One rare study comparing the Insula structure between a handful of different primate species subjects demonstrated that the anterior portion of the Insula, in which Von Economo neurons were found, displayed a human unique left asymmetry [151]. Further studies with a larger sample size are needed to investigate whether the Insula and particularly its anterior portion is structurally lateralised in nonhuman primates in relation to planification of communication, especially gesture.

6. Cingulate Cortex/Sulcus

The Cingulate Sulcus is in the medial part of the cerebral cortex delimitated ventrally by the Cingulate cortex and dorsally by the paracentral lobe and the superior frontal cortex. Its anterior part is considered a hub for domain-general cognitive processing, such as counterfactual thinking, mentalizing, and language, including cognitive control on signals production [153–156]. Pioneering studies regarding language processes have shown that for the Anterior and Midcingulate cortex, (1) stimulations evoke orofacial and tongue movements [157] and (2) lesioned patients experience akinetic mutism, associated with an absent motivation to speak [158]. Little is known about the Cingulate cortex concerning direct language development. Rare results come from Lœvenbruck et al. [147] highlighting, in adults, the functional neuroanatomical activation of the left anterior Cingulate cortex, besides Broca's area and the Insula in communicative pointing gestures [147]. However, the anterior Cingulate cortex was prominently shown to be important for joint attention in both adults and infants [159]. Joint attention is considered a prerequisite of the theory of mind as well as a prelinguistic communication act [159,160]. Interestingly, the hand, mouth, and tongue motor representations are grouped together around the caudal end of each vertical sulci departing from the Cingulate Sulcus [161], suggesting its key implication for a multimodal language system. In addition, 50% of human subjects present a Paracingulate Sulcus, located more rostrally above the anterior and Midcingulate Sulcus [162].

The presence of the Paracingulate Sulcus is lateralised in the human brain with nearly 70% located in the left hemisphere [14,162]. This lateralisation is influenced by genetic factors and the in-womb environment [163]. At the functional level, this human asymmetry has been shown to be correlated with the involvement of the left Cingulate cortex in language tasks in right-handed subjects [153,164].

Some authors have suggested that the anterior and Midcingulate cortices might have also played a role in language evolution. Loh et al. [155] hypothesised the existence of an evolutionary conserved ventrolateral frontal (around Broca's region) and dorsomedial frontal (roughly the Midcingulate cortex) network, which enables cognitive control of vocalisations. In fact, it is known that innate reflexive vocalisations such as 'screams' and 'shrieks', are associated with the 'cingulate vocalization pathway' [165]. In fact, the 'cingulate vocalisation area' in the anterior and Midcingulate cortex are connected (1) to the periaque–ductal gray, which directly projects to premotor nuclei in the brainstem and controls laryngeal motoneurons, which elicits vocalisations; and (2) to the facial motor nuclei to also produce affective facial movements [155]. In addition, in nonhuman primates, innate calls can be evoked by anterior and Midcingulate cortex stimulations [166] and lesions impair the production of calls [167] similar as in humans. Additionally, as in the Insula cortex (see above), large spindle shaped Von Economo neurons are present in the anterior Cingulate cortex in humans, apes, and other mammals [152,168,169]. Together with the presence of Von Economo neurons in the anterior Insula, it has been proposed that Von Economo neurons may be implicated in primate communication [150]. In contrast to the Insula cortex, no Von Economo neurons have been found in the monkey's Cingulate cortex yet. According to a recent comparative study including macaques, baboons, chimpanzees, and humans, the Cingulate Sulcus shows a highly conserved morphological antero-posterior organisation of vertical sulci or their precursor 'dimples' [14].

While, surprisingly, no structural asymmetries data on the Cingulate Sulcus are available so far in humans, significant population-level leftward asymmetries were found in the anterior portion of the Cingulate Sulcus in chimpanzee, whereas significant rightward biases were found in its posterior portion [170]. In contrast to humans, no population-level lateralisation for the presence of the Paracingulate Sulcus was found in chimpanzees [14]. Nevertheless, interindividual variation of the presence or absence of a Paracingulate Sulcus and variability of the intralimbic sulcus was associated with the production of attention-getting sounds and right handedness for gestural communication in chimpanzees [170]. It is, therefore, not to be excluded that the Cingulate Sulcus could be linked to precursors of human language such as communicative gesture and vocalisations.

7. Superior Temporal Sulcus (STS)

The Superior Temporal sulcus [STS] is a long fold separating the superior from the middle temporal lobe. It is important for a variety of social cognition tasks important for communication [171,172] containing both specialised regions for a particular task and regions that respond to several tasks simultaneously [171]. In fact, specialised areas of the STS have been shown to be implicated in the perception of voices in the right hemisphere [172], faces [173], biological motion [174], audiovisual integration [175], and in the theory of the mind [176,177]. Regarding speech perception in particular, three temporal voice areas are dispatched symmetrically in both hemispheres along an anterio-posterior gradient [178]. Distinct areas of the STS have also been implicated in language processing in hearing and deaf participants, especially in the left hemisphere [2,47,179]. In children, responses to voices in the STS and STG are strongly right-lateralised, an asymmetry which decreased with age [114,180]. This finding suggested that newborns rely for speech processing more on prosodic information, known to be processes in the right hemisphere, than for phonological information, processed in the left hemisphere (see also [181] for a right hemispheric STG temporal primacy in children in contrast to adults). Interestingly, together with Broca's area, the posterior STS is proposed to constitute the neural network supporting syntactic processes [79], which, until the age of 10, does not process syntax and semantics independently [182].

Regarding structural asymmetries, the left STS has been shown to be longer than the right, but matures later, as seen in preterm newborn infants [55,56,58,183,184]. In addition, the right STS was found to be deeper than the left STS in a portion ventral to Heschl's gyrus, called the STAP (Superior Temporal Asymmetrical Pit, [53,178,183]). This robust asymmetry is irrespective of age, handedness, and language lateralisation, suggesting a strong genetic influence [53,180,183]. Functional correspondence was found between its deepest point and location of the voice sensitive peak [178]. Several sulcus interruptions 'plis de passage' are more present in the left STS, probably resulting from stronger white matter fibres passing underneath the STS [185]. The aforementioned results highlight the STS' implication in communication through the direct link of language perception and the indirect link with diverse social cognition tasks, which are crucial for complex language processing. This makes the STS and its asymmetric structure a promising key area in comparative studies in order to search for language prerequisites shared between primate relatives.

The STS was intensively studied in macaque monkeys and shown to be one main hub in the social interaction processing network, where it was mapped onto a fine-grain pattern of object, body, and face selectivity [186]. Indeed, its implication in a variety of social cognition tasks was demonstrated. This includes gaze following and joint attention [187] as well as facial movements in the upper STS [188]. In addition, the mid-STS in macaques was recently hypothesised to be equivalent of the human TPJ area, important for the theory of mind, because it was active for predicting social situations [189] and because macaques living in larger groups demonstrate an increased grey matter of this area [190]. Especially regarding communication, similar to humans, voice and face patches were also found in the monkey's STS [178,191–194]. In fact, recently it was demonstrated that audiovisual integration occurred in a distinct region in the anterior fundus [195]. The Superior Temporal Asymmetrical Pit 'STAP' was not robustly shown to be present in chimpanzees, suggesting a human specific landmark of perisylvian organisation, which was related to human specific social cognition and communication [183]. According to results presented at conferences, structural rightward STS depth asymmetries were also found present in adult and juvenile baboons, within a portion that may overlap with the STAP in humans. Interestingly, preliminary results in baboons suggest that the strength of this rightward STS asymmetry varies according to social cognition proxies such as social group size and gestural communication's right-handedness [196,197].

8. Inferior Parietal Lobe

The Inferior Parietal Lobe or Geschwind's territory comprising the angular and the supramarginal gyrus was demonstrated as essential in language processing, connecting indirectly to primary language areas (such as the Planum Temporale or Broca's area). For example, it was shown to be involved in episodic memory retrieval of words [198] or verbal working memory [199], but also in tool-use [200,201]. The Inferior Parietal Lobe is one of the latest to myelinate in development [202] and was related to the emergence of language in evolution [203] and in development [204]. The Inferior Parietal Lobe includes the Planum Parietale, which lies in the supramarginal gyrus, in the ascending portion of the Sylvian fissure, next to the Planum Temporale [62]. The Planum Parietale is implicated in dyslexia and communication disorders [69] and processing voice spectral information [205]. Adaptation during primate evolution of this area and its connectivity may provide the capacity of enhanced visual analysis of moving images that is important for tool handling and control [206–208]. However, due to its anatomical position, this region is a zone of convergence and integration of sensory and motor information via the fronto-parietal network [209]. In fact, the connection between Broca's area and the Inferior Parietal Lobe is right-lateralised, in contrast to the left lateralisation of the Arcuate Fasciculus (see above). In addition, the Inferior Parietal Lobe inhabits mirror neurons that fire not only during motor execution, but also when observing actions performed by others, and might therefore lead to action understanding and language evolution [210,211].

Due to the aforementioned data, the Inferior Parietal Lobe is a key-region for investigating the lateralised links between actions, tool-use, and language evolution. A structural asymmetry of this area was found in the Planum Parietale. It showed a rightward asymmetry in both right-handed males and left-handed females, which was not correlated with the Planum Temporale asymmetry, indicating functional independence [62]. Another structural asymmetry was found for the parietal operculum, which constitutes the gyrus directly above the Planum Temporale as part of the supramarginal gyrus [212]. It shows a leftward asymmetry, especially for right-handers [213]. Regarding connectivity, tool-making skills elicited plastic remodeling of fronto-parietal white matter projections from the right Inferior Parietal Lobe into the right Broca's area [201].

Because of the potential overlapping of brain circuits for tool-use and language, nonhuman primate brain studies have focussed on the Inferior Parietal Lobe. A potential interaction might lie in the semantic knowledge important for both language and tools to acquire the skill necessary to perform these actions [200,214,215]. In this view, modifications of circuits that subserve gestures and imitations may have paved the evolutionary way for language and tool-use in humans [215], with the Inferior Parietal Lobe in its heart. Recently, Cheng et al. [216] found that leftward rostral- and rightward caudal inferior parietal structural asymmetries connecting to several areas related to tool-use in humans [215] were similar in chimpanzees and humans but not present in macaques. This finding could be related to human-like leftward asymmetries of the parietal operculum, which were also reported in chimpanzees and related to tool-use but not non-tool-use motor actions [76]. To learn to use tools, humans need a certain level of body awareness to match variations in kinematic details with the desired outcome during practice. Similar awareness is required for other animals on the Mirror Self-Recognition test [217], which also some chimpanzee subjects pass. In fact, chimpanzee subjects that passed the test (to recognize themself in a mirror) also possessed a more right lateralised fronto-parietal connection, exactly as in humans during tool-making learning [82,201,218,219]. This rightward asymmetry of connectivity could be related to a human-like rightward asymmetry initially found for the Planum Parietale in apes [220], a finding which was replicated in chimpanzees, particularly in females in relation to handedness [221]. Future studies in nonhuman primates should continue investigating potential links in lateralisation of the Inferior Parietal Lobe and behaviour. It would help clarifying whether left or right lateralised brain circuits for tool-use paved the way for language brain circuits.

9. The Central Sulcus (CS)

The Central Sulcus (CS) is a major landmark in the brain, dividing the parietal from the frontal lobe and is one of the primary sulci developed in the brain [57]. It also divides the primary motor cortex from the primary somatosensory cortex, where topographic sensory and motor representations of human body parts are organised [222]. Within this topographic organisation, a morphological landmark of hand and finger representations has been documented across the dorsal-ventral plane of the CS, known as KNOB or motor hand area, which has an omega-like shape [223].

Regardless of the hand, the direction of handedness was found associated with contralateral asymmetries of the motor hand area (humans: [92,224–226]. In fact, the portion of the CS that delimits the motor hand area was found deeper in the hemisphere contralateral to the preferred hand of the subjects [92]. This feature seems, however, not to be related with language lateralisation. In fact, it has recently been demonstrated that the neural substrates of typical handedness measures and language brain organisation might be not related, but rather independent from each other [27,29,227].

Within an evolutionary framework, hemispheric specialisation and handedness have been historically considered unique to human language evolution [228,229]. However, many primates, such as baboons or chimpanzees, also present right-handed dominance for manipulative actions (wild subjects: [230]), and even stronger right-handed dominance for communicative gestures (captive subjects: [30]). Just as in humans, the nonhuman primates' direction of handedness for object manipulation was found associated with contralateral asymmetries of the motor hand area within the Central Sulcus (Chimpanzees: [231,232]; Baboons: [233]; Capuchin monkeys: [234]; Squirrel monkeys: [235]). It is notable that the contralateral hand area effect in the CS was found exclusively for manipulative actions [233] but not for communicative gesture handedness. As mentioned in the previous section, handedness for gestural communication was exclusively found related to Broca's homologue [106], suggesting a potential independent neural substrate of handedness and language organisation in evolution. Preliminary results in juvenile baboons reported a similar neuroanatomical manifestation of early handedness, highlighting that structural asymmetries in the Central Sulcus appear early in development with the emergence of handedness behaviour [236]. Interestingly, the motor hand area of the Central Sulcus was also shown to be related to more complex hierarchical organisation of behaviours as it was related to tool-use handedness in chimpanzees [97].

10. Conclusions

Within an evolutionary framework across primates, we compared data on interhemispheric structural asymmetries in most of the key brain structures that are known to be associated with hemispheric specialisation for language processing in humans.

The results of this comparative neuroanatomical approach conducted between humans and mostly chimpanzees, and to a lesser extent baboons, are quite straightforward and challenge the historical view that hemispheric specialisation is a human specific feature of language evolution [228,229]. It becomes clear that we share the structural lateralised patterns for most language-related regions with other primate species, and even new-born monkeys, especially for perisylvian regions, including the Planum Temporale and Broca's area [24,69,71,73,74,95]. Although less documented in comparison to those two historical lateralised perisylvian regions, there is some evidence that other important regions of interest within the large human language network might share the same hemispheric structural lateralisation across primate species. According to rare recent comparative results, mostly conducted in chimpanzees and to a lesser extend in baboons, those shared features includes the leftward lateralised white matter connectivity tract between Broca's area and the Planum Temporale-the Arcuate Fasciculus [134], leftward lateralised parietal operculum, rightward lateralised Planum Parietale and fronto-parietal projection [76,220,221], the rightward posterior section of the STS [196,197], as well as the presence of paracingulate sulcus in the left hemisphere [170]. These collective findings suggest an important shared biological ancestral encoding between Old World monkeys, great apes, and humans, that were initially considered as neuroanatomical landmarks of brain lateralisation for language. Their presence in non-linguistic primate species clearly states that these landmarks are not human- or language- specific.

Nevertheless, it remains unclear for which shared 'domain-general' cognitive function between human and nonhuman primate species, that is also at the heart of language processing in modern humans, these structural hemispheric asymmetry patterns are related to and have evolved. While reviewing the nonhuman primate literature that reported clear links between anatomical asymmetries for language area homologues and hand preferences for tool-use or for gestural communication [71,75,106], some functional roads become plausible. For instance, it is thus not excluded that both 'syntactic' hierarchical sequential processing (such as the one involved in tool making and use in great apes), and the intentional communication properties (shared with the communicative gestural system in primates) might thus constitute a functional candidate to have paved the way for such brain asymmetric organisation within the evolutionary framework of human language. We demonstrate, therefore, that nonhuman primates such as chimpanzees, and even more phylogenetic distant species such as baboons among the Old World monkeys, offer compelling comparative models for the evolution of human brains and behaviours. In particular, a sulci and region-of-interest approach seems fruitful for comparing brain structures. In the same vein, handedness measurements for different behaviours seem fruitful for linking the lateralised brain anatomy to lateralised behaviours. Interesting future work could now be in transferring this handedness knowledge to humans, to clarify, for example, whether communicative gesture handedness might be a marker for language lateralisation.

To interpret brain asymmetry and its evolution, we highlighted the importance of investigating the lateralisation of behaviours such as gestural communication, tool use, and bimanual coordination within a comparative approach [38]. Indeed, a blending of the two is necessary in order to gain a holistic view on how language lateralisation has evolved, which ultimately also provides an initial platform for its emergence.

Regarding specifically language evolution, we hypothesise that asymmetries for language areas may not have initially evolved for language, if language is seen as a cognitive module. Rather, each asymmetry could have evolved independently for independent cognitive functions to adapt to unknown environmental pressures. This could explain the unclear relationship between structural and functional asymmetries related to language areas. The structural asymmetry would here be a more ancient fossil of other cognitive specialisations on which the functional language asymmetry got piggybacked. The functional and structural specialisation of these structures may next have been important nests for developing intentional communicative behaviour in nonhuman primates, which later evolved to language processing in our species. In other words, 'perisylvian language asymmetries' are not specific to language but could rather be exaptations of pre-existing specialisations for other cognitive functions, which together make up what we call 'language'. Therefore, 'domain-general' language-related brain architecture associated with intentional communicative or syntactic behaviours might rather be shared between humans, apes, and at least baboons of the Old World monkey family. Such an asymmetric brain organisation might have, thus, emerged from their common ancestor around 25 million years ago and later increased during hominin evolution.

Author Contributions: Conceptualization, Y.B. and A.M.; writing—original draft preparation, Y.B.; writing—review and editing, Y.B. and A.M.; supervision, A.M.; project administration, A.M.; funding acquisition, A.M. All authors have read and agreed to the published version of the manuscript.

Funding: The project has received funding from the European Research Council under the European Union's Horizon 2020 research and innovation program grant agreement No. 716931-GESTIMAGE-ERC-2016-STG (P.I. Adrien Meguerditchian), as well as from grants ANR-16-CONV-0002 (ILCB), ANR-17-EURE-0029 (NeuroSchool) and the Excellence Initiative of Aix-Marseille University A*MIDEX AMX-19-IET-004 (NeuroMarseille).

Institutional Review Board Statement: Not applicable.

Informed Consent Statement: Not applicable.

Conflicts of Interest: The authors declare no conflict of interest.

References

1. Berwick, R.C.; Chomsky, N. *Why Only Us: Language and Evolution*; MIT Press: Cambridge, MA, USA, 2016.
2. Fedorenko, E. The role of domain-general cognitive control in language comprehension. *Front. Psychol.* **2014**, *5*, 335. [CrossRef] [PubMed]
3. Fitch, W.T. *The Evolution of Language*; Cambridge University Press: Cambridge, UK, 2010.
4. Liebal, K.; Waller, B.M.; Slocombe, K.E.; Burrows, A.M. *Primate Communication: A Multimodal Approach*; Cambridge University Press: Cambridge, UK, 2014.
5. Fagot, J.; Boë, L.-J.; Berthomier, F.; Claidière, N.; Malassis, R.; Meguerditchian, A.; Rey, A.; Montant, M. The baboon: A model for the study of language evolution. *J. Hum. Evol.* **2019**, *126*, 39–50. [CrossRef]
6. Geschwind, N. The Organization of Language and the Brain. *Science* **1970**, *170*, 940–944. [CrossRef]
7. Toga, A.W.; Thompson, P.M. Mapping brain asymmetry. *Nat. Rev. Neurosci.* **2003**, *4*, 37–48. [CrossRef] [PubMed]
8. Biduła, S.P.; Kroliczak, G. Structural asymmetry of the insula is linked to the lateralization of gesture and language. *Eur. J. Neurosci.* **2015**, *41*, 1438–1447. [CrossRef] [PubMed]
9. Becker, Y.; Loh, K.K.; Coulon, O.; Meguerditchian, A. The Arcuate Fasciculus and language origins: Disentangling existing conceptions that influence evolutionary accounts. *Neurosci. Biobehav. Rev.* **2022**, *134*, 104490. [CrossRef] [PubMed]
10. Catani, M.; Mesulam, M. The arcuate fasciculus and the disconnection theme in language and aphasia: History and current state. *Cortex* **2008**, *44*, 953–961. [CrossRef] [PubMed]
11. Duffau, H. The error of Broca: From the traditional localizationist concept to a connectomal anatomy of human brain. *J. Chem. Neuroanat.* **2018**, *89*, 73–81. [CrossRef] [PubMed]
12. Catani, M.; Dawson, M. Language Processing, Development and Evolution. *Conn's Transl. Neurosci.* **2017**, 679–692. [CrossRef]
13. Keller, S.S.; Roberts, N.; García-Fiñana, M.; Mohammadi, S.; Ringelstein, E.B.; Knecht, S.; Deppe, M. Can the Language-dominant Hemisphere Be Predicted by Brain Anatomy? *J. Cogn. Neurosci.* **2010**, *23*, 2013–2029. [CrossRef] [PubMed]
14. Amiez, C.; Sallet, J.; Hopkins, W.D.; Meguerditchian, A.; Hadj-Bouziane, F.; Ben Hamed, S.; Wilson, C.R.E.; Procyk, E.; Petrides, M. Sulcal organization in the medial frontal cortex provides insights into primate brain evolution. *Nat. Commun.* **2019**, *10*, 3437. [CrossRef]
15. Gerrits, R.; Verhelst, H.; Dhollander, T.; Xiang, L.; Vingerhoets, G. Structural perisylvian asymmetry in naturally occurring atypical language dominance. *Anat. Embryol.* **2021**, *227*, 573–586. [CrossRef] [PubMed]
16. Tzourio-Mazoyer, N.; Crivello, F.; Mazoyer, B. Is the planum temporale surface area a marker of hemispheric or regional language lateralization? *Anat. Embryol.* **2018**, *223*, 1217–1228. [CrossRef]
17. Dehaene-Lambertz, G.; Dehaene, S.; Hertz-Pannier, L. Functional Neuroimaging of Speech Perception in Infants. *Science* **2002**, *298*, 2013–2015. [CrossRef] [PubMed]

18. Catani, M.; Allin, M.P.G.; Husain, M.; Pugliese, L.; Mesulam, M.M.; Murray, R.M.; Jones, D.K. Symmetries in human brain language pathways correlate with verbal recall. *Proc. Natl. Acad. Sci. USA* **2007**, *104*, 17163–17168. [CrossRef]
19. Darwin, C. *The Descent of Man and Selection in Relation to Sex*; Princeton University Press: Princeton, NJ, USA, 1981; Volume 1, [Originally Work Published 1871].
20. Hopkins, W.D.; Misiura, M.; Pope, S.M.; Latash, E.M. Behavioral and brain asymmetries in primates: A preliminary evaluation of two evolutionary hypotheses. *Ann. N. Y. Acad. Sci.* **2015**, *1359*, 65–83. [CrossRef] [PubMed]
21. Poirier, C.; Ben Hamed, S.; Garcia-Saldivar, P.; Kwok, S.C.; Meguerditchian, A.; Merchant, H.; Rogers, J.; Wells, S.; Fox, A.S. Beyond MRI: On the scientific value of combining non-human primate neuroimaging with metadata. *NeuroImage* **2021**, *228*, 117679. [CrossRef] [PubMed]
22. Milham, M.; Petkov, C.I.; Margulies, D.S.; Schroeder, C.E.; Basso, M.A.; Belin, P.; Fair, D.A.; Fox, A.; Kastner, S.; Mars, R.; et al. Accelerating the Evolution of Nonhuman Primate Neuroimaging. *Neuron* **2020**, *105*, 600–603. [CrossRef] [PubMed]
23. Hopkins, W.D.; Taglialatela, J.P.; Meguerditchian, A.; Nir, T.; Schenker, N.M.; Sherwood, C.C. Gray matter asymmetries in chimpanzees as revealed by voxel-based morphometry. *NeuroImage* **2008**, *42*, 491–497. [CrossRef]
24. Marie, D.; Roth, M.; Lacoste, R.; Nazarian, B.; Bertello, A.; Anton, J.-L.; Hopkins, W.D.; Margiotoudi, K.; Love, S.A.; Meguerditchian, A. Left Brain Asymmetry of the Planum Temporale in a Nonhominid Primate: Redefining the Origin of Brain Specialization for Language. *Cereb. Cortex* **2018**, *28*, 1808–1815. [CrossRef] [PubMed]
25. Knecht, S.; Dräger, B.; Deppe, M.; Bobe, L.; Lohmann, H.; Flöel, A.; Ringelstein, E.-B.; Henningsen, H. Handedness and hemispheric language dominance in healthy humans. *Brain* **2000**, *123*, 2512–2518. [CrossRef]
26. Amunts, K.; Jäncke, L.; Mohlberg, H.; Steinmetz, H.; Zilles, K. Interhemispheric asymmetry of the human motor cortex related to handedness and gender. *Neuropsychologia* **2000**, *38*, 304–312. [CrossRef]
27. Groen, M.A.; Whitehouse, A.J.O.; Badcock, N.A.; Bishop, D.V.M. Associations between Handedness and Cerebral Lateralisation for Language: A Comparison of Three Measures in Children. *PLoS ONE* **2013**, *8*, e64876. [CrossRef]
28. Mazoyer, B.; Zago, L.; Jobard, G.; Crivello, F.; Joliot, M.; Perchey, G.; Mellet, E.; Petit, L.; Tzourio-Mazoyer, N. Gaussian Mixture Modeling of Hemispheric Lateralization for Language in a Large Sample of Healthy Individuals Balanced for Handedness. *PLoS ONE* **2014**, *9*, e101165. [CrossRef] [PubMed]
29. Ocklenburg, S.; Beste, C.; Arning, L.; Peterburs, J.; Güntürkün, O. The ontogenesis of language lateralization and its relation to handedness. *Neurosci. Biobehav. Rev.* **2014**, *43*, 191–198. [CrossRef]
30. Meguerditchian, A.; Vauclair, J.; Hopkins, W.D. On the origins of human handedness and language: A comparative review of hand preferences for bimanual coordinated actions and gestural communication in nonhuman primates. *Dev. Psychobiol.* **2013**, *55*, 637–650. [CrossRef] [PubMed]
31. Hewes, G.W.; Andrew, R.J.; Carini, L.; Choe, H.; Gardner, R.A.; Kortlandt, A.; Krantz, G.S.; McBride, G.; Nottebohm, F.; Pfeiffer, J.; et al. Primate Communication and the Gestural Origin of Language [and Comments and Reply]. *Curr. Anthr.* **1973**, *14*, 5–24. [CrossRef]
32. Tomasello, M. *Origins of Human Communication*; The MIT Press: Cambridge, UK, 2008.
33. Meguerditchian, A.; Vauclair, J. Communicative Signaling, Lateralization and Brain Substrate in Nonhuman Primates: Toward a Gestural or a Multimodal Origin of Language? *Hum. Mente J. Philos. Stud.* **2014**, *7*, 135–160.
34. Molesti, S.; Meguerditchian, A.; Bourjade, M. Gestural communication in olive baboons [Papio anubis]: Repertoire and intentionality. *Anim. Cogn.* **2019**, *23*, 19–40. [CrossRef]
35. Greenfield, P.M. Language, tools and brain: The ontogeny and phylogeny of hierarchically organized sequential behavior. *Behav. Brain Sci.* **1991**, *14*, 531–551. [CrossRef]
36. Corballis, M.C. From mouth to hand: Gesture, speech and the evolution of right- handedness. *Brain Sci.* **2004**, *26*, 199–260. [CrossRef] [PubMed]
37. Stout, D.; Toth, N.; Schick, K.; Chaminade, T. Neural correlates of Early Stone Age toolmaking: Technology, language and cognition in human evolution. *Philos. Trans. R. Soc. B Biol. Sci.* **2008**, *363*, 1939–1949. [CrossRef] [PubMed]
38. Forrester, G.S.; Quaresmini, C.; Leavens, D.; Mareschal, D.; Thomas, M.S. Human handedness: An inherited evolutionary trait. *Behav. Brain Res.* **2013**, *237*, 200–206. [CrossRef] [PubMed]
39. Higuchi, S.; Chaminade, T.; Imamizu, H.; Kawato, M. Shared neural correlates for language and tool use in Broca's area. *NeuroReport* **2009**, *20*, 1376–1381. [CrossRef]
40. Thibault, S.; Py, R.; Gervasi, A.M.; Salemme, R.; Koun, E.; Lövden, M.; Boulenger, V.; Roy, A.C.; Brozzoli, C. Tool use and language share syntactic processes and neural patterns in the basal ganglia. *Science* **2021**, *374*, eabe0874. [CrossRef]
41. Galaburda, A.M.; LeMay, M.; Kemper, T.L.; Geschwind, N. Right-Left Asymmetries in the Brain. *Science* **1978**, *199*, 852–856. [CrossRef]
42. Mesulam, M.M. From sensation to cognition. *Brain* **1998**, *121*, 1013–1052. [CrossRef]
43. Wernicke, C. *Der Aphasische Symptomencomplex: Eine Psychologische Studie auf Anatomischer Basis*; Cohn: Breslau, Poland, 1874.
44. Dronkers, N.F.; Wilkins, D.P.; Van Valin, R.D.; Redfern, B.B.; Jaeger, J.J. Lesion analysis of the brain areas involved in language comprehension. *Cognition* **2004**, *92*, 145–177. [CrossRef]
45. Borovsky, A.; Saygin, A.P.; Bates, E.; Dronkers, N. Lesion correlates of conversational speech production deficits. *Neuropsychologia* **2007**, *45*, 2525–2533. [CrossRef]

46. Shapleske, J.; Rossell, S.; Woodruff, P.; David, A. The planum temporale: A systematic, quantitative review of its structural, functional and clinical significance. *Brain Res. Rev.* **1999**, *29*, 26–49. [CrossRef]
47. Vigneau, M.; Beaucousin, V.; Hervé, P.; Duffau, H.; Crivello, F.; Houdé, O.; Mazoyer, B.; Tzourio-Mazoyer, N. Meta-analyzing left hemisphere language areas: Phonology, semantics, and sentence processing. *NeuroImage* **2006**, *30*, 1414–1432. [CrossRef] [PubMed]
48. Dehaene-Lambertz, G.; Montavont, A.; Jobert, A.; Allirol, L.; Dubois, J.; Hertz-Pannier, L.; Dehaene, S. Language or music, mother or Mozart? Structural and environmental influences on infants' language networks. *Brain Lang.* **2010**, *114*, 53–65. [CrossRef]
49. Mahmoudzadeh, M.; Dehaene-Lambertz, G.; Fournier, M.; Kongolo, G.; Goudjil, S.; Dubois, J.; Grebe, R.; Wallois, F. Syllabic discrimination in premature human infants prior to complete formation of cortical layers. *Proc. Natl. Acad. Sci. USA* **2013**, *110*, 4846–4851. [CrossRef] [PubMed]
50. Geschwind, N.; Levitsky, W. Human Brain: Left-Right Asymmetries in Temporal Speech Region. *Science* **1968**, *161*, 186–187. [CrossRef]
51. Wada, J.A.; Clarke, R.; Hamm, A. Cerebral Hemispheric Asymmetry in Humans. *Arch. Neurol.* **1975**, *32*, 239–246. [CrossRef] [PubMed]
52. Witelson, S.F.; Pallie, W. Left hemisphere Spezialisation for Language in the Newborn. *Brain* **1973**, *96*, 641–646. [CrossRef]
53. Glasel, H.; Leroy, F.; Dubois, J.; Hertz-Pannier, L.; Mangin, J.; Dehaene-Lambertz, G. A robust cerebral asymmetry in the infant brain: The rightward superior temporal sulcus. *NeuroImage* **2011**, *58*, 716–723. [CrossRef]
54. Hill, J.; Inder, T.; Neil, J.; Dierker, D.; Harwell, J.; Van Essen, D. Similar patterns of cortical expansion during human development and evolution. *Proc. Natl. Acad. Sci. USA* **2010**, *107*, 13135–13140. [CrossRef]
55. Dubois, J.; Benders, M.; Cachia, A.; Lazeyras, F.; Leuchter, R.H.-V.; Sizonenko, S.V.; Borradori-Tolsa, C.; Mangin, J.F.; Hüppi, P.S. Mapping the Early Cortical Folding Process in the Preterm Newborn Brain. *Cereb. Cortex* **2008**, *18*, 1444–1454. [CrossRef]
56. Dubois, J.; Benders, M.; Lazeyras, F.; Borradori-Tolsa, C.; Leuchter, R.H.-V.; Mangin, J.; Hüppi, P. Structural asymmetries of perisylvian regions in the preterm newborn. *NeuroImage* **2010**, *52*, 32–42. [CrossRef]
57. Chi, J.G.; Dooling, E.C.; Gilles, F.H. Gyral development of the human brain. *Ann. Neurol.* **1977**, *1*, 86–93. [CrossRef] [PubMed]
58. Chi, J.G.; Dooling, E.C.; Gilles, F.H. Left-Right Asymmetries of the Temporal Speech Areas of the Human Fetus. *Arch. Neurol.* **1977**, *34*, 346–348. [CrossRef] [PubMed]
59. Josse, G.; Hervé, P.-Y.; Crivello, F.; Mazoyer, B.; Tzourio-Mazoyer, N. Hemispheric specialization for language: Brain volume matters. *Brain Res.* **2006**, *1068*, 184–193. [CrossRef]
60. Gauger, L.M.; Lombardino, L.J.; Leonard, C.M. Brain Morphology in Children with Specific Language Impairment. *J. Speech Lang. Hear. Res.* **1997**, *40*, 1272–1284. [CrossRef] [PubMed]
61. Altarelli, I.; Leroy, F.; Monzalvo, K.; Fluss, J.; Billard, C.; Dehaene-Lambertz, G.; Galaburda, A.M.; Ramus, F. Planum temporale asymmetry in developmental dyslexia: Revisiting an old question. *Hum. Brain Mapp.* **2014**, *35*, 5717–5735. [CrossRef]
62. Jäncke, L.; Schlaug, G.; Huang, Y.; Steinmetz, H. Asymmetry of the planum parietale. *NeuroReport* **1994**, *5*, 1161–1163. [CrossRef]
63. Dorsaint-Pierre, R.; Penhune, V.B.; Watkins, K.; Neelin, P.; Lerch, J.P.; Bouffard, M.; Zatorre, R.J. Asymmetries of the planum temporale and Heschl's gyrus: Relationship to language lateralization. *Brain* **2006**, *129*, 1164–1176. [CrossRef]
64. Eckert, M.A.; Leonard, C.M.; Possing, E.T.; Binder, J.R. Uncoupled leftward asymmetries for planum morphology and functional language processing. *Brain Lang.* **2006**, *98*, 102–111. [CrossRef]
65. Greve, D.N.; Van der Haegen, L.; Cai, Q.; Stufflebeam, S.; Sabuncu, M.R.; Fischl, B.; Brysbaert, M. A Surface-based Analysis of Language Lateralization and Cortical Asymmetry. *J. Cogn. Neurosci.* **2013**, *25*, 1477–1492. [CrossRef]
66. Kolinsky, R.; Morais, J.; Cohen, L.; Dehaene-Lambertz, G.; Dehaene, S. The impact of literacy on the language brain areas. *Rev. Neuropsychol.* **2014**, *6*, 173–181. [CrossRef]
67. Tzourio-Mazoyer, N.; Mazoyer, B. Variations of planum temporale asymmetries with Heschl's Gyri duplications and association with cognitive abilities: MRI investigation of 428 healthy volunteers. *Anat. Embryol.* **2017**, *222*, 2711–2726. [CrossRef] [PubMed]
68. Ocklenburg, S.; Friedrich, P.; Fraenz, C.; Schlüter, C.; Beste, C.; Güntürkün, O.; Genç, E. Neurite architecture of the planum temporale predicts neurophysiological processing of auditory speech. *Sci. Adv.* **2018**, *4*, eaar6830. [CrossRef] [PubMed]
69. Gannon, P.J.; Holloway, R.L.; Broadfield, D.C.; Braun, A.R. Asymmetry of Chimpanzee Planum Temporale: Humanlike Pattern of Wernicke's Brain Language Area Homolog. *Science* **1998**, *279*, 220–222. [CrossRef] [PubMed]
70. Hopkins, W.D.; Marino, L.; Rilling, J.K.; MacGregor, L.A. Planum temporale asymmetries in great apes as revealed by magnetic resonance imaging [MRI]. *NeuroReport* **1998**, *9*, 2913–2918. [CrossRef] [PubMed]
71. Hopkins, W.D.; Nir, T.M. Planum temporale surface area and grey matter asymmetries in chimpanzees [Pan troglodytes]: The effect of handedness and comparison with findings in humans. *Behav. Brain Res.* **2010**, *208*, 436–443. [CrossRef]
72. Cantalupo, C.; Pilcher, D.L.; Hopkins, W.D. Are planum temporale and sylvian fissure asymmetries directly related? A MRI study in great apes. *Neuropsychologia* **2003**, *41*, 1975–1981. [CrossRef]
73. Becker, Y.; Sein, J.; Velly, L.; Giacomino, L.; Renaud, L.; Lacoste, R.; Anton, J.-L.; Nazarian, B.; Berne, C.; Meguerditchian, A. Early Left-Planum Temporale Asymmetry in newborn monkeys [Papio anubis]: A longitudinal structural MRI study at two stages of development. *NeuroImage* **2021**, *227*, e117575. [CrossRef]
74. Becker, Y.; Phelipon, R.; Sein, J.; Velly, L.; Renaud, L.; Meguerditchian, A. Planum temporale grey matter volume asymmetries in newborn monkeys [Papio anubis]. *Brain Struct. Funct.* **2022**, *227*, 463–468. [CrossRef]

75. Meguerditchian, A.; Gardner, M.J.; Schapiro, S.J.; Hopkins, W.D. The sound of one-hand clapping: Handedness and perisylvian neural correlates of a communicative gesture in chimpanzees. *Proc. R. Soc. B Boil. Sci.* **2012**, *279*, 1959–1966. [CrossRef]
76. Gilissen, E.P.; Hopkins, W.D. Asymmetries of the Parietal Operculum in Chimpanzees [Pan troglodytes] in Relation to Handedness for Tool Use. *Cereb. Cortex* **2013**, *23*, 411–422. [CrossRef]
77. Broca, P. *Remarques sur le siège de la Faculté du Langage Articulé, Suivies d'une Observation d'aphémie [perte de la parole]*; Bulletin et Memoires de la Societe Anatomique de Paris: Paris, France, 1861; Volume 6, pp. 330–357.
78. Hickok, G.; Poeppel, D. The cortical organization of speech processing. *Nat. Rev. Neurosci.* **2007**, *8*, 393–402. [CrossRef] [PubMed]
79. Friederici, A.D. *Language in Our Brain: The Origins of a Uniquely Human Capacity*; The MIT Press: Cambridge, MA, USA, 2017.
80. Gentilucci, M.; Volta, R.D. Spoken Language and arm Gestures are Controlled by the same Motor Control System. *Q. J. Exp. Psychol.* **2008**, *61*, 944–957. [CrossRef] [PubMed]
81. Koechlin, E.; Jubault, T. Broca's Area and the Hierarchical Organization of Human Behavior. *Neuron* **2006**, *50*, 963–974. [CrossRef]
82. Stout, D.; Hecht, E.E. Evolutionary neuroscience of cumulative culture. *Proc. Natl. Acad. Sci. USA* **2017**, *114*, 7861–7868. [CrossRef]
83. Emmorey, K.; Grabowski, T.; McCullough, S.; Damasio, H.; Ponto, L.; Hichwa, R.; Bellugi, U. Motor-iconicity of sign language does not alter the neural systems underlying tool and action naming. *Brain Lang.* **2004**, *89*, 27–37. [CrossRef]
84. Campbell, R.; Macsweeney, M.; Waters, D. Sign Language and the Brain: A Review. *J. Deaf Stud. Deaf Educ.* **2007**, *13*, 3–20. [CrossRef] [PubMed]
85. Imada, T.; Zhang, Y.; Cheour, M.; Taulu, S.; Ahonen, A.; Kuhl, P.K. Infant speech perception activates Broca's area: A developmental magnetoencephalography study. *NeuroReport* **2006**, *17*, 957–962. [CrossRef]
86. Dehaene-Lambertz, G.; Hertz-Pannier, L.; Dubois, J.; Mériaux, S.; Roche, A.; Sigman, M.; Dehaene, S. Functional organization of perisylvian activation during presentation of sentences in preverbal infants. *Proc. Natl. Acad. Sci. USA* **2006**, *103*, 14240–14245. [CrossRef]
87. Keller, S.S.; Crow, T.; Foundas, A.; Amunts, K.; Roberts, N. Broca's area: Nomenclature, anatomy, typology and asymmetry. *Brain Lang.* **2009**, *109*, 29–48. [CrossRef]
88. Sprung-Much, T.; Eichert, N.; Nolan, E.; Petrides, M. Broca's area and the search for anatomical asymmetry: Commentary and perspectives. *Anat. Embryol.* **2022**, *227*, 441–449. [CrossRef]
89. Keller, S.S.; Highley, J.R.; Garcia-Finana, M.; Sluming, V.; Rezaie, R.; Roberts, N. Sulcal variability, stereological measurement and asymmetry of Broca's area on MR images. *J. Anat.* **2007**, *211*, 534–555. [CrossRef] [PubMed]
90. Scheibel, A.B. A dendritic correlate of human speech. In *Cerebral Dominance: The Biological Foundations*; Harvard University Press: Cambridge, MA, USA, 1984; pp. 43–52.
91. Amunts, K.; Schleicher, A.; Mohlberg, H.; Uylings, H.B.; Zilles, K. Broca's region revisited: Cytoarchitecture and intersubject variability. *J. Comp. Neurol.* **1999**, *412*, 319–341. [CrossRef]
92. Amunts, K.; Schleicher, A.; Ditterich, A.; Zilles, K. Broca's region: Cytoarchitectonic asymmetry and developmental changes. *J. Comp. Neurol.* **2003**, *465*, 72–89. [CrossRef]
93. Kurth, F.; Cherbuin, N.; Luders, E. Speaking of aging: Changes in gray matter asymmetry in Broca's area in later adulthood. *Cortex* **2020**, *129*, 133–140. [CrossRef] [PubMed]
94. Simonds, R.J.; Scheibel, A.B. The postnatal development of the motor speech area: A preliminary study. *Brain Lang.* **1989**, *37*, 42–58. [CrossRef]
95. Cantalupo, C.; Hopkins, W.D. Asymmetric Broca's area in great apes. *Nature* **2001**, *414*, 505. [CrossRef] [PubMed]
96. Keller, S.S.; Deppe, M.; Herbin, M.; Gilissen, E. Variability and asymmetry of the sulcal contours defining Broca's area homologue in the chimpanzee brain. *J. Comp. Neurol.* **2012**, *520*, 1165–1180. [CrossRef]
97. Hopkins, W.D. Motor and Communicative Correlates of the Inferior Frontal Gyrus [Broca's Area] in Chimpanzees. In *Origins of Human Language: Continuities and Splits with Nonhuman Primates*; Boë, L.-J., Fagot, J., Perrier, P., Schwartz, J.-L., Eds.; Peter Lang: Oxford, UK, 2017; pp. 153–186.
98. Schenker-Ahmed, N.; Hopkins, W.D.; Spocter, M.; Garrison, A.R.; Stimpson, C.D.; Erwin, J.M.; Hof, P.R.; Sherwood, C.C. Broca's Area Homologue in Chimpanzees [Pan troglodytes]: Probabilistic Mapping, Asymmetry, and Comparison to Humans. *Cereb. Cortex* **2010**, *20*, 730–742. [CrossRef]
99. Graïc, J.-M.; Peruffo, A.; Corain, L.; Centelleghe, C.; Granato, A.; Zanellato, E.; Cozzi, B. Asymmetry in the Cytoarchitecture of the Area 44 Homolog of the Brain of the Chimpanzee Pan troglodytes. *Front. Neuroanat.* **2020**, *14*. [CrossRef]
100. Petrides, M. Lateral prefrontal cortex: Architectonic and functional organization. *Philos. Trans. R. Soc. B Biol. Sci.* **2005**, *360*, 781–795. [CrossRef]
101. Petrides, M.; Cadoret, G.; Mackey, S. Orofacial somatomotor responses in the macaque monkey homologue of Broca's area. *Nature* **2005**, *435*, 1235–1238. [CrossRef] [PubMed]
102. Belmalih, A.; Borra, E.; Contini, M.; Gerbella, M.; Rozzi, S.; Luppino, G. Multimodal architectonic subdivision of the rostral part [area F5] of the macaque ventral premotor cortex. *J. Comp. Neurol.* **2009**, *512*, 183–217. [CrossRef] [PubMed]
103. Hage, S.R.; Nieder, A. Single neurons in monkey prefrontal cortex encode volitional initiation of vocalizations. *Nat. Commun.* **2013**, *4*, 2409. [CrossRef] [PubMed]
104. Taglialatela, J.P.; Russell, J.L.; Schaeffer, J.A.; Hopkins, W.D. Communicative Signaling Activates 'Broca's' Homolog in Chimpanzees. *Curr. Biol.* **2008**, *18*, 343–348. [CrossRef]

105. Hopkins, W.D.; Meguerditchian, A.; Coulon, O.; Misiura, M.; Pope, S.; Mareno, M.C.; Schapiro, S.J. Motor skill for tool-use is associated with asymmetries in Broca's area and the motor hand area of the precentral gyrus in chimpanzees [Pan troglodytes]. Behav. *Brain Res.* **2017**, *318*, 71–81. [CrossRef]
106. Becker, Y.; Claidière, N.; Margiotoudi, K.; Marie, D.; Roth, M.; Nazarian, B.; Anton, J.-L.; Coulon, O.; Meguerditchian, A. Broca's cerebral asymmetry reflects gestural communication's lateralisation in monkeys [Papio anubis]. *eLife* **2022**, *11*, e70521. [CrossRef]
107. Catani, M.; Schotten, M.T. *Atlas of Human Brain Connections*; OUP Oxford: Oxford, UK, 2012.
108. Metellus, P.; Boussen, S.; Guye, M.; Trebuchon, A. Successful Insular Glioma Removal in a Deaf Signer Patient during an Awake Craniotomy Procedure. *World Neurosurg.* **2017**, *98*, 883-e1. [CrossRef]
109. Dubois, J.; Poupon, C.; Thirion, B.; Simonnet, H.; Kulikova, S.; Leroy, F.; Hertz-Pannier, L.; Dehaene-Lambertz, G. Exploring the Early Organization and Maturation of Linguistic Pathways in the Human Infant Brain. *Cereb. Cortex* **2016**, *26*, 2283–2298. [CrossRef]
110. Brauer, J.; Anwander, A.; Perani, D.; Friederici, A.D. Dorsal and ventral pathways in language development. *Brain Lang.* **2013**, *127*, 289–295. [CrossRef]
111. Brauer, J.; Anwander, A.; Friederici, A.D. Neuroanatomical Prerequisites for Language Functions in the Maturing Brain. *Cereb. Cortex* **2011**, *21*, 459–466. [CrossRef]
112. Dubois, J.; Hertz-Pannier, L.; Dehaene-Lambertz, G.; Cointepas, Y.; Le Bihan, D. Assessment of the early organization and maturation of infants' cerebral white matter fiber bundles: A feasibility study using quantitative diffusion tensor imaging and tractography. *NeuroImage* **2006**, *30*, 1121–1132. [CrossRef] [PubMed]
113. Dubois, J.; Hertz-Pannier, L.; Cachia, A.; Mangin, J.F.; Le Bihan, D.; Dehaene-Lambertz, G. Structural Asymmetries in the Infant Language and Sensori-Motor Networks. Cereb. *Cortex* **2009**, *19*, 414–423. [CrossRef] [PubMed]
114. Perani, D.; Saccuman, M.C.; Scifo, P.; Anwander, A.; Spada, D.; Baldoli, C.; Poloniato, A.; Lohmann, G.; Friederici, A.D. Neural language networks at birth. *Proc. Natl. Acad. Sci. USA* **2011**, *108*, 16056–16061. [CrossRef] [PubMed]
115. Friederici, A.D. Language Development and the Ontogeny of the Dorsal Pathway. Front. *Evol. Neurosci.* **2012**, *4*, 3. [CrossRef]
116. Büchel, C.; Raedler, T.; Sommer, M.; Sach, M.; Weiller, C.; Koch, M. White Matter Asymmetry in the Human Brain: A Diffusion Tensor MRI Study. *Cereb. Cortex* **2004**, *14*, 945–951. [CrossRef] [PubMed]
117. Nucifora, P.G.P.; Verma, R.; Melhem, E.R.; Gur, R.E.; Gur, R.C. Leftward asymmetry in relative fiber density of the arcuate fasciculus. *NeuroReport* **2005**, *16*, 791–794. [CrossRef]
118. Powell, H.R.; Parker, G.J.; Alexander, D.C.; Symms, M.R.; Boulby, P.A.; Wheeler-Kingshott, C.A.; Barker, G.J.; Noppeney, U.; Koepp, M.J.; Duncan, J.S. Hemispheric asymmetries in language-related pathways: A combined functional MRI and tractography study. *NeuroImage* **2006**, *32*, 388–399. [CrossRef] [PubMed]
119. Takaya, S.; Kuperberg, G.R.; Liu, H.; Greve, D.N.; Makris, N.; Stufflebeam, S.M. Asymmetric projections of the arcuate fasciculus to the temporal cortex underlie lateralized language function in the human brain. *Front. Neuroanat.* **2015**, *9*, 119. [CrossRef]
120. Thiebaut de Schotten, M.; Ffytche, D.H.; Bizzi, A.; Dell'Acqua, F.; Allin, M.; Walshe, M.; Murray, R.; Williams, S.C.; Murphy, D.G.M.; Catani, M. Atlasing location, asymmetry and inter-subject variability of white matter tracts in the human brain with MR diffusion tractography. *NeuroImage* **2011**, *54*, 49–59. [CrossRef] [PubMed]
121. Liu, Y.; Balériaux, D.; Kavec, M.; Metens, T.; Absil, J.; Denolin, V.; Pardou, A.; Avni, F.; Van Bogaert, P.; Aeby, A. Structural asymmetries in motor and language networks in a population of healthy preterm neonates at term equivalent age. A diffusion tensor imaging and probabilistic tractography study. *NeuroImage* **2010**, *51*, 783–788. [CrossRef]
122. Song, J.W.; Mitchell, P.D.; Kolasinski, J.; Ellen Grant, P.; Galaburda, A.M.; Takahashi, E. Asymmetry of White Matter Pathways in Developing Human Brains. *Cereb. Cortex* **2015**, *25*, 2883–2893. [CrossRef] [PubMed]
123. Lebel, C.; Beaulieu, C. Lateralization of the arcuate fasciculus from childhood to adulthood and its relation to cognitive abilities in children. *Hum. Brain Mapp.* **2009**, *30*, 3563–3573. [CrossRef] [PubMed]
124. López-Barroso, D.; Catani, M.; Ripollés, P.; Dell'Acqua, F.; Rodríguez-Fornells, A.; de Diego-Balaguer, R. Word learning is mediated by the left arcuate fasciculus. *Proc. Natl. Acad. Sci. USA* **2013**, *110*, 13168–13173. [CrossRef]
125. Salvan, P.; Tournier, J.D.; Batalle, D.; Falconer, S.; Chew, A.; Kennea, N.; Aljabar, P.; Dehaene-Lambertz, G.; Arichi, T.; Edwards, A.D.; et al. Language ability in preterm children is associated with arcuate fasciculi microstructure at term. *Hum. Brain Mapp.* **2017**, *38*, 3836–3847. [CrossRef] [PubMed]
126. Propper, R.E.; O'Donnell, L.J.; Whalen, S.; Tie, Y.; Norton, I.H.; Suarez, R.O.; Zollei, L.; Radmanesh, A.; Golby, A.J. A combined fMRI and DTI examination of functional language lateralization and arcuate fasciculus structure: Effects of degree versus direction of hand preference. *Brain Cogn.* **2010**, *73*, 85–92. [CrossRef] [PubMed]
127. Zhu, L.; Fan, Y.; Zou, Q.; Wang, J.; Gao, J.-H.; Niu, Z. Temporal Reliability and Lateralization of the Resting-State Language Network. *PLoS ONE* **2014**, *9*, e85880. [CrossRef] [PubMed]
128. Silva, G.; Citterio, A. Hemispheric asymmetries in dorsal language pathway white-matter tracts: A magnetic resonance imaging tractography and functional magnetic resonance imaging study. *Neuroradiol. J.* **2017**, *30*, 470–476. [CrossRef]
129. Verhelst, H.; Dhollander, T.; Gerrits, R.; Vingerhoets, G. Fibre-specific laterality of white matter in left and right language dominant people. *NeuroImage* **2021**, *230*, 117812. [CrossRef]
130. Schmahmann, J.D.; Pandya, D.N. *Fiber Pathways of the Brain*; Oxford University Press: Oxford, UK, 2006.
131. Petrides, M.; Pandya, D.N. Distinct Parietal and Temporal Pathways to the Homologues of Broca's Area in the Monkey. *PLoS Biol.* **2009**, *7*, e1000170. [CrossRef]

132. Petrides, M. *Neuroanatomy of Language Regions of the Human Brain*, 1st ed.; Elsevier/AP, Academic Press is an imprint of Elsevier: Amsterdam, The Netherlands, 2014.
133. Rilling, J.K.; Glasser, M.F.; Preuss, T.M.; Ma, X.; Zhao, T.; Hu, X.; Behrens, T.E.J. The evolution of the arcuate fasciculus revealed with comparative DTI. *Nat. Neurosci.* **2008**, *11*, 426–428. [CrossRef]
134. Rilling, J.K.; Glasser, M.F.; Jbabdi, S.; Andersson, J.; Preuss, T.M. Continuity, Divergence, and the Evolution of Brain Language Pathways. *Front. Evol. Neurosci.* **2012**, *3*, 11. [CrossRef]
135. Frey, S.; Mackey, S.; Petrides, M. Cortico-cortical connections of areas 44 and 45B in the macaque monkey. *Brain Lang.* **2014**, *131*, 36–55. [CrossRef]
136. Eichert, N.; Verhagen, L.; Folloni, D.; Jbabdi, S.; Khrapitchev, A.A.; Sibson, N.R.; Mantini, D.; Sallet, J.; Mars, R. What is special about the human arcuate fasciculus? Lateralization, projections, and expansion. *Cortex* **2019**, *118*, 107–115. [CrossRef] [PubMed]
137. Barrett, R.L.C.; Dawson, M.; Dyrby, T.B.; Krug, K.; Ptito, M.; D'Arceuil, H.; Croxson, P.L.; Johnson, P.J.; Howells, H.; Forkel, S.; et al. Differences in Frontal Network Anatomy Across Primate Species. *J. Neurosci.* **2020**, *40*, 2094–2107. [CrossRef] [PubMed]
138. Balezeau, F.; Wilson, B.; Gallardo, G.; Dick, F.; Hopkins, W.; Anwander, A.; Friederici, A.D.; Griffiths, T.D.; Petkov, C.I. Primate auditory prototype in the evolution of the arcuate fasciculus. *Nat. Neurosci.* **2020**, *23*, 611–614. [CrossRef] [PubMed]
139. Rocchi, F.; Oya, H.; Balezeau, F.; Billig, A.; Kocsis, Z.; Jenison, R.L.; Nourski, K.; Kovach, C.; Steinschneider, M.; Kikuchi, Y.; et al. Common fronto-temporal effective connectivity in humans and monkeys. *Neuron* **2021**, *109*, 852–868.e8. [CrossRef]
140. Dronkers, N.F. A new brain region for coordinating speech articulation. *Nature* **1996**, *384*, 159–161. [CrossRef]
141. Oh, A.; Duerden, E.G.; Pang, E.W. The role of the insula in speech and language processing. *Brain Lang.* **2014**, *135*, 96–103. [CrossRef] [PubMed]
142. Bamiou, D.-E.; Musiek, F.E.; Luxon, L.M. The insula [Island of Reil] and its role in auditory processing: Literature review. *Brain Res. Rev.* **2003**, *42*, 143–154. [CrossRef]
143. Ackermann, H.; Riecker, A. The contribution of the insula to motor aspects of speech production: A review and a hypothesis. *Brain Lang.* **2004**, *89*, 320–328. [CrossRef]
144. Friederici, A.D. The neural basis for human syntax: Broca's area and beyond. *Curr. Opin. Behav. Sci.* **2018**, *21*, 88–92. [CrossRef]
145. Zaccarella, E.; Friederici, A.D. Merge in the Human Brain: A Sub-Region Based Functional Investigation in the Left Pars Opercularis. *Front. Psychol.* **2015**, *6*, 1818. [CrossRef]
146. Lœvenbruck, H.; Baciu, M.; Segebarth, C.; Abry, C. The left inferior frontal gyrus under focus: An fMRI study of the production of deixis via syntactic extraction and prosodic focus. *J. Neurolinguistics* **2005**, *18*, 237–258. [CrossRef]
147. Lœvenbruck, H.; Vilain, C.; Dohen, M. From gestural pointing to vocal pointing in the brain. *Rev. Fr. Linguist. Appl.* **2008**, *13*, 23–33. [CrossRef]
148. Chiarello, C.; Vazquez, D.; Felton, A.; Leonard, C.M. Structural asymmetry of anterior insula: Behavioral correlates and individual differences. *Brain Lang.* **2013**, *126*, 109–122. [CrossRef] [PubMed]
149. Economo, C.V. Eine neue art spezialzellen des lobus cinguli und lobus insulae. *Z. Die Gesamte Neurol. Und Psychiatr.* **1926**, *100*, 706–712. [CrossRef]
150. Evrard, H.C. Von Economo and fork neurons in the monkey insula, implications for evolution of cognition. *Curr. Opin. Behav. Sci.* **2018**, *21*, 182–190. [CrossRef]
151. Bauernfeind, A.L.; de Sousa, A.A.; Avasthi, T.; Dobson, S.D.; Raghanti, M.A.; Lewandowski, A.H.; Zilles, K.; Semendeferi, K.; Allman, J.M.; Craig, A.D.; et al. A volumetric comparison of the insular cortex and its subregions in primates. *J. Hum. Evol.* **2013**, *64*, 263–279. [CrossRef]
152. Nimchinsky, E.A.; Gilissen, E.; Allman, J.M.; Perl, D.P.; Erwin, J.M.; Hof, P.R. A neuronal morphologic type unique to humans and great apes. *Proc. Natl. Acad. Sci. USA* **1999**, *96*, 5268–5273. [CrossRef]
153. Toga, A.W.; Thompson, P.M. Temporal Dynamics of Brain Anatomy. *Annu. Rev. Biomed. Eng.* **2003**, *5*, 119–145. [CrossRef] [PubMed]
154. Amodio, D.M.; Frith, C.D. Meeting of minds: The medial frontal cortex and social cognition. *Nat. Rev. Neurosci.* **2006**, *7*, 268–277. [CrossRef] [PubMed]
155. Loh, K.K.; Petrides, M.; Hopkins, W.D.; Procyk, E.; Amiez, C. Cognitive control of vocalizations in the primate ventrolateral-dorsomedial frontal [VLF-DMF] brain network. *Neurosci. Biobehav. Rev.* **2017**, *82*, 32–44. [CrossRef] [PubMed]
156. Loh, K.K.; Procyk, E.; Neveu, R.; Lamberton, F.; Hopkins, W.D.; Petrides, M.; Amiez, C. Cognitive control of orofacial motor and vocal responses in the ventrolateral and dorsomedial human frontal cortex. *Proc. Natl. Acad. Sci. USA* **2020**, *117*, 4994–5005. [CrossRef] [PubMed]
157. Talairach, J.; Bancaud, J.; Geier, S.; Bordas-Ferrer, M.; Bonis, A.; Szikla, G.; Rusu, M. The cingulate gyrus and human behaviour. *Electroencephalogr. Clin. Neurophysiol.* **1973**, *34*, 45–52. [CrossRef]
158. Ackermann, H.; Ziegler, W. Akinetischer Mutismus-eine Literaturübersicht. Fortschr. *Neurol. Psychiatr.* **1995**, *63*, 59–67. [CrossRef]
159. Benga, O. Intentional communication and the anterior cingulate cortex. *Interact. Stud.* **2005**, *6*, 201–221. [CrossRef]
160. Mundy, P. A review of joint attention and social-cognitive brain systems in typical development and autism spectrum disorder. *Eur. J. Neurosci.* **2018**, *47*, 497–514. [CrossRef]
161. Amiez, C.; Petrides, M. Neuroimaging Evidence of the Anatomo-Functional Organization of the Human Cingulate Motor Areas. *Cereb. Cortex* **2014**, *24*, 563–578. [CrossRef] [PubMed]

162. Ide, A.; Dolezal, C.; Fernández, M.; Labbé, E.; Mandujano, R.; Montes, S.; Segura, P.; Verschae, G.; Yarmuch, P.; Aboitiz, F. Hemispheric differences in variability of fissural patterns in parasylvian and cingulate regions of human brains. *J. Comp. Neurol.* **1999**, *410*, 235–242. [CrossRef]
163. Amiez, C.; Wilson, C.R.E.; Procyk, E. Variations of cingulate sulcal organization and link with cognitive performance. *Sci. Rep.* **2018**, *8*, 13988. [CrossRef] [PubMed]
164. Paus, T.; Tomaiuolo, F.; Otaky, N.; Macdonald, D.; Petrides, M.; Atlas, J.; Morris, R.; Evans, A.C. Human Cingulate and Paracingulate Sulci: Pattern, Variability, Asymmetry, and Probabilistic Map. *Cereb. Cortex* **1996**, *6*, 207–214. [CrossRef] [PubMed]
165. Jürgens, U.; Pratt, R. The cingular vocalization pathway in the squirrel monkey. *Exp. Brain Res.* **1979**, *34*, 499–510. [CrossRef] [PubMed]
166. Jürgens, U.; Ploog, D. Cerebral representation of vocalization in the squirrel monkey. *Exp. Brain Res.* **1970**, *10*, 532–554. [CrossRef] [PubMed]
167. Aitken, P.G. Cortical control of conditioned and spontaneous vocal behavior in rhesus monkeys. *Brain Lang.* **1981**, *13*, 171–184. [CrossRef]
168. Allman, J.M.; Tetreault, N.A.; Hakeem, A.Y.; Manaye, K.F.; Semendeferi, K.; Erwin, J.M.; Park, S.; Goubert, V.; Hof, P.R. The von Economo neurons in frontoinsular and anterior cingulate cortex in great apes and humans. *Anat. Embryol.* **2010**, *214*, 495–517. [CrossRef] [PubMed]
169. Hakeem, A.Y.; Sherwood, C.C.; Bonar, C.J.; Butti, C.; Hof, P.R.; Allman, J.M. Von Economo Neurons in the Elephant Brain. *Anat. Rec.* **2009**, *292*, 242–248. [CrossRef] [PubMed]
170. Hopkins, W.D.; Procyk, E.; Petrides, M.; Schapiro, S.J.; Mareno, M.C.; Amiez, C. Sulcal Morphology in Cingulate Cortex is Associated with Voluntary Oro-Facial Motor Control and Gestural Communication in Chimpanzees [Pan troglodytes]. *Cereb. Cortex* **2021**, *31*, 2845–2854. [CrossRef]
171. Deen, B.; Koldewyn, K.; Kanwisher, N.; Saxe, R. Functional Organization of Social Perception and Cognition in the Superior Temporal Sulcus. *Cereb. Cortex* **2015**, *25*, 4596–4609. [CrossRef] [PubMed]
172. Belin, P.; Zatorre, R.J.; Lafaille, P.; Ahad, P.A.; Pike, B. Voice-selective areas in human auditory cortex. *Nature* **2000**, *403*, 309–312. [CrossRef] [PubMed]
173. Pitcher, D.; Dilks, D.D.; Saxe, R.R.; Triantafyllou, C.; Kanwisher, N. Differential selectivity for dynamic versus static information in face-selective cortical regions. *NeuroImage* **2011**, *56*, 2356–2363. [CrossRef]
174. Pelphrey, K.A.; Morris, J.P.; Michelich, C.R.; Allison, T.; McCarthy, G. Functional Anatomy of Biological Motion Perception in Posterior Temporal Cortex: An fMRI Study of Eye, Mouth and Hand Movements. *Cereb. Cortex* **2005**, *15*, 1866–1876. [CrossRef] [PubMed]
175. Taylor, K.I.; Moss, H.E.; Stamatakis, E.A.; Tyler, L.K. Binding crossmodal object features in perirhinal cortex. *Proc. Natl. Acad. Sci. USA* **2006**, *103*, 8239–8244. [CrossRef] [PubMed]
176. Ciaramidaro, A.; Adenzato, M.; Enrici, I.; Erk, S.; Pia, L.; Bara, B.; Walter, H. The intentional network: How the brain reads varieties of intentions. *Neuropsychologia* **2007**, *45*, 3105–3113. [CrossRef]
177. Wyk, B.C.V.; Hudac, C.M.; Carter, E.J.; Sobel, D.M.; Pelphrey, K.A. Action Understanding in the Superior Temporal Sulcus Region. *Psychol. Sci.* **2009**, *20*, 771–777. [CrossRef] [PubMed]
178. Bodin, C.; Takerkart, S.; Belin, P.; Coulon, O. Anatomo-functional correspondence in the superior temporal sulcus. *Anat. Embryol.* **2018**, *223*, 221–232. [CrossRef]
179. Moreno, A.; Limousin, F.; Dehaene, S.; Pallier, C. Brain correlates of constituent structure in sign language comprehension. *NeuroImage* **2018**, *167*, 151–161. [CrossRef]
180. Bonte, M.; Frost, M.A.; Rutten, S.; Ley, A.; Formisano, E.; Goebel, R. Development from childhood to adulthood increases morphological and functional inter-individual variability in the right superior temporal cortex. *NeuroImage* **2013**, *83*, 739–750. [CrossRef]
181. Brauer, J.; Neumann, J.; Friederici, A.D. Temporal dynamics of perisylvian activation during language processing in children and adults. *NeuroImage* **2008**, *41*, 1484–1492. [CrossRef]
182. Skeide, M.A.; Brauer, J.; Friederici, A.D. Syntax gradually segregates from semantics in the developing brain. *NeuroImage* **2014**, *100*, 106–111. [CrossRef]
183. Leroy, F.; Cai, Q.; Bogart, S.L.; Dubois, J.; Coulon, O.; Monzalvo, K.; Fischer, C.; Glasel, H.; Van der Haegen, L.; Bénézit, A.; et al. New human-specific brain landmark: The depth asymmetry of superior temporal sulcus. *Proc. Natl. Acad. Sci. USA* **2015**, *112*, 1208–1213. [CrossRef]
184. Leroy, F.; Glasel, H.; Dubois, J.; Hertz-Pannier, L.; Thirion, B.; Mangin, J.-F.; Dehaene-Lambertz, G. Early Maturation of the Linguistic Dorsal Pathway in Human Infants. *J. Neurosci.* **2011**, *31*, 1500–1506. [CrossRef] [PubMed]
185. Le Guen, Y.; Leroy, F.; Auzias, G.; Riviere, D.; Grigis, A.; Mangin, J.-F.; Coulon, O.; Dehaene-Lambertz, G.; Frouin, V. The chaotic morphology of the left superior temporal sulcus is genetically constrained. *NeuroImage* **2018**, *174*, 297–307. [CrossRef] [PubMed]
186. Sliwa, J.; Freiwald, W.A. A dedicated network for social interaction processing in the primate brain. *Science* **2017**, *356*, 745–749. [CrossRef] [PubMed]
187. Marciniak, K.; Atabaki, A.; Dicke, P.W.; Thier, P. Disparate substrates for head gaze following and face perception in the monkey superior temporal sulcus. *eLife* **2014**, *3*, e03222. [CrossRef] [PubMed]

188. Fisher, C.; Freiwald, W.A. Contrasting Specializations for Facial Motion within the Macaque Face-Processing System. *Curr. Biol.* **2015**, *25*, 261–266. [CrossRef] [PubMed]
189. Roumazeilles, L.; Eichert, N.; Bryant, K.L.; Folloni, D.; Sallet, J.; Vijayakumar, S.; Foxley, S.; Tendler, B.C.; Jbabdi, S.; Reveley, C.; et al. Longitudinal connections and the organization of the temporal cortex in macaques, great apes, and humans. *PLoS Biol.* **2020**, *18*, e3000810. [CrossRef] [PubMed]
190. Sallet, J.; Mars, R.B.; Noonan, M.P.; Andersson, J.L.; O'Reilly, J.X.; Jbabdi, S.; Croxson, P.L.; Jenkinson, M.; Miller, K.L.; Rushworth, M.F.S. Social Network Size Affects Neural Circuits in Macaques. *Science* **2011**, *334*, 697–700. [CrossRef]
191. Ghazanfar, A.A.; Chandrasekaran, C.; Logothetis, N.K. Interactions between the Superior Temporal Sulcus and Auditory Cortex Mediate Dynamic Face/Voice Integration in Rhesus Monkeys. *J. Neurosci.* **2008**, *28*, 4457–4469. [CrossRef]
192. Petkov, C.I.; Kayser, C.; Steudel, T.; Whittingstall, K.; Augath, M.; Logothetis, N.K. A voice region in the monkey brain. *Nat. Neurosci.* **2008**, *11*, 367–374. [CrossRef]
193. Belin, P.; Bodin, C.; Aglieri, V. A "voice patch" system in the primate brain for processing vocal information? *Hear. Res.* **2018**, *366*, 65–74. [CrossRef]
194. Bodin, C.; Trapeau, R.; Nazarian, B.; Sein, J.; Degiovanni, X.; Baurberg, J.; Rapha, E.; Renaud, L.; Giordano, B.L.; Belin, P. Functionally homologous representation of vocalizations in the auditory cortex of humans and macaques. *Curr. Biol.* **2021**, *31*, 4839–4844.e4. [CrossRef] [PubMed]
195. Khandhadia, A.P.; Murphy, A.P.; Romanski, L.M.; Bizley, J.K.; Leopold, D.A. Audiovisual integration in macaque face patch neurons. *Curr. Biol.* **2021**, *31*, 1826–1835.e3. [CrossRef] [PubMed]
196. Meguerditchian, A. "Human-specific" brain lateralization landmarks found in monkeys and their socio-cognitive correlates in both adults and infants [Papio anubis]. In Proceedings of the Communication at NeuroFrance, Strasbourg, France, 19–21 May 2021.
197. Meguerditchian, A.; Marie, D.; Love, S.A.; Margiotoudi, K.; Bertello, A.; Lacoste, R.; Roth, M.; Nazarian, B.; Anton, J.-L.; Coulon, O. Human-Like Brain Specialization in Baboons: An in vivo Anatomical Mri Study of Language Areas Homologs in 96 Subjects. In *The Evolution of Language, Proceedings of the 11th International Conference [EVOLANG11], New Oleans, LA, USA, 20–24 March 2016*; Roberts, S.G., Cuskley, C., McCrohon, L., Barceló-Coblijn, L., Feher, O., Verhoef, T., Eds.; EvoLang Scientific Committee: New Orleans, Tulane, 2016. [CrossRef]
198. Vilberg, K.L.; Rugg, M.D. Memory retrieval and the parietal cortex: A review of evidence from a dual-process perspective. *Neuropsychologia* **2008**, *46*, 1787–1799. [CrossRef] [PubMed]
199. Jacquemot, C.; Scott, S.K. What is the relationship between phonological short-term memory and speech processing? *Trends Cogn. Sci.* **2006**, *10*, 480–486. [CrossRef] [PubMed]
200. Stout, D.; Chaminade, T. Stone tools, language and the brain in human evolution. *Philos. Trans. R. Soc. B Biol. Sci.* **2012**, *367*, 75–87. [CrossRef] [PubMed]
201. Hecht, E.E.; Gutman, D.A.; Khreisheh, N.; Taylor, S.V.; Kilner, J.; Faisal, A.A.; Bradley, B.A.; Chaminade, T.; Stout, D. Acquisition of Paleolithic toolmaking abilities involves structural remodeling to inferior frontoparietal regions. *Anat. Embryol.* **2015**, *220*, 2315–2331. [CrossRef]
202. Flechsig, P. Developmental [myelogenetic] localisation of the cerebral cortex in the human subject. *Lancet* **1901**, *2*, 1027–1029. [CrossRef]
203. Geschwind, N. Disconnexion syndromes in animals and man. *Brain* **1965**, *88*, 237. [CrossRef]
204. Catani, M.; Bambini, V. A model for Social Communication and Language Evolution and Development [SCALED]. *Curr. Opin. Neurobiol.* **2014**, *28*, 165–171. [CrossRef] [PubMed]
205. Lattner, S.; Meyer, M.E.; Friederici, A.D. Voice perception: Sex, pitch, and the right hemisphere. *Hum. Brain Mapp.* **2005**, *24*, 11–20. [CrossRef]
206. Vanduffel, W.; Fize, D.; Peuskens, H.; Denys, K.; Sunaert, S.; Todd, J.T.; Orban, G.A. Extracting 3D from Motion: Differences in Human and Monkey Intraparietal Cortex. *Science* **2002**, *298*, 413–415. [CrossRef] [PubMed]
207. Grefkes, C.; Fink, G.R. REVIEW: The functional organization of the intraparietal sulcus in humans and monkeys. *J. Anat.* **2005**, *207*, 3–17. [CrossRef]
208. Orban, G.A.; Claeys, K.; Nelissen, K.; Smans, R.; Sunaert, S.; Todd, J.T.; Wardak, C.; Durand, J.-B.; Vanduffel, W. Mapping the parietal cortex of human and non-human primates. *Neuropsychologia* **2006**, *44*, 2647–2667. [CrossRef]
209. Budisavljevic, S.; Castiello, U.; Begliomini, C. Handedness and White Matter Networks. *Neuroscientist* **2021**, *27*, 88–103. [CrossRef] [PubMed]
210. Arbib, M.A. From monkey-like action recognition to human language: An evolutionary framework for neurolinguistics. *Behav. Brain Sci.* **2005**, *28*, 105–124. [CrossRef] [PubMed]
211. Rizzolatti, G.; Sinigaglia, C. The functional role of the parieto-frontal mirror circuit: Interpretations and misinterpretations. *Nat. Rev. Neurosci.* **2010**, *11*, 264–274. [CrossRef] [PubMed]
212. Eickhoff, S.B.; Schleicher, A.; Zilles, K.; Amunts, K. The Human Parietal Operculum. I. Cytoarchitectonic Mapping of Subdivisions. *Cereb. Cortex* **2006**, *16*, 254–267. [CrossRef]
213. Habib, M.; Robichon, F.; Levrier, O.; Khalil, R.; Salamon, G. Diverging Asymmetries of Temporo-parietal Cortical Areas: A Reappraisal of Geschwind/Galaburda Theory. *Brain Lang.* **1995**, *48*, 238–258. [CrossRef] [PubMed]

214. Johnson-Frey, S.H.; Newman-Norlund, R.; Grafton, S.T. A Distributed Left Hemisphere Network Active during Planning of Everyday Tool Use Skills. *Cereb. Cortex* **2005**, *15*, 681–695. [CrossRef] [PubMed]
215. Lewis, J.W. Cortical Networks Related to Human Use of Tools. *Neurosci* **2006**, *12*, 211–231. [CrossRef]
216. Cheng, L.; Zhang, Y.; Li, G.; Wang, J.; Hopkins, W.D.; Sherwood, C.C.; Gong, G.; Fan, L.; Jiang, T. Divergent Connectional Asymmetries of the Inferior Parietal Lobule Shape Hemispheric Specialization in Humans, Chimpanzees, and Macaque Monkeys. *Neuroscience* **2021**. [CrossRef]
217. Anderson, J.R.; Gallup, G.G. Mirror self-recognition: A review and critique of attempts to promote and engineer self-recognition in primates. *Primates* **2015**, *56*, 317–326. [CrossRef] [PubMed]
218. Hecht, E.; Mahovetz, L.M.; Preuss, T.M.; Hopkins, W.D. A neuroanatomical predictor of mirror self-recognition in chimpanzees. *Soc. Cogn. Affect. Neurosci.* **2017**, *12*, 37–48. [CrossRef]
219. Stout, D.; Hecht, E.; Khreisheh, N.; Bradley, B.; Chaminade, T. Cognitive Demands of Lower Paleolithic Toolmaking. *PLoS ONE* **2015**, *10*, e0121804. [CrossRef]
220. Gannon, P.J.; Kheck, N.M.; Braun, A.R.; Holloway, R.L. Planum parietale of chimpanzees and orangutans: A comparative resonance of human-like planum temporale asymmetry. *Anat. Rec. Part A* **2005**, *287*, 1128–1141. [CrossRef] [PubMed]
221. Taglialatela, J.P.; Dadda, M.; Hopkins, W.D. Sex differences in asymmetry of the planum parietale in chimpanzees [Pan troglodytes]. *Behav. Brain Res.* **2007**, *184*, 185–191. [CrossRef] [PubMed]
222. Penfield, W.; Boldrey, E. somatic motor and sensory representation in the cerebral cortex of man as studied by electrical stimulation. *Brain* **1937**, *60*, 389–443. [CrossRef]
223. Yousry, T. Localization of the motor hand area to a knob on the precentral gyrus. *A New Landmark. Brain* **1997**, *120*, 141–157. [CrossRef] [PubMed]
224. Amunts, K.; Schlaugb, G.; Schleichera, A.; Steinmetzb, H.; Dabringhausa, A.; Roland, P.E.; Zilles, K. Asymmetry in the Human Motor Cortex and Handedness. *NeuroImage* **1996**, *4*, 216–222. [CrossRef]
225. Cykowski, M.D.; Coulon, O.; Kochunov, P.V.; Amunts, K.; Lancaster, J.L.; Laird, A.R.; Glahn, D.C.; Fox, P.T. The Central Sulcus: An Observer-Independent Characterization of Sulcal Landmarks and Depth Asymmetry. *Cereb. Cortex* **2008**, *18*, 1999–2009. [CrossRef] [PubMed]
226. Sun, Z.Y.; Klöppel, S.; Rivière, D.; Perrot, M.; Frackowiak, R.; Siebner, H.; Mangin, J.-F. The effect of handedness on the shape of the central sulcus. *NeuroImage* **2012**, *60*, 332–339. [CrossRef] [PubMed]
227. Häberling, I.S.; Corballis, P.M.; Corballis, M.C. Language, gesture, and handedness: Evidence for independent lateralized networks. *Cortex* **2016**, *82*, 72–85. [CrossRef]
228. Crow, T. Directional asymmetry is the key to the origin of modern Homo sapiens [the Broca-Annett axiom]: A reply to Rogers' review of The Speciation of Modern Homo Sapiens. *Laterality Asymmetries Body Brain Cogn.* **2004**, *9*, 233–242. [CrossRef]
229. Warren, J.M. Handedness and laterality in humans and other animals. *Physiol. Psychol.* **1980**, *8*, 351–359. [CrossRef]
230. Hobaiter, C.; Byrne, R.W. Laterality in the gestural communication of wild chimpanzees. *Ann. N. Y. Acad. Sci.* **2013**, *1288*, 9–16. [CrossRef]
231. Hopkins, W.D.; Cantalupo, C. Handedness in Chimpanzees [Pan troglodytes] Is Associated with Asymmetries of the Primary Motor Cortex but Not With Homologous Language Areas. *Behav. Neurosci.* **2004**, *118*, 1176–1183. [CrossRef] [PubMed]
232. Dadda, M.; Cantalupo, C.; Hopkins, W.D. Further evidence of an association between handedness and neuroanatomical asymmetries in the primary motor cortex of chimpanzees [Pan troglodytes]. *Neuropsychologia* **2006**, *44*, 2582–2586. [CrossRef] [PubMed]
233. Margiotoudi, K.; Marie, D.; Claidière, N.; Coulon, O.; Roth, M.; Nazarian, B.; Lacoste, R.; Hopkins, W.D.; Molesti, S.; Fresnais, P.; et al. Handedness in monkeys reflects hemispheric specialization within the central sulcus. An in vivo MRI study in right- and left-handed olive baboons. *Cortex* **2019**, *118*, 203–211. [CrossRef] [PubMed]
234. Phillips, K.A.; Sherwood, C.C. Primary motor cortex asymmetry is correlated with handedness in capuchin monkeys [cebus apella]. *Behav. Neurosci.* **2005**, *119*, 1701–1704. [CrossRef]
235. Nudo, R.; Jenkins, W.; Merzenich, M.; Prejean, T.; Grenda, R. Neurophysiological correlates of hand preference in primary motor cortex of adult squirrel monkeys. *J. Neurosci.* **1992**, *12*, 2918–2947. [CrossRef]
236. Bouziane, S.; Loh, K.K.; Becker, Y.; Brunschvig, S.; Picchiottino, A.; Sein, J.; Coulon, O.; Velly, L.; Renaud, L.; Meguerditchian, A. Early structural asymmetry in the central sulcus is associated with handedness in infant baboons. In Proceedings of the OHBM 2021: 27th Annual Meeting of the Organization for Human Brain Mapping, Monday, Virtual, 21–25 June 2021.

MDPI

Article

Brain Size Associated with Foot Preferences in Australian Parrots

Gisela Kaplan * and Lesley J. Rogers *

School of Science and Technology, University of New England, Armidale, NSW 2351, Australia
* Correspondence: gkaplan@une.edu.au (G.K.); lrogers@une.edu.au (L.J.R.)

Abstract: Since foot preference of cockatoos and parrots to hold and manipulate food and other objects has been associated with better ability to perform certain tasks, we predicted that either strength or direction of foot preference would correlate with brain size. Our study of 25 psittacine species of Australia found that species with larger absolute brain mass have stronger foot preferences and that percent left-footedness is correlated positively with brain mass. In a sub-sample of 11 species, we found an association between foot preference and size of the nidopallial region of the telencephalon, an area equivalent to the mammalian cortex and including regions with executive function and other higher-level functions. Our analysis showed that percent left-foot use correlates positively and significantly with size of the nidopallium relative to the whole brain, but not with the relative size of the optic tecta. Psittacine species with stronger left-foot preferences have larger brains, with the nidopallium making up a greater proportion of those brains. Our results are the first to show an association between brain size and asymmetrical limb use by parrots and cockatoos. Our results support the hypothesis that limb preference enhances brain capacity and higher (nidopallial) functioning.

Keywords: parrots; footedness; brain mass; body mass; nidopallium; optic tectum; optic tecta; Wulst; lateral asymmetry

Citation: Kaplan, G.; Rogers, L.J. Brain Size Associated with Foot Preferences in Australian Parrots. *Symmetry* **2021**, *13*, 867. https://doi.org/10.3390/sym13050867

Academic Editors: Sebastian Ocklenburg and Onur Güntürkün

Received: 14 April 2021
Accepted: 10 May 2021
Published: 12 May 2021

1. Introduction

Hand and foot preferences (footedness) have often been used as proxy measures of brain lateralization although there is little evidence that these preferences correlate with structural differences in the brain [1]. In humans, however, non-right handedness has been associated with particular dysfunctional conditions, although not consistently [2–5]. In non-human species, absence of hand or limb preference has been considered to indicate weak or absent asymmetry of the brain [6,7]. However, although hand-preference, or limb-preference, reflects which hemisphere is in control of motor behaviour [8], its presence or absence cannot not necessarily be used as a measure of the strength or direction of asymmetry in the brain itself, either at the individual level or the population level. For example, in any group of marmosets, approximately half the individuals have a left-hand preference and the other half a right-hand preference [9]. Nevertheless, almost all individuals have the same eye preference for viewing particular stimuli [10]. Furthermore, left-handed marmosets have a negative cognitive bias, whereas right-handed marmosets have a positive cognitive bias [11]: a result explained by specialised processing of the hemisphere contralateral to the preferred hand. In marmosets, hand preference is an individual characteristic, whereas eye preference has a population bias. Even when no consistent limb preference is present, a species may still have population asymmetry for processing sensory information in the brain.

Some researchers adhere to the hypothesis that hand preference in humans is stronger than any hand or limb preference in non-human species [12]. By extrapolation, it has also been argued that brain lateralization is stronger in humans, thereby adhering to a view of a significant discontinuity of brain function between humans and other animals [13,14] but see [15]. Although there is some support for this idea when only the evidence for hand preference in primates is considered, that too has been contested [14,16]. In fact, we now know that limb preferences present as population-level asymmetries are quite common in vertebrate species [12]. They occur in some amphibians [17,18], and footedness has been reported for several avian orders (wildfowl and waders [19], yellow-bellied tits, *Pardaliparus venustulus*, [20] and many species of parrots [21,22]). Cockatoos also display foot preferences [21,23–25] and, in some species, foot preference is as strong as hand preference in humans [21]. Furthermore, the well-studied laterality of a broad range of perceptual functions in chickens and pigeons is as strong as laterality in humans [26–29].

Having a lateralized brain has been shown to increase the processing capacity of the brain, to permit complementary and parallel processing of sensory information and to improve motor control [27]. Consistent with this, psittacine species that display foot preferences have better ability to perform certain tasks than species with weak on no foot preference [30,31]. This raises questions about potential associations between brain size and foot preference in different psittacine species. Do species with footedness have larger brains, or is footedness a way of compensating for having a smaller brain?

In this paper we are concerned with species-level population lateralization of foot use in a range of species of Australian parrots and cockatoos. Psittaciformes are usually subdivided into three superfamilies Cacatuoidea (cockatoos), Psittacoidea (true parrots) and Strigopoidea [32]. Species in Strigopoidea were not included here because they are extant New Zealand parrots. Worldwide, there are about 375 species of parrots, of which about 56 species (depending on taxonomical consideration of counting some birds as subspecies or separate species) are native to Australia. Modern extant parrots and songbirds are of particular interest for evolutionary reasons [33]. Both orders have their origin in East-Gondwana, now Australia [34]. While radiations and departures from the continent eventually occurred, the evolution of the two superfamilies from ancestral surviving lineages has been uninterrupted to this day, despite the mass extinction events of 65 million years ago [33,35].

Large-brained psittacines, as cockatoos are, have a unique cerebrotype compared to large-brained songbirds: they have a relatively larger subpallium within the telencephalon, containing more telencephalic neurons [36]. The subpallium is responsible for neural regulation of feeding, reproduction, voluntary movement, and agonistic and stress behaviours. It is also associated with reward, memory and learning [37].

Parrots are an anomaly amongst avian species in that their brains are lavishly equipped with nuclei for vocalizations, having the same seven nuclei of the song control system as songbirds, although they are not songbirds. Parrots are well-known for their extraordinary ability to mimic and to retain the memory of a large number of sounds that are not species-specific. Why this capacity has developed is not clear because it has so far not been confirmed that parrots use mimicry in the wild on a regular or even just occasional basis. Even more puzzling, from a functional point of view, is, as Chakraborty and colleagues [38,39] discovered, that parrots have a core and shell song systems, i.e., an additional set of nuclei not present in songbirds. It seems to be a structure unique to the parrot brain but its function is still not entirely clear. We now know, however, that within the song control system the magnocellular nucleus of the medial striatum (MMSt) is a prime target for somato-motor outputs from the hyperpallium apicale of the rostral Wulst, the avian equivalent of the mammalian motor cortex [40]. This projection may be significant in parrots as it potentially mediates control of the body and limbs along with vocalization during elaborate, ritualized visual displays [41]. One notes that the song nuclei are largely located in the nidopallium, the part of the forebrain that is involved in cognition.

We report associations between foot preference (footedness) in species of Australian parrots and brain size, measured as whole brain mass and as whole brain mass relative to body mass, and between footedness and the size of two regions of the brain, the nidopallium, including the primary visual centre (entopallium), and the optic tectum. The avian nidopallium (see Figure 1), an analogue to the mammalian cortex, is an important area of the cortical telencephalon of the avian forebrain. Some of its sub-regions, such as the caudal nidopallium, the nidopallium caudocentral (NCC), caudomedial (NCM) and caudolateral (NCL), are considered vital for many complex, higher order cognitive functions in birds [42–44]. Indeed, the NCL is the seat of executive function, functionally equivalent to the prefrontal cortex of mammals [44,45] and, amongst other connections, it is reciprocally connected to sensory areas in all modalities and to basal ganglia and premotor areas [46]. It plays a key role in cognitive control of a number of functions, including roles in reward systems [47] and choice behaviour [48]. Although no studies of NCL have yet been conducted using parrots, this region of the nidopallium is almost certainly involved in feeding using the feet.

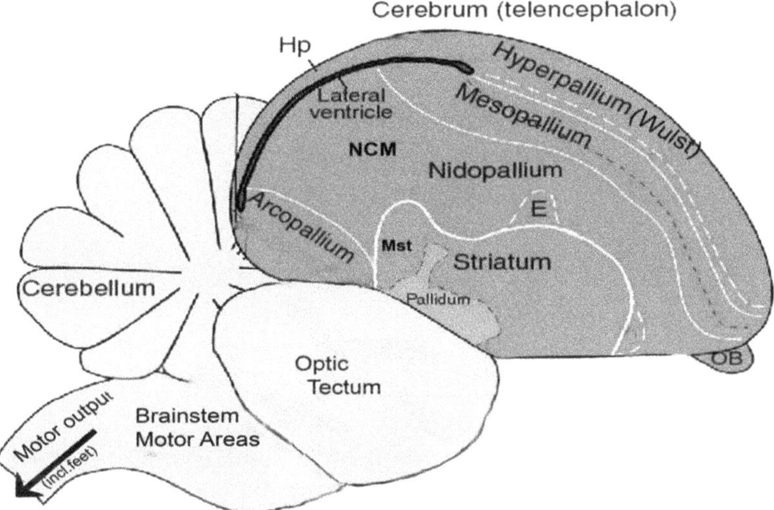

Figure 1. Brain regions of the parrot brain. Our text focusses on the Nidopallium and Optic tectum, as marked. Note that the optic tectum is located laterally on each side of the brain and in this Figure it obscures the midbrain and most of the hypothalamus. The darker section indicates the forebrain (telencephalon). Abbreviations: Hp = Hippocampus, NCM = Caudiomedial nidopallium; MSt = medial striatum; E = Entopallium; OB = Olfactory bulb. Adapted from Chakraborty and Jarvis [39], Kuenzel et al. [49] and Reiner et al. [50]. The Nidopallium includes NCM, E and Arcoplallium, as well as other regions discussed in the text.

The optic tectum is part of the main visual input system of birds and must be involved in foot/eye coordination. However, Niederleitner et al. [51] discovered a relay nucleus between the inferior colliculus and the optic tectum in the chicken, providing a solid basis for demonstrating visual–auditory integration.

We were interested in testing whether limb preference might be associated with the size of these of the optic tectum and the nidopallium because eye preferences and foot preferences are linked, as Brown and Magat showed in 16 species of Australian parrots [52].

2. Materials and Methods

2.1. Subjects

Psittaciformes are subdivided into the superfamilies Strigopoidea (New Zealand parrots), Cacatuoidea (cockatoos) and Psittacoidea (true parrots). Species in our samples, which did not include the New Zealand parrots, belonged to the two superfamilies of cockatoos (Cacatuoidea) and true parrots (Psittacoidea). Those examined here are exclusively native Australian species, excluding closely related cockatoo species endemic to islands north of Australia (such as New Guinea, including the Bismarck Archipelago, the Solomon

Islands, or Tanimbar Islands Archipelago (Indonesia). The subjects used were those for which published data on brain mass, body mass and foot preference are available (see next section) but for which no previous study exists that has tested a potential association between these variables.

2.2. Source of Data

Data on body mass, brain mass and volumes of brain regions were obtained from Franklin et al. [53] and Iwaniuk et al. [54]. We compared published data on foot preferences (see below) with published data on brain mass, relative brain mass and the size of two regions of the forebrain, the nidopallium and the optic tectum.

Foot preference in 25 species of Australian parrots was obtained from three publications. Data for nine species were obtained from a previous study by the co-author of this paper, Rogers [21]. Scores for the budgerigar came from Rogers and Workman [55] and for another 15 species from Brown and Magat [22] (see Table 1). We used data for two measures of footedness: (1) percent use of the left foot and (2) strength of foot preference, regardless of whether the left or right foot is used.

Table 1. Psittacine species used. No = number of individuals scored, % Left = (Left/Left + Right) ×100, and the ratio of brain mass/body mass × 1000.

Common Name	Latin Species Name	No	% Left	Strength	Brain Mass /Body Mass × 1000
Galah	*Eolophus roseicapella*	58	89	39	21.3509
Sulphur-crested cockatoo	*Cacatua galerita*	98	87	37	20.2646
Little corella	*Cacatua sanguinea*	14	93	43	19.5567
Long-billed corella	*Cacatua tenuirostris*	17	89	39	19.5098
Pink cockatoo	*Cacatua leadbeateri*	24	100	50	22.6742
Yellow-tailed black cockatoo	*Calyptorhynchus funereus*	7	100	50	23.6710
Gang-gang cockatoo	*Callocephalon fimbriatum*	38	100	50	30.9971
Yellow rosella	*Platycercus f. flaveolus*	6	100	50	30.6296
Crimson rosella	*Platycercus elegans*	17	23	27	31.5813
Budgerigar	*Melopsittacus undulatus*	9	51	1	59.7692
Cockatiel	*Nymphicus hollandicus*	20	90	40	28.0889
Red-tailed black cockatoo	*Calyptorhynchus banksii*	20	93	43	18.4779
King parrot	*Alisterus scapularis*	20	8.5	41.5	22.4216
Palm cockatoo	*Probosciger aterrimus*	5	80	30	26.3499
Eclectus parrot	*Eclectus roratus*	20	26	24	15.0394
Turquoise parrot	*Neophema pulchella*	10	45	5	33.4110
Red-winged parrot	*Aprosmictus erythropterus*	10	10	40	25.7329
Australian ringneck	*Barnardius zonarius*	5	20	30	28.6071
Red-capped parrot	*Purpureicephalus spurius*	5	72	22	31.4403
Superb parrot	*Plytelis swainsonii*	20	27.5	22.5	23.9352
Red-rumped parrot	*Psephotus haematonotus*	20	72	22	32.6547
Little lorikeet	*Glossopsitta pusilla*	15	49.3	0.7	39.6925
Varied lorikeet	*Psitteuleles versicolor*	5	48	2	38.6567
Rainbow lorikeet	*Trichoglossus moluccanus*	20	46	4	30.3341
Bourke's parrot	*Neopsephotus bourkii*	20	49.5	0.5	28.7778

The number of individuals scored varied considerably between species (see Table 1) due to differences in availability of birds to test. Both caged and wild birds were tested and, for all species apart from the budgerigar, data were collected from multiple locations in order to make the scores representative of the species.

The behavioural score was percent left-foot use to hold food. Brown and Magat [22] determined foot preferences by scoring the foot used to grasp food items, with 10 trials per individual bird. Rogers [21] scored the foot used to hold food while eating (Figure 2), the number of scores per individual varying from 1 to 6. Since budgerigars rarely hold food in a foot, in this species preferred foot was determined by placing a small piece of adhesive tape on the dorsal surface of the beak and then scoring the foot used in attempts to remove the tape (10 scores per bird) [55].

Figure 2. Two left-footed cockatoos. **Left**: female red-tailed black cockatoo (*Calyptorhynchus banksia*). **Right**: male sulphur-crested cockatoo (*Cacatua galerita*). Note that the left foot grasps the food item and the entire leg is lifted to the beak (Photo credits: **left**: B. Machini, **right**: G. Kaplan).

The formula (L/L + R) × 100 was used to determine % Left, where L refers to the number of times the left foot was used and R to the number of times the right foot was used. At least five subjects per species were assessed (Table 1). The scores determined were mean percentages for each species; hence, they represented the %L for each species, considered as a group or population.

Strength of foot preference was determined as the difference between the scored % Left and 50% (no preference). Hence, it was an absolute score, not taking into account the direction of the foot preference. These scores ranged from 0% to 50%.

Out of our main group of species, we selected for more detailed examination 11 species for which the volume of various brain regions, relative to the volume of the whole brain, had been determined using histological sections and Nissl staining by Iwaniuk and Hurd [56] (see Table 2). We selected to compare % Left and strength of foot preference with two brain regions: viz., the nidopallium (N) and the optic tectum. Data for these two regions were given as proportions of the total brain volume.

2.3. Statistical Tests

Pearson correlations (in Excel) were performed between % Left and brain mass relative to body mass (Brain mass/body mass × 1000) and, in the smaller group, between % Left and the volumes of the two brain regions relative to total brain volume. Where needed, due to multiple comparisons, Bonferroni corrections were applied. Correlations were also made using scores for the strength of footedness (absolute value of difference between score and no preference, 50%).

Table 2. A list of the species in the sub-group tested for correlation between % Left foot use and the volumes of the nidopallium and the optic tecta relative to the volume of the whole brain (i.e., the scores are proportions). The figures for the two brain regions are given as proportions of the whole brain, sourced from Iwaniuk and Hurd [56].

Common Name	Scientific Name	% Left	Nidopallium	Optic Tectum
Galah	*Eolophus roseicapella*	89	0.3618	0.0314
Yellow-tailed black cockatoo	*Calyptorhynchus funereus*	100	0.3887	0.0196
Cockatiel	*Nymphicus hollandicus*	90	0.3571	0.0350
Crimson rosella	*Platycerus elegans*	23	0.3401	0.0429
Budgerigar	*Melopsittacus undulatus*	53	0.3210	0.0514
Superb parrot	*Plytelis swainsonii*	27.5	0.3200	0.0556
Red-rumped parrot	*Psephotus haematonotus*	72	0.3540	0.0387
Rainbow lorikeet	*Trichoglossus haematodus*	46	0.3370	0.0339
Bourke's parrot	*Neopsephotus bourkii*	49.5	0.3586	0.0482
Australian king parrot	*Alisterus scapularis*	8.5	0.3098	0.0424
Eclectus parrot	*Eclectus roratus*	26	0.3534	0.0351

3. Results

3.1. Association between Foot Preference and Body Mass

First, we tested our data to see whether they showed results similar to those of Brown and Magat [22] and, consistent with that report, strength of foot preference correlated significantly with body mass (R (25) = 0.5339, $p = 0.0059$, Bonferroni correction $\alpha = 0.0166$) and strongly with $\log_{10}$ Body mass (R(25) = 0.7166, $p = 0.000056$). The larger the bird, the stronger the foot preference.

3.2. Association between Foot Preference and Brain Mass

Strength of foot preference correlated significantly with brain mass (R (25) = 0.5259, $p = 0.0069$, Bonferroni correction $\alpha = 0.0166$) and with $\log_{10}$ Brain mass (R (25) = 0.7166, $p = 0.00005$; Figure 3A). Species with larger brains have stronger foot preferences. There was also a significant association between % Left and Brain mass (R (25) = 0.4857, $p = 0.01384$) and $\log_{10}$ Brain mass (R (25) = 0.4525, $p = 0.02111$). Hence, left-footedness is stronger in species with a larger brain.

3.3. Association between Foot Preference and Brain Mass Relative to Body Mass

There was no significant correlation between % Left and brain mass/body mass (R(25) = −0.1655, $p = 0.4291$). However, the correlation between strength of foot preference and brain mass/body mass was significant (R (25) = −0.6032, $p = 0.0014$, Bonferroni correction $\alpha = 0.0166$: Figure 3B). The larger the size (or mass) of the brain relative to body size (or mass), the weaker the foot preference. This result comes about because body mass increases across species at a greater rate than does brain weight. Hence, species with larger bodies, and larger brains per se, but not relative to body mass, have stronger foot preferences.

3.4. Association between Foot Preference and Brain Mass in the Smaller Sample

First, this subset of 11 species was tested for Pearson correlation between the strength of foot preference and brain mass. As for the larger sample, this set showed a significant positive correlation (R (11) = 0.6410, $p = 0.0335$). This result shows that the subset was representative of the larger sample.

A Pearson correlation was applied to % Left scores versus relative volume of the nidopallial region and a positive association was found (R (11) = 0.7674, $p = 0.0058$, Bonferroni corrected α = 0.025; Figure 4A). Hence, the larger the volume of the nidopallium compared to the whole brain, the stronger the left foot preference.

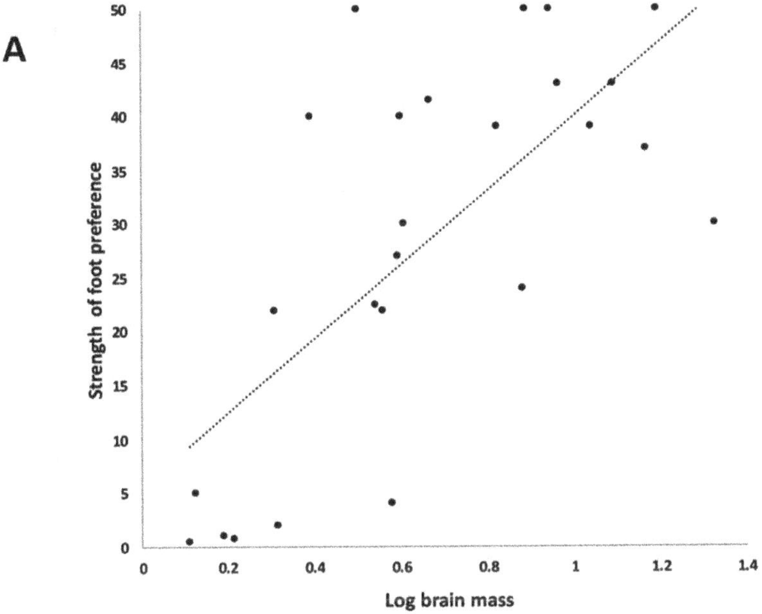

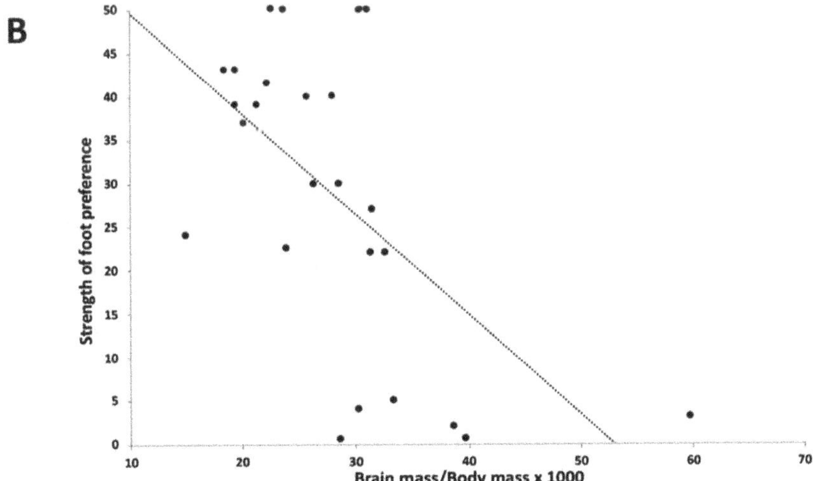

Figure 3. Strength of foot preference for the larger sample (n = 25) plotted against (**A**). Log_{10} of Brain mass, and (**B**). Brain mass/Body mass × 1000. The correlation between Strength of foot preference and brain mass is significant and positive (see text) and between strength of foot preference and Brain mass/Body mass is negative and significant (see text).

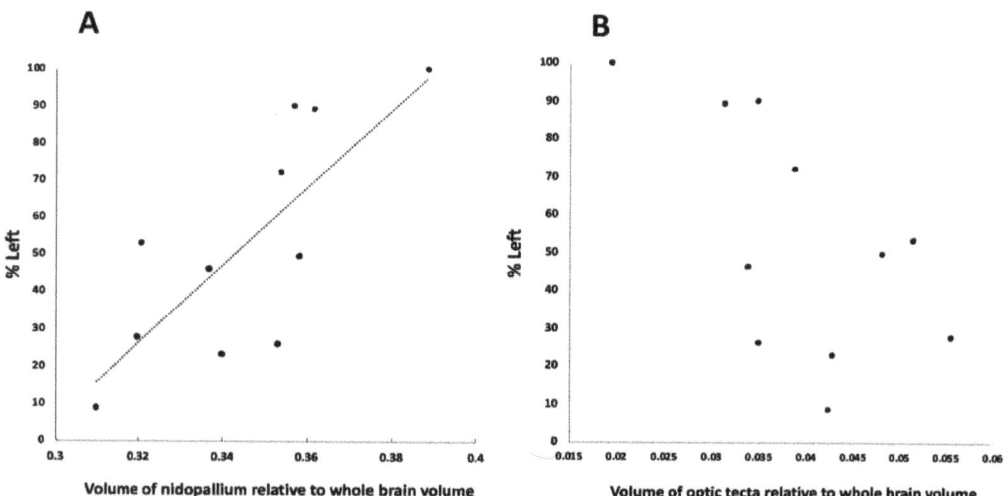

Figure 4. Percent left foot preference plotted against volume of the nidopallium relative to volume of the whole brain (**A**), and volume of the optic tecta relative to the volume of the whole brain (**B**). The positive correlation in A is significant: the nidopallium is larger, relative to the whole brain, in left footed parrots, and it is smaller in right-footed parrots. There is a trend for the opposite association between foot preference and size of the optic tecta but it is not significant after Bonferroni correction (see text).

A negative correlation was found between % Left and the relative volume of the optic tecta but it failed to be significant once the Bonferroni correction was applied ($R(11) = -0.6129$, $p = 0.0449$, Bonferroni correction $\alpha = 0.025$; Figure 4B).

Strength of foot preference did not correlate significantly with the volumes of either the nidopallium or optic tecta regions (nidopallium, $R(11) = 0.3393$, $p = 0.3072$; optic tecta, $R(8) = -0.5908$, $p = 0.0556$, Bonferroni correction $\alpha = 0.025$).

4. Discussion

First, we analysed the data to see whether they supported the finding of Brown and Margat [22] that, in species of Australian parrots, strength of foot preference correlates positively with body length, the latter being taken to indicate body size. As we were able to examine strength of foot preference versus body mass, this gave a better indication of the relationship between strength of foot preference and body size. By doing so, we found a significant positive correlation, which supports the findings of Magat and Brown [22]: the larger the species of parrot, the stronger is the foot preference. Magat and Brown [22] provided some evidence that this relationship may be due to larger parrots feeding on larger seeds. Such feeding would require more eye-foot manipulation. In fact, larger species often hold a whole seed pod in one foot and manipulate it carefully in order to extract the seeds (see Figure 1 in [57], p. 27, showing a yellow-tailed black cockatoo feeding on a large Banksia pod).

We also found a positive correlation between strength of foot preference and brain mass (Figure 3). The larger the brain size, the stronger the preference to hold food or objects in a preferred foot. In turn, this reflects control by the hemisphere opposite the preferred foot and use of the specific specialised functions of that hemisphere. However, the association between strength of foot-preference and brain mass relative to body mass is negative (Figure 3). To summarise the results so far, although larger birds have stronger foot preferences and larger brains, they have smaller brain mass relative to body mass: this means that, across species, the body size increases at a greater rate than does brain size.

These measures may also correlate with other factors; for example, with genome size and, hence, neuronal complexity of the brain. In fact, Andrews and Gregory [58] found that, in 54 species of parrots and cockatoos, genome size correlated negatively with relative brain size. They attributed this to species differences in neuronal complexity: the larger the genome, the more complex are the neural connections in the brain [58]. It would now be worth investigating whether a similar association between brain size and genome size holds for the species we tested. Since we have found a negative correlation between strength of footedness and relative brain mass, strength of foot preference may correlate positively with genome size, and hence, complexity of brain structure. This is merely a thought for future research. We are aware that whole brain size is a rough measure to associate with behaviour [59] but it is a starting point leading to investigation of more specific brain regions associated with particular patterns of behaviour.

Our findings also illustrate the difference between using absolute brain size compared to relative brain size when drawing links between brain and behaviour [60]. We suggest that absolute brain size (or mass) is a better measure to associate with foot preference, and possibly also with cognitive behaviour, than is brain mass relative to body mass. As Herculano-Houzel [61] emphasised, body mass is very variable and not tightly correlated with the number of neurones in the brain. Nevertheless, although it is preferrable not to use body mass as a measure related to behaviour or cognition, within Psittaciformes, Herculano-Houzel [61] did report a positive relationship between the number of neurones in the brain and body mass and, more specifically, between the number of neurones in the pallium and body mass. To extrapolate to our data, since foot preferences are stronger in larger parrots with larger brains, these features may go along with more neurones in the brain and with higher computational capacity or "cognitive power" [36].

The positive relationship between strength of foot preference and brain mass was also present in our subgroup of 11 species selected in order to examine correlations between foot preference and regional areas of the brain. In this representative subgroup, we also found that % Left correlated positively with volume of the nidopallium relative to whole brain volume (Figure 4). Parrots with proportionately larger nidopallial regions express stronger left-foot preferences, and hence use of the right hemisphere. Put simply, the nidopallium is larger in left-footed species than it is in right-footed species.

The nidopallial region measured by Iwaniuk and Hurd [56] included all of the subregions (nucleus basorostralis pallii, entopallium and arcopallium) as well as area temporo-parieto-occipitalis. The entopallium is a primary visual centre, receiving inputs from the retina, via the optic tectum and nucleus rotundus, and sending outputs to the arcopallium (see Figure 1), which in turn sends outputs to the brain stem and controls motor function (summarised in [28]). It is a visuo–motor system clearly involved in eye–foot co-ordination. It is not known whether these regions of the nidopallial complex differ in size between the hemispheres, and so may be associated with asymmetry of limb use, but research on pigeons has shown that asymmetry is present in the visual pathway sending inputs to the entopallium: the entopallium in the left hemisphere receives strong inputs from both eyes, whereas the same region in the right hemisphere receives inputs mainly from the left eye [28,62,63]. In left-footed birds, therefore, the right hemisphere is being used rather exclusively to carry out eye-foot co-ordination needed to hold and manipulate food items [52].

Recently, Morandi-Raikova et al. [64] reported asymmetry in entopallial neurones in domestic chicks: they found more parvalbumin-expressing neurones, most likely GABAergic inhibitory neurones, in the entopallium of right hemisphere than in the same region of the left hemisphere. This suggests that primary visual inputs are processed differently in the left and right hemispheres. Furthermore, as shown in the pigeon, there is asymmetry in the arcopallial regions, due to differences in left-to-right versus right-to-left exchange of visual information via the anterior commissure [65].

We found a trend for a negative correlation between % Left and the relative size of the optic tecta (Figure 4) but it was not significant in our sample, perhaps due to our small sample size. Nevertheless, the larger nidopallium, as we found in species with stronger foot preferences, is not matched by any difference between species in size of the optic tecta. It seems, therefore, that visual processing in the optic tecta may well be similar across species. Hence, there is regional specificity in size increase of the nidopallium related to foot preference. Our data provide an example, across species, of brain structures contributing differently to whole brain size, a consideration discussed in detail by Willemet [60].

It is not known whether evolution of the bird brain involved coordination between the telencephalon and non-telencephalic regions [54] but this seems highly unlikely. The latter is not necessarily related to cognition. From Striedter and Charvet's work [66], we have evidence that certain areas of the brain differ in size relative to the rest of the bird's brain in different species from embryonic stage onwards. In a morphometric comparison between budgerigars, *Melopsittacus undulatus*, and quails, *Colinus virginianus*, the researchers showed [66] that species differences in telencephalon and optic tectum size occur right from the beginning of development of the embryo: the optic tectum being much larger in the quail than in the budgerigar, whereas the telencephalon occupies more than 60% of the budgerigar brain and slightly more than 40% in the quail [66]. These differences reflect the cognitive differences between budgerigars and quails.

The evidence is already clear that parrots have brain/body ratios and encephalization quotients similar to those of primates [54]. However, in most research on the size of the telencephalon and sub-structures, there has been no consideration of differences between the left and the right hemispheres. Apart from a recent study of cichlid fish, showing associations between lateralized behaviour and brain structures, as well as gene expression [67], research on asymmetry in non-human species has focused on left–right differences in function rather than structure.

In a range of species, the left hemisphere has been shown to be responsible for top-down control (such as routine behaviour) and the right hemisphere for environmentally elicited behaviour such as response to threats and social partners [68]. Social play behaviour in parrots and other clades is correlated positively with brain size [69] but, for this behaviour, there has been no study of differences between the hemispheres. By contrast, social recognition is known to be largely right-hemisphere controlled [70,71] and, as Yamazaki at al. [72] concluded, cognition overall may be largely asymmetrical [73].

The significant results shown here concern the relationship between footedness and the size of the nidopallium, a region of the forebrain with executive and other higher-level functions. Our finding, coupled with the evidence of parrots' exceptional vocal and cognitive abilities, indicates that the nidopallium deserves more and detailed attention. Our finding of a significant positive correlation between % Left foot use and the size of the nidopallium further suggests that it would be worth looking in future for asymmetries in the size of the nidopallial regions of the hemispheres and, particularly, the areas that are included in that region, the entopallium, arcopallium and NCL. We hope that our findings might encourage future research on left–right differences, not only of gross structure, but also of neural processing [74], subcellular structures [64] and gene expression [67].

Author Contributions: Conceptualization, G.K. and L.J.R.; Data curation, G.K. and L.J.R.; Formal analysis, G.K. and L.J.R.; Investigation, G.K. and L.J.R.; Methodology, G.K. and L.J.R.; Project administration, G.K. and L.J.R.; Writing—original draft, G.K. and L.J.R.; Writing—review & editing, G.K. and L.J.R. All authors have read and agreed to the published version of the manuscript.

Funding: No funding was requested for this research.

Institutional Review Board Statement: Not applicable.

Informed Consent Statement: Not applicable.

Data Availability Statement: The data have been published previously in papers cited in the Methods.

Acknowledgments: We thank the authors of papers from which we could select data for analysis.

Conflicts of Interest: The authors declare no conflict of interest.

References

1. Ocklenburg, S.; Metzen, D.; Schlüter, C.; Fraenz, C.; Arning, L.; Streit, F.; Güntürkün, O.; Kumsta, R.; Genc, E. Polygenic scores for handedness and their association with asymmetries in brain structure. *Res. Sq.* **2021**. [CrossRef]
2. Dragovic, M.; Hammond, G. Handedness in schizophrenia: A quantitative review of evidence. *Acta Psychiatr. Scand.* **2005**, *111*, 410–419. [CrossRef]
3. Ocklenburg, S.; Güntürkün, O.; Hugdahl, K.; Hirnstein, M. Laterality and mental disorders in the postgenomic age—A closer look at schizophrenia and language lateralization. *Neurosci. Biobehav. Rev.* **2015**, *59*, 100–110. [CrossRef] [PubMed]
4. Packheiser, J.; Schmitz, J.; Stein, C.C.; Pfeifer, L.S.; Berretz, G.; Papadatou-Pastou, M.; Peterburs, J.; Ocklenburg, S. Handedness and depression: A meta-analysis across 87 studies. *PsyArXiv* **2021**. [CrossRef]
5. Sommer, I.E.C.; Aleman, A.; Ramsey, N.; Bouma, A.; Kahn, R. Handedness, language lateralisation and anatomical asymmetry in schizophrenia—Meta-analysis. *Br. J. Psychiatr.* **2001**, *178*, 344–351. [CrossRef]
6. Chance, S.A.; Crow, T.J. Distinctively human: Cerebral lateralisation and language in *Homo sapiens*. *J. Anthropol. Sci.* **2007**, *85*, 83–100.
7. McGrew, W.C.; Marchant, L.F. On the other hand: Current issues in and meta-analysis of the behavioral laterality of hand function in nonhuman primates. *Yearb. Phys. Anthropol.* **1997**, *40*, 201–232. [CrossRef]
8. Hopkins, W.D.; Cantalupo, C. Handedness in chimpanzees (*Pan troglodytes*) is associated with asymmetries of the primary motor cortex but not with homologous language areas. *Behav. Neurosci.* **2004**, *118*, 1176–1183. [CrossRef]
9. Hook, M.A.; Rogers, L.J. Visuospatial reaching preferences of common marmosets: An assessment of individual biases across a variety of tasks. *J. Comp. Psychol.* **2008**, *122*, 41–51. [CrossRef]
10. Hook-Costigan, M.A.; Rogers, L.J. Eye preferences in common marmosets (*Callithrix jacchus*): Influence of age, stimulus and hand preference. *Laterality* **1998**, *3*, 109–130. [CrossRef] [PubMed]
11. Gordon, D.J.; Rogers, L.J. Cognitive bias, hand preference and welfare in common marmosets. *Behav. Brain Res.* **2015**, *287*, 100–108. [CrossRef] [PubMed]
12. Ströckens, F.; Güntürkün, O.; Ocklenburg, S. Limb preferences in non-human vertebrates. *Laterality* **2013**, *18*, 536–575. [CrossRef]
13. Crow, T.J. Why cerebral asymmetry is the key to the origin of *Homo sapiens*: How to find the gene of eliminate the theory. *Curr. Psychol. Cogn.* **1998**, *17*, 1237–1277.
14. Corballis, M.C. Cerebral asymmetry and human uniqueness. In *The Evolution of Hemispheric Specialization in Primates*; Special Topics in Primatology; Hopkins, W.D., Ed.; Elsevier: Amsterdam, The Netherlands, 2007; Volume 5, pp. 1–21.
15. Corballis, M.C. Bilaterally symmetrical: To be or not to be? *Symmetry* **2020**, *12*, 326. [CrossRef]
16. Corballis, M.C. Humanity and the left hemisphere: The story of half a brain. *Laterality* **2020**, *26*. [CrossRef]
17. Malashichev, Y.B.; Wassersug, R.J. Left and right in the amphibian world: Which way to develop and where to turn? *BioEssays* **2004**, *26*, 512–522. [CrossRef] [PubMed]
18. Stancher, G.; Sovrano, V.A.; Vallortigara, G. Motor asymmetries in fishes, amphibians, and reptiles. *Proc. Brain Res.* **2018**, *238*, 33–56.
19. Randler, C. Foot preferences during resting in wildfowl and waders. *Laterality* **2007**, *12*, 191–197. [CrossRef]
20. Yu, G.; Guo, J.; Xie, W.; Wang, J.; Wu, Y.; Zhang, J.; Xu, J.; Li, J. Footedness predicts escape performance in a passerine bird. *Ecol. Evol.* **2020**, *10*, 4251–4260. [CrossRef] [PubMed]
21. Rogers, L.J. Lateralisation in the avian brain. *Bird Behav.* **1980**, *2*, 1–12. [CrossRef]
22. Brown, C.; Magat, M. The evolution of lateralized foot use in parrots: A phylogenetic approach. *Behav. Ecol.* **2011**, *22*, 1201–1208. [CrossRef]
23. Rogers, L.J. Laterality in animals. *Int. J. Comp. Psychol.* **1989**, *3*, 5–25.
24. Magrath, D.I. Footedness in the glossy black-cockatoo: Some observations and a review of the literature with a note on the husking of Allocasuarina cones by this species. *Corella* **1994**, *18*, 21–24.
25. Woodall, P.F.; Woodall, L.B. Little and long-billed Corellas feeding on hoop pine seeds, and their 'footedness'. *Sunbird J. Qld. Ornithol. Soc.* **2001**, *31*, 30–32.
26. Rogers, L.J.; Vallortigara, G.; Andrew, R.J. *Divided Brains: The Biology and Behaviour of Brain Asymmetries*; Cambridge University Press: Cambridge, UK, 2013.
27. Vallortigara, G.; Rogers, L.J. A function for the bicameral mind. *Cortex* **2020**, *124*, 274–285. [CrossRef] [PubMed]
28. Güntürkün, O.; Strökens, F.; Ocklenburg, S. Brain lateralization: A comparative perspective. *Physiol. Rev.* **2020**, *100*, 1019–1063. [CrossRef] [PubMed]
29. Ocklenburg, S.; Güntürkün, O. *The Lateralized Brain: The Neuroscience and Evolution of Hemispheric Asymmetries*; Academic Press: London, UK, 2018.
30. Magat, M.; Brown, C. Laterality enhances cognition in Australian parrots. *Proc. R. Soc. B* **2009**, *276*, 4155–4162. [CrossRef] [PubMed]
31. Cussen, V.A.; Mench, J.A. Performance on the Hamilton search task, and the influence of lateralization, in captive orange-winged Amazon parrots (*Amazona amazonica*). *Anim. Cogn.* **2014**, *17*, 901–909. [CrossRef] [PubMed]
32. Joseph, L.; Toon, A.; Schirtzinger, R.E.; Wright, T.F.; Schodde, R. A revised nomenclature and classification for family-group taxa of parrots (Psittaciformes). *Zootaxa* **2012**, *40*, 26–40. [CrossRef]

33. Barker, F.K.; Cibois, A.; Schikler, P.A.; Feinstein, J.; Cracraft, J. Phylogeny and diversification of the largest avian radiation. *Proc. Nat. Acad. Sci. USA* **2004**, *101*, 11040–11045. [CrossRef]
34. Edwards, S.V.; Bowles, W.E. Out of Gondwana: The origin of passerine birds. *Trends Ecol. Evol.* **2002**, *17*, 347–349. [CrossRef]
35. Wright, T.F.; Schirtzinger, E.E.; Matsumoto, T.; Eberhard, J.R.; Graves, G.R.; Sanchez, J.J.; Capelli, S.; Müller, H.; Scharpegge, J.; Chambers, G.K.; et al. A multilocus molecular phylogeny of the parrots (Psittaciformes): Support for a Gondwanan origin during the Cretaceous. *Mol. Biol. Evol.* **2008**, *25*, 2141–2156. [CrossRef] [PubMed]
36. Olkowicz, S.; Kocourek, M.; Lučan, R.K.; Porteš, M.; Fitch, W.T.; Herculano-Houzel, S.; Němec, P. Birds have primate-like numbers of neurons in the forebrain. *Proc. Natl. Acad. Sci. USA* **2016**, *113*, 7255–7260. [CrossRef]
37. Kuenzel, W.J. The Avian subpallium and autonomic nervous system. In *Sturkie's Avian Physiology*, 6th ed.; Elsevier: Amsterdam, The Netherlands, 2014; pp. 135–163.
38. Chakraborty, M.; Jarvis, E.D. Brain evolution by brain pathway duplication. *Philos. Trans. R. Soc. Lond. B* **2015**, *370*, 20150056. [CrossRef]
39. Chakraborty, M.; Walløe, S.; Nedergaard, S.; Fridel, E.E.; Dabelsteen, T.; Pakkenberg, B.; Bertelsen, M.F.; Dorrestein, G.M.; Brauth, S.E.; Durand, S.; et al. Core and shell song systems unique to the parrot brain. *PLoS ONE* **2015**, *10*, e0118496. [CrossRef]
40. Wild, J.M.; Williams, M.N. Rostral Wulst in passerine birds. 1. Origin, course, and terminations of an avian pyramidal tract. *J. Comp. Neurol.* **2000**, *416*, 429–450. [CrossRef]
41. Wild, J.M. Neural pathways for the control of birdsong production. *J. Neurobiol.* **1997**, *33*, 653–670. [CrossRef]
42. Güntürkün, O. The avian 'prefrontal cortex' and cognition. *Curr. Opin. Neurobiol.* **2005**, *15*, 686–693. [CrossRef] [PubMed]
43. Güntürkün, O. The convergent evolution of neural substrates for cognition. *Psychol. Res.* **2012**, *76*, 212–219. [CrossRef]
44. Herold, C.; Palomero-Gallagher, N.; Hellmann, B.; Kröner, S.; Theiss, C.; Güntürkün, O.; Zilles, K. The receptor architecture of the pigeons' nidopallium caudolaterale: An avian analogue to the mammalian prefrontal cortex. *Brain Struct. Funct.* **2011**, *216*, 239–254. [CrossRef]
45. Von Eugen, K.; Tabrik, S.; Güntürkün, O.; Ströckens, F. A comparative analaysis of the dopaminergic innervation of the executive caudal nidopallium in pigeon, chicken, zebra finch, and carrion crow. *J. Comp. Neurol.* **2020**, *528*, 2929–2955. [CrossRef]
46. Kröner, S.; Güntürkün, O. Afferent and efferent connections of the caudolateral neostriatum in the pigeon (*Columba livia*): A retro-and antero-grade pathway tracing study. *J. Comp. Neurol.* **1999**, *407*, 228–260. [CrossRef]
47. Dykes, M.; Klarer, A.; Porter, B.; Rose, J.; Colombo, M. Neurons in the pigeon nidopallium caudolaterale display value-related activity. *Sci. Rep.* **2018**, *8*, 5377. [CrossRef] [PubMed]
48. Kalenscher, T.; Windmann, S.; Diekamp, B.; Rose, J.; Güntürkün, O.; Colombo, M. Single units in the pigeon brain integrate reward amount and time-to-reward in an impulsive choice task. *Curr. Biol.* **2005**, *15*, 594–602. [CrossRef]
49. Kuenzel, W.J.; Medina, L.; Csillag, A.; Perkel, D.J.; Reiner, A. The avian subpallium: New insights into structural and functional subdivisions occupying the lateral subpallial wall and their embryological origins. *Brain Res.* **2011**, *1424*, 67–101. [CrossRef]
50. Reiner, A.; Perkel, D.J.; Bruce, L.L.; Butler, A.B.; Csillag, A.; Kuenzel, W.; Medina, L.; Paxinos, G.; Shimizu, T.; Striedter, G.; et al. Revised nomenclature for avian telencephalon and some related brainstem nuclei. *J Comp. Neurol.* **2004**, *473*, 377–414. [CrossRef]
51. Niederleitner, B.; Gutierrez-Ibanez, C.; Krabichler, Q.; Weigel, S.; Luksch, H. A novel relay nucleus between the inferior colliculus and the optic tectum in the chicken (*Gallus gallus*). *J. Comp. Neurol.* **2017**, *525*, 513–534. [CrossRef]
52. Brown, C.; Magat, M. Cerebral lateralization determines hand preferences in Australian parrots. *Biol. Letts* **2011**, *7*, 496–498. [CrossRef] [PubMed]
53. Franklin, D.C.; Garnett, S.T.; Luck, G.W.; Gutierrez-Ibanez, C.; Iwaniuk, A.N. Relative brain size in Australian birds. *Emu* **2014**, *114*. [CrossRef]
54. Iwaniuk, A.N.; Dean, K.M.; Nelson, J.E. Interspecific allometry of the brain and brain regions in parrots (Psittaciformes): Comparisons with other birds and primates. *Brain Behav. Evol.* **2005**, *65*, 40–59. [CrossRef] [PubMed]
55. Rogers, L.J.; Workman, L. Footedness in birds. *Anim. Behav.* **1993**, *45*, 409–411. [CrossRef]
56. Iwaniuk, A.N.; Hurd, P.L. The evolution of cerebrotypes in birds. *Brain Behav. Evol.* **2005**, *65*, 215–230. [CrossRef] [PubMed]
57. Rogers, L.J. Lateralization in its many forms, and its evolution and development. In *The Evolution of Hemispheric Specialization in Primates;* Special Topics in Primatology; Hopkins, W.D., Ed.; Elsevier: Amsterdam, The Netherlands, 2007; Volume 5, pp. 23–56.
58. Andrews, C.B.; Gregory, T.R. Genome size is inversely correlated with relative brain size in parrots and cockatoos. *Genome* **2009**, *52*, 261–267. [CrossRef] [PubMed]
59. Healy, S.D.; Rowe, C. A critique of comparative brain studies of brain size. *Proc. R. Soc. B* **2007**, *274*, 453–464. [CrossRef]
60. Willemet, R. Reconsidering the evolution of brain, cognition, and behavior in birds and mammals. *Front. Psychol.* **2013**, *4*, 396. [CrossRef] [PubMed]
61. Herculano-Houzel, S. Numbers of neurons as biological correlates of cognitive capability. *Curr. Opin. Behav. Sci.* **2017**, *16*, 1–7. [CrossRef]
62. Güntürkün, O.; Hellmann, B.; Melsbach, G.; Prior, H. Asymmetries of representation in the visual system of pigeons. *Neuroreport* **1998**, *9*, 4127–4130. [CrossRef]
63. Güntürkün, O.; Stüttgen, M.C.; Manns, M. Pigeons as a model species for cognitive neuroscience. *Neuroforum* **2014**, *5*, 86–92. [CrossRef]

64. Morandi-Raikova, A.; Danieli, K.; Lorenzi, E.; Rosa-Salva, O.; Mayer, U. Anatomical asymmetries in the tectofugal pathway of dark-incubated domestic chicks: Rightwards lateralization of parvalbumin neurons in the entopallium. *Laterality* **2021**, *26*. [CrossRef] [PubMed]
65. Xiao, Q.; Güntürkün, O. The commissura anterior compensates asymmetries of visual representation in pigeons. *Laterality* **2021**, *26*. [CrossRef]
66. Striedter, G.F.; Charvet, C.J. Developmental origins of species differences in telencephalon and tectum size: Morphometric comparisons between a par-akeet (*Melopsittacus undulatus*) and a quail (*Colinus virgianus*). *J. Comp. Neurol.* **2008**, *507*, 1663–1675. [CrossRef]
67. Lee, H.J.; Schneider, R.F.; Manousaki, T.; Kang, J.H.; Lein, E.; Franchini, P.; Meyer, A. Lateralized feeding behavior is associated with asymmetrical neuroanatomy and lateralized gene expressions in the brain in scale-eating cichlid fish. *Genome Biol. Evol.* **2017**, *9*, 3122–3136. [CrossRef]
68. MacNeilage, P.; Rogers, L.J.; Vallortigara, G. Origins of the left and right brain. *Sci. Am.* **2009**, *301*, 60–67. [CrossRef]
69. Kaplan, G. Play behaviour, not tool using, relates to brain mass in a sample of birds. *Sci. Rep.* **2020**, *10*, 20437. [CrossRef]
70. Zucca, P.; Sovrano, V.A. Animal lateralization and social recognition: Quails use their left visual hemifield when approaching a companion and their right visual hemifield when approaching a stranger. *Cortex* **2008**, *44*, 13–20. [CrossRef] [PubMed]
71. Salva, O.R.; Regolin, L.; Mascalzoni, E.; Vallortigara, G. Cerebral and behavioural asymmetries in animal social recognition. *Comp. Cogn. Behav. Rev.* **2012**, *7*, 110–138. [CrossRef]
72. Yamazaki, Y.; Aust, U.; Huber, L.; Hausmann, M.; Güntürkün, O. Lateralized cognition: Asymmetrical and complementary strategies of pigeons during discrimination of the "human concept". *Cognition* **2007**, *104*, 315–344. [CrossRef]
73. Kaplan, G. Audition and hemispheric specialization in songbirds and new evidence from Australian magpies. *Symmetry* **2017**, *9*, 99. [CrossRef]
74. Costalunga, G.; Kobylkov, D.; Rosa-Salva, O.; Vallortigara, G.; Mayer, U. Light-Incubation effects on lateralisation of single unit responses in the visual Wulst of domestic chicks. *Brain Struct. Funct.* **2021**. [CrossRef] [PubMed]

Article

Hemispheric and Sex Differences in Mustached Bat Primary Auditory Cortex Revealed by Neural Responses to Slow Frequency Modulations

Stuart D. Washington [1,2,*], Dominique L. Pritchett [3], Georgios A. Keliris [4] and Jagmeet S. Kanwal [2]

1 Department of Radiology, Howard University Hospital, 2041 Georgia Ave NW, Washington, DC 20060, USA
2 Laboratory of Auditory Communication and Cognition, Department of Neurology, Georgetown University, 3700 O St. NW, Washington, DC 20057, USA; kanwalj@georgetown.edu
3 Department of Biology, EE Just Hall Building, Howard University, 415 College St. NW, Washington, DC 20059, USA; dominique.pritchett@howard.edu
4 Bio-Imaging Lab, Department of Biomedical Sciences, University of Antwerp, Universiteitsplein 1, B-2610 Antwerp, Belgium; Georgios.Keliris@uantwerpen.be
* Correspondence: sdw4@georgetown.edu

Abstract: The mustached bat (*Pteronotus parnellii*) is a mammalian model of cortical hemispheric asymmetry. In this species, complex social vocalizations are processed preferentially in the left Doppler-shifted constant frequency (DSCF) subregion of primary auditory cortex. Like hemispheric specializations for speech and music, this bat brain asymmetry differs between sexes (i.e., males>females) and is linked to spectrotemporal processing based on selectivities to frequency modulations (FMs) with rapid rates (>0.5 kHz/ms). Analyzing responses to the long-duration (>10 ms), slow-rate (<0.5 kHz/ms) FMs to which most DSCF neurons respond may reveal additional neural substrates underlying this asymmetry. Here, we bilaterally recorded responses from 176 DSCF neurons in male and female bats that were elicited by upward and downward FMs fixed at 0.04 kHz/ms and presented at 0–90 dB SPL. In females, we found inter-hemispheric latency differences consistent with applying different temporal windows to precisely integrate spectrotemporal information. In males, we found a substrate for asymmetry less related to spectrotemporal processing than to acoustic energy (i.e., amplitude). These results suggest that in the DSCF area, (1) hemispheric differences in spectrotemporal processing manifest differently between sexes, and (2) cortical asymmetry for social communication is driven by spectrotemporal processing differences and neural selectivities for amplitude.

Keywords: primary auditory cortex (A1); Doppler-shifted constant frequency (DSCF); mustached bat; sex differences; amplitude; spectral; temporal; hemispheric specialization; social communication; frequency modulation (FM)

Citation: Washington, S.D.; Pritchett, D.L.; Keliris, G.A.; Kanwal, J.S. Hemispheric and Sex Differences in Mustached Bat Primary Auditory Cortex Revealed by Neural Responses to Slow Frequency Modulations. *Symmetry* **2021**, *13*, 1037. https://doi.org/10.3390/sym13061037

Academic Editor: Giorgio Vallortigara

Received: 15 April 2021
Accepted: 4 June 2021
Published: 8 June 2021

1. Introduction

A left-hemispheric advantage for receptive language in general and especially speech perception [1,2] is characteristic of the human auditory cortex. Numerous studies of healthy [3–9] and clinical [10–15] human populations report that the left auditory cortex (AC) displays high temporal resolution relative to the right. This enhanced temporal resolution enables left AC to better process speech sounds containing rapid formant transitions, which are comparable to frequency modulations (FMs) [16,17]. Conversely, these [6–8,14,17] and other [18–22] studies report that the right AC has enhanced spectral resolution relative to the left. The right AC has a greater contribution to pitch discrimination [14,18,19,21] and musical processing [22–25] along with the detection of speaker identity and prosodic variation [26,27] than the left due to this higher spectral resolution. Multiple domain-general hypotheses [28,29] attribute these findings to the *acoustic uncertainty principle*, which states that there is an inverse relationship between temporal and spectral resolution governed by the same mathematics underlying Heisenberg's quantum uncertainty principle [30,31]. Some caveats to these domain-general explanations include: (1) this asymmetry for speech and language processing is often reported to be less pronounced in females than in males [9,32–35], and (2) these explanations do not preclude additional perceptual underpinnings for this asymmetry.

Hemispheric specialization for speech and language was traditionally considered to be unique either to humans [36] or to mammals with large brains [37]. However, there is substantial evidence of hemispheric specialization for conspecific communication sounds (i.e., social calls) in relatively large and small non-human primates [38–43]. Further, the number of studies reporting hemispheric specialization for social calls across avian [44–49] and small mammalian [50–53] species is growing [54–57]. Indeed, there is evidence that sea lions [58] and frogs [59] display hemispheric specialization for conspecific social calls. There is even evidence that such lateralization can occur for non-conspecific social communication since domesticated dogs display hemispheric biases for processing human speech [60,61]. Studies of dogs and other domesticated animals report that the right and left hemispheres process vocalizations differently based on their emotional valence and their acoustic structure [62,63]. An increasing number of studies also provide evidence for hemispheric differences in temporal and spectral processing in the non-human mammalian auditory cortex [52,53] or its avian homologues [64,65]. There are even reports of a sex-dependent asymmetry for temporal processing in rodents [66] and other mammals [67].

Processing of social calls in the primary AC (A1) of the mustached bat (*Pteronotus parnellii*) is known to be lateralized to the left side [50]. Previously, we provided evidence for sex-dependent hemispheric asymmetries for processing constant frequencies (CF or tone-bursts) and FMs in the Doppler-shifted constant frequency (DSCF) processing area [67], a subfield encompassing ~46% of mustached bat A1 [68]. These results must be contextualized within the broader scope of mustached bat echolocation and social communication to fully grasp its implications. During echolocation, mustached bats emit biosonar pulses composed of a fundamental CF, a downward frequency modulation (FM), and three harmonics thereof (Figure 1). Subfields of AC in this species have evolved to extract orientation and environmental information from the pulse, echo, CF ($CF_{1–4}$), and FM ($FM_{1–4}$) components of these biosonar signals [69]. For example, neurons in the FM-FM processing area use the delay between the pulse-FM_1 and echo-$FM_{2–4}$ to compute the target range [70,71]. In contrast, neurons in the DSCF area compute relative target velocity [68] and/or aid in maintaining distance from background objects during foraging [72] by firing in response to the returning echo-CF_2 but remaining unresponsive to the emitted pulse-CF_2. DSCF neural responses are facilitated when CFs in the echo-CF_2 range (60–63 kHz in *P.p. parnellii* [68] and 57.5–60 kHz in *P.p. rubiginosus* [73]) are paired at the onset with CFs in the pulse-FM_1 range (23–27 kHz) [74,75]. Interestingly, neurons in these same subfields also process conspecific social calls during communication [76–80]. The conspecific social calls of mustached bats are characterized by a phonetic-like syntax and have high acoustic complexity relative to the calls of most other mammalian species [81]. Despite their long-established specialization for processing echolocation, neurons in the DSCF area, FM-FM area, and other mustached bat auditory cortical subfields are responsive to complete social calls and their acoustic components. Furthermore, cortical FM-FM neurons are selective for the natural phonetic syntax in social calls [76], and DSCF neurons have a directional preference for upward FMs that exist primarily in social calls [80].

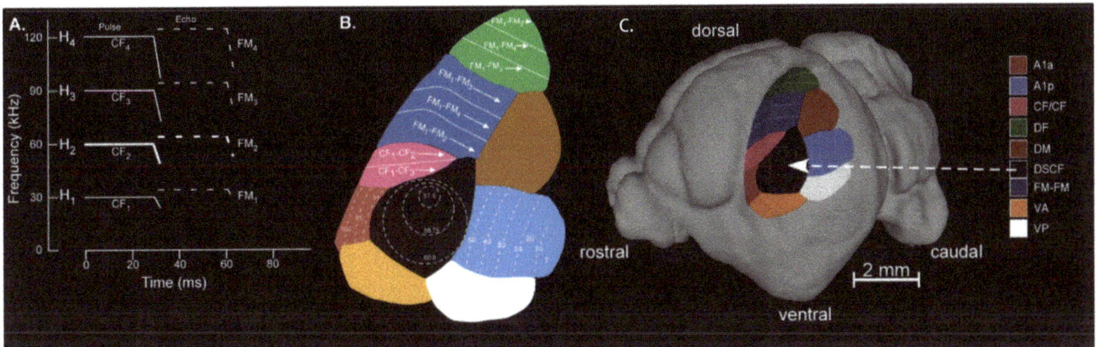

Figure 1. (**A**) Schematized spectrogram of the mustached bat's echolocation signal. $H_{1–4}$ refers to harmonics 1–4 of the echolocation pulse and/or echo. Note that the $H_{1–4}$ signal is composed of constant frequency ($CF_{1–4}$) and frequency-modulated ($FM_{1–4}$) components present in the pulse and echo (e.g., pulse-FM_1 or echo-CF_2). (**B**) The organization of functionally defined subdivisions of the mustached bat auditory cortex. Regions showed here include A1-anterior (A1a), A1-posterior (A1p), CF/CF area, dorsal fringe (DF) area, dorsal medial (DM), Doppler-shifted constant frequency (DSCF), and FM-FM areas. Map of functional areas was adapted from Suga (1985) [69] and Fitzpatrick et al. (1998) [82]. (**C**) Map of mustached bat auditory cortex superimposed on a lateral view of an MRI-based 3D reconstruction of the bat's brain. All recordings in the present study were obtained from the DSCF area, which occupies nearly 50% of A1 and represents narrow ranges of frequencies centered around the echo-CF_2 (57.5–60 kHz) and pulse-FM_1 (23–27 kHz). 3D-rendering adapted from Washington et al. (2018) [83].

Left DSCF neurons are generally more responsive to conspecific mustached bat social calls than those on the right [50]. To understand the neuro-acoustic basis of this finding, we investigated how DSCF neural selectivities for CFs and linear FMs differ between hemispheres, taking into account the sex differences commonly observed in hemispheric asymmetries in songbirds, rodents, and humans [67]. Left DSCF neurons in male bats were generally selective for shorter linear FMs with faster rates and responded to CFs and FMs with shorter latencies than those on the right, suggesting relatively higher temporal resolution amongst left DSCF neurons. Conversely, right DSCF neurons in male bats are selective for FMs with longer durations and relatively narrow bandwidths, suggesting a higher spectral resolution. Left DSCF neurons in female bats selected for shorter FMs and responded to them with shorter latencies than those on the right, but otherwise displayed fewer significant hemispheric differences than males [67]. The evolutionary pressures and underlying mechanisms for this sex difference remain elusive. However, acoustic uncertainty represents a potential evolutionary pressure for the asymmetry amongst DSCF neurons overall [84]. Specifically, the temporal resolution required for DSCF neurons to process mustached bat social calls likely conflicts with the refined spectral resolution they need to calculate Doppler shifts during echolocation. Separating temporally and spectrally refined DSCF neurons into different cerebral hemispheres could alleviate this conflict.

Here, we focus on asymmetrical processing of long, slow (<0.5 kHz/ms) FMs to which neurons in the DSCF areas in both hemispheres are highly responsive [67,80]. These types of FMs are commonly present within CF-like whistling sounds (termed long, quasi-CF or QCFl calls) as well as True CF or TCF call types that are frequently produced by male bats within a colony [81,85]. We, therefore, hypothesized that latencies and other characteristics of DSCF neural responses to slow FMs would differ between hemispheres and/or sexes. To test this hypothesis, we compared DSCF neural responses to linear upward and downward FMs with rates of 0.04 kHz/ms (duration = 131.25 ms; bandwidth = 5.25 kHz) across sound levels (i.e., amplitudes) in both males and females.

2. Methods

2.1. Surgery and Electrophysiological Recordings

The Georgetown University Animal Care and Use Committee (GUACUC Protocol #04-075) approved all methods presented here. These surgical and electrophysiological recording procedures have been described previously [74,79,80]. Six (four male) wild-caught mustached bats (*Pteronotus parnellii rubiginosus*) were used in these experiments. Bats were housed in one of two humidity (>60%) and temperature (~20–30 °C) controlled flight rooms with dimensions of either 6.6 or 4.3 m^2 (ceiling height = 3 m). Bats were fed a daily diet of nutrient-enriched mealworms. Under an anesthesia mixture of isoflurane/air (medical grade, Anaquest, Murray Hill, NJ, USA), we made an incision in the skin along the midline of each bat's head and glued a 2-mm-diameter metal post caudal to the sagittal-coronal suture intersection. Each bat was allowed >3 days to recover before electrophysiological recordings began. Bats were awake and restrained throughout recordings. Restraints entailed clamping the metal headpost while allowing the body to hang in a Styrofoam mold stabilized by rubber bands in a sound-proof and echo-attenuated chamber (IAC 400A) heated to 31 °C. Bats sat undisturbed in this recording set up for a day or two prior to recording so that they could be acclimated to the experimental environment. Careful to avoid the recording site, we treated the skin and muscle of the wound area with medetomidine (Domitor) during the acclimation period and in subsequent experiments. We used sharpened, vinyl-coated tungsten-microelectrodes (>1 MΩ) to perform electrophysiological recordings from the AC at a depth of 300–650 μm through a small (50μm) hole bored into the skull. We placed another microelectrode (< 1 MΩ) onto the dura mater of a non-auditory region of the cortex to establish a reference for differential recording. Electric signals acquired through the recording electrode were then amplified and band-pass filtered between 300 and 3000 Hz before being converted to digital format via SciWorks 3.0 software (Data Wave, Sequim, WA, USA).

2.2. Acoustic Stimuli

We used constant frequencies (CFs or "tone burst") and frequency modulations (FMs) to study neural responses within the DSCF processing area of the mustached bat primary auditory cortex (A1). CFs were created using analog function generators. A customized SIGNAL 3.0 script (Engineering Design) was used to generate FMs [86]. All CFs were 30 ms in duration and tapered (0.5 ms) at both ends. FM duration ranged between 0.4 and 131.5 ms and was tapered only when their durations were greater than 2 ms. CFs were presented via loudspeaker, and FMs were presented via the leaf-tweeter speaker.

Constant Frequencies: We first presented CFs to determine the frequencies that elicited peak responses from each neuron in order (A) to determine if the neuron was a DSCF neuron and (B) to determine the best frequency on which to center FMs. We classified a neuron as a "DSCF neuron" if it had a peak response to a CF between 57.5 and 60 kHz (best high frequency, or BF_{high}, within the echo-CF_2 range), and this peak response was facilitated when paired at the onset with a CF between 23 and 28 kHz (best low frequency, or BF_{low}, within the pulse-FM_1 range). Neurons generally showed only a small response to CFs at BF_{low} alone, and the facilitation criteria were as described previously [74,75]. CFs in the 57.5–60 kHz ranged paired with those in the 23–28 kHz were presented to facilitate responses per established DSCF neural tuning properties [74,75]. Amplitudes of CFs were also adjusted to obtain the best frequencies at their best amplitudes (BA).

Frequency Modulations: FMs were linear modulations of frequency (f) in the echo-CF_2 (57.5–60 kHz) range. We detailed the procedure for studying FM response characteristics elsewhere [80]. Linear FMs are defined by the following four parameters: duration (Δt) in ms, bandwidth (Δf) in kHz, rate of modulation ($\Delta f/\Delta t$) in kHz/ms, and the central frequency (f) of an FM in kHz. We generated 14 FM stimuli (or an FM rate array), changed the rates of those FMs between 0.04 and 4.0 kHz/ms, kept FM bandwidth constant at 5.25 kHz, and allowed FM duration to co-vary with rate. Each FM rate array was presented 100 times and had its amplitude decreased by 10 dB SPL every 10 repetitions (i.e., from

90 to 0 dB SPL). All FMs in the rate arrays were paired at the onset with a CF at BF_{low} in order to facilitate response magnitude. We presented two types of FM arrays, one where all the FMs were upward and a second where all FMs were downward, to determine the FM directional preference for each neuron [86]. The magnitudes of peak responses (10 ms bin) to FMs in the rate array were used as the criterion for determining the "best-FM rate" for a given neuron. Other arrays were generated to assess the "best-FM bandwidth" and "best-FM central frequency" for each neuron [67,80].

The present study focuses primarily on the responses of 176 DSCF neurons (Left Male = 43; Left Female = 40; Right Male = 35; Right Female = 58) to upward and downward FMs within the FM rate array modulated at 0.04 kHz/ms, repeated 100 times, from 0–90 dB SPL. Neural responses of 64 DSCF neurons to 200 repetitions of FMs modulated at 0.04 kHz/ms were also measured. These 64 neurons (Left Male = 4; Left Female = 10; Right Male = 17; Right Female = 33) all had best-FM rates equal to 0.04 kHz/ms and were presented at their respective BAs, best-FM bandwidths, best-FM central frequencies, and best-FM directions. These "best-FMs" were paired at the onset with a 30 ms CF at the BF_{low} of the respective neuron.

2.3. Data Analysis

Each FM rate array was a series of 14 FM stimuli totaling 3750 ms in duration. Specifically, the FM rate arrays were composed of an initial 250 ms period without a stimulus ("null" stimulus period) in the echo-CF_2 range followed by 14 linear FMs (presentation rate 4/s or one presentation every 250 ms) centered on the BF_{high} of the neuron under study and increasing in modulation rate from 0.04–4.0 kHz/ms. Here, we extracted and analyzed only neural responses to the 250–500 ms section of each FM rate array. This 250–500 ms section corresponded to the presentation of FMs with modulation rates of 0.04 kHz/ms. Sound-level (i.e., amplitude) decreased by 10 dB SPL every 10 trials (100 trials total). All stimuli in the FM rate array were paired at the onset with a 30 ms CF at the BF_{low} of the neuron under study. An example of a DSCF neuron's responses to the14 FM stimuli is provided in Figure 2A. Data corresponding to 0.04 kHz/ms (250–500 ms range) are highlighted to emphasize that this is the focus of the present study. Note that in the expanded view of this range in Figure 2B, the example neuron's response latency increases as the sound level decreases, a typical "latency shift" common to most auditory neurons [87].

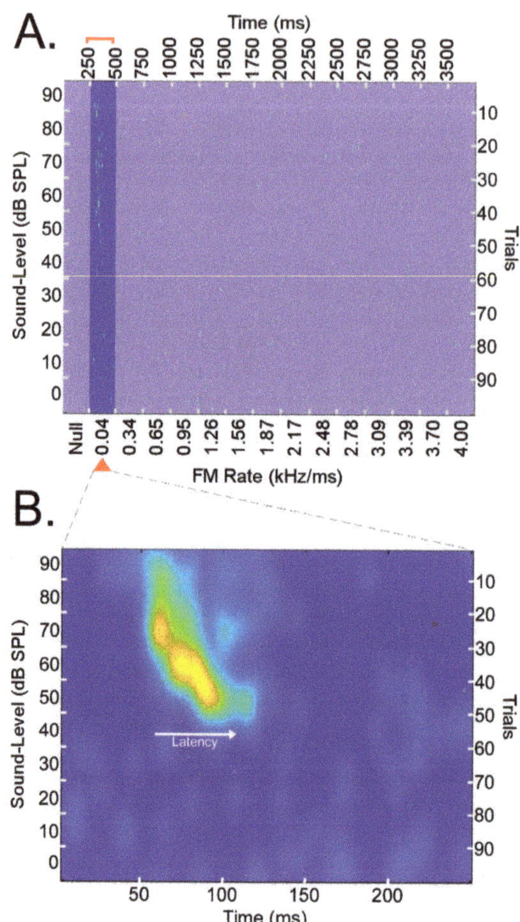

Figure 2. (**A**) A matrix of responses to an array of 14 linear FMs presented for 100 trials each (right vertical axis). The presentation rate was 4/s or one presentation every 250 ms for a total of 3750 ms (top horizontal axis). FMs had bandwidths of 5.25 kHz, durations of 131.25 ms, and increased in rate from 0.04 to 4.0 kHz/ms (bottom horizontal axis). The sound level (i.e., amplitude) decreased by 10 dB SPL every 10 trials from 90 to 0 dB SPL (left vertical axis). A 250 ms period with no stimulus ("null") preceded the presentation of the 14 FM for a series of 15 stimuli total. Presentations of linear FMs were performed in the upward and downward directions such that, for each neuron, there was a matrix of responses to 14 upward FMs and another to 14 downward FMs, both sets repeated 100 times at different sound levels. Rasters shown above correspond to the responses of a right-hemispheric DSCF neuron from a male bat elicited by a series of 14 upward FMs. Above, the 0–250 ms and 500–3750 ms time periods corresponding to the presentation of a "null" stimulus control and FMs with rates >0.04 kHz/ms are deemphasized to highlight this study's focus on neural responses to the 0.04 kHz/ms FMs presented during the 250–500 ms time period. (**B**) Spike density function generated by performing a 2D convolution between the 250–500 ms time period of the response matrix in "A" with a rotationally symmetric Gaussian lowpass filter (size = 25 × 25, sigma = 5). Spike density functions were generated for the 250–500 ms time periods for upward and downward FM rate arrays of each of 176 neurons. Spike density functions were then grouped as upward (N = 176), downward (N = 176), left (N = 93), right (N = 83), male (N = 78), and female (N = 98) prior to any analyses.

Single unit recording data were converted from SciWorks to MATLAB format and then sorted into matrices corresponding to responses elicited by either upward or downward FMs (i.e., upward and downward response matrices). Response matrix dimensions were $100 \times 250 \times 176$, corresponding to "number of trials" × "duration in ms" × "number of neurons." Every 10 trials corresponded to the same sound level, so we grouped the 100 trials into 10 bins corresponding to 10 sound levels ($10 \times 250 \times 176$ = sound level × duration × neuron). We generated a rotationally symmetric Gaussian lowpass filter (size = 25×25, sigma = 5) and then performed a 2D convolution between the Gaussian filter and the response matrices of each neuron. Convolved response matrices (i.e., spike density functions) were then grouped by hemisphere and sex (Left Male, Left Female, Right Male, and Right Female) and statistically compared via two-sample Kolmogorov–Smirnov goodness-of-fit tests. Absolute maxima in the spike density functions were used to find peak response latencies (duration/time axis), best amplitude (sound level), and response magnitude (firing intensity). Effect sizes for significant findings were assessed using *Cohen's d*, which we symbolize here with " $|d|$ " because the directionality of the *Cohen's d* statistic is irrelevant for our purposes.

Peak response magnitudes and latencies of DSCF neural responses to "best FMs" were elicited by CFs at BF_{low}, and best FMs paired at onset (BF_{low} + best FM) and presented 200 times at BA. Responses to these 200 repetitions of BF_{low} + best FM were recorded in the form of peri-stimulus time histograms (PSTHs). These histograms, calculated online by summation of spike trains over repeated trials, were used to measure the neuronal response that represents a stimulus-locked change in peak response magnitude and latency. Here, we selected 64 neurons that had best-FMs rates of 0.04 kHz/ms.

3. Results

The data presented here are a reanalyzed subset of previously reported data [67,80]. Our previous results demonstrated that there are sex-dependent hemispheric differences for processing FMs in the DSCF neural population [67]. Thus, we organized our data into four groups (Left Male, Right Male, Left Female, and Right Female) to test a prior hypothesis based on the sex-dependent asymmetries and combined these groups whenever the need to test other related hypotheses arose. Responses to downward and upward FMs were analyzed separately before being analyzed in their "best directions" (i.e., the direction of the FM that elicited the greatest neural response magnitude). Figure 3 displays mean spike density functions for DSCF neural responses elicited by FMs with rates of 0.04 kHz/ms in the upward and downward FM directions in male bats. Figure 4 displays the corresponding data in female bats.

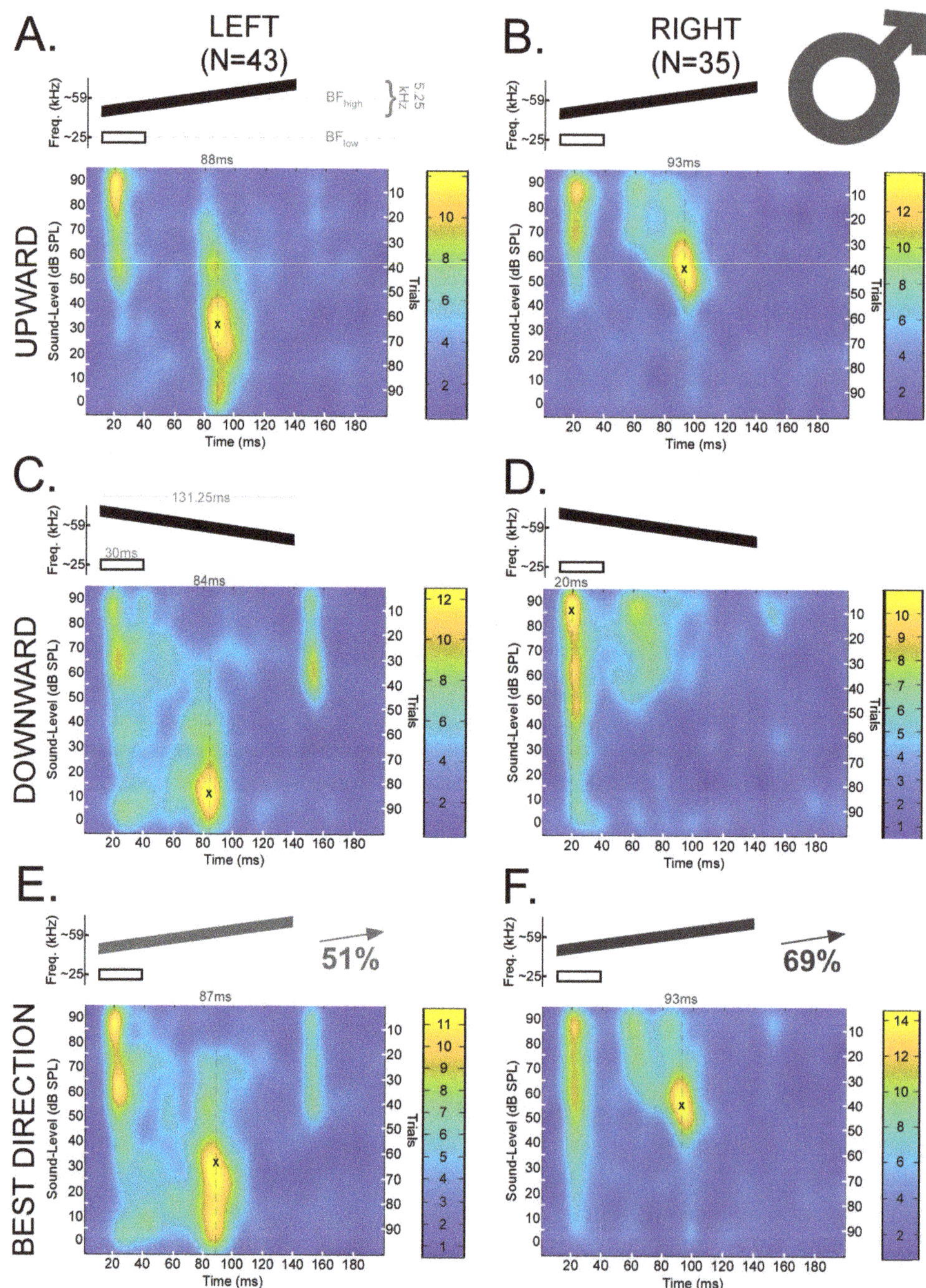

Figure 3. Mean spike density functions based on DSCF neural responses elicited by 100 trials of FMs with rates of 0.04 kHz/ms presented at 0–90 dB SPL and recorded from male mustached bats. **(A)** Top: Schematic of an upward FM (black) paired at the onset with a 30 ms CF (white) at BF$_{low}$ (23–27 kHz). The upward FM had a bandwidth of 5.25 kHz, a duration of 131.25 ms, and a central frequency equal to the BF$_{high}$ (57.5–60 kHz) of the individual DSCF neuron under study.

The FM and CF had an onset delay of 10 ms. Bottom left: Mean spike density function of DSCF neural responses to FMs with rates of 0.04 kHz/ms in the left hemispheres of male bats (N = 43). "X" marks the location of peak firing (88 ms, 30 dB SPL) in the spike density function, and the hatched line corresponds to the horizontal coordinate. Bottom right: Color bar encoding the firing rate intensity in the spike density function at left. (**B**) Top: Schematic of a CF and upward FM identical to those depicted in "A." Bottom left: Mean spike density function of DSCF neural responses to upward FMs with rates of 0.04 kHz/ms in the right hemispheres of male bats (N = 35). "X" marks the location of peak firing (93 ms, 50–60 dB SPL) in the spike density function, and the hatched line corresponds to the horizontal coordinate. Bottom right: Color bar encoding the firing rate intensity in the spike density function at left. (**C**) Top: Schematic of a downward FM (black) paired at the onset with a 30 ms CF at BF_{low} (white). All other parameters identical to "A." Bottom left: Mean spike density function of DSCF neural responses to downward FMs with rates of 0.04 kHz/ms in the left hemispheres of male bats (N = 43). "X" marks the location of peak firing (84 ms, 10 dB SPL) in the spike density function, and the hatched line corresponds to the horizontal coordinate. Bottom right: Color bar encoding the firing rate intensity in the spike density function at left. (**D**) Top: Schematic of a CF and downward FM identical to those depicted in "C." Bottom left: Mean spike density function of DSCF neural responses to downward FMs with rates of 0.04 kHz/ms in the right hemispheres of male bats (N = 35). "X" marks the location of peak firing (20 ms, 80–90 dB SPL) in the spike density function, and the hatched line corresponds to the horizontal coordinate. Bottom right: Color bar encoding the firing rate intensity in the spike density function at left. (**E**) Top: Schematic of an FM (gray or black at 51% opacity) paired at the onset with a 30 ms CF (white) at BF_{low} (23–27 kHz). Opacity denotes that the "best direction" for 51% of the neurons in this sample (22/43) was upward. Otherwise, stimuli are identical to those depicted in "A" and "C." Bottom left: Mean spike density function of DSCF neural responses to 0.04 kHz/ms FMs modulated in the "best directions" of each neuron in the left hemispheres of male bats. "X" marks the location of peak firing (87 ms, 30 dB SPL) in the spike density function, and the hatched line corresponds to the horizontal coordinate. Bottom right: Color bar encoding the firing rate intensity in the spike density function at left. (**F**) Top: Schematic of an FM (gray or black at 69% opacity) paired at the onset with a 30 ms CF (white) at BF_{low} (23–27 kHz). Opacity denotes that the "best direction" for 69% of the neurons in this sample (24/35) was upward. Otherwise, stimuli are identical to those depicted in "A" and "C." Bottom left: Mean spike density function of DSCF neural responses to 0.04 kHz/ms FMs modulated in the "best directions" of each neuron in the right hemispheres of male bats. "X" marks the location of peak firing (93 ms, 50–60 dB SPL) in the spike density function, and the hatched line corresponds to the horizontal coordinate. Bottom right: Color bar encoding the firing rate intensity in the spike density function at left. No responses occur in the last 50 ms, so they are omitted to provide greater detail.

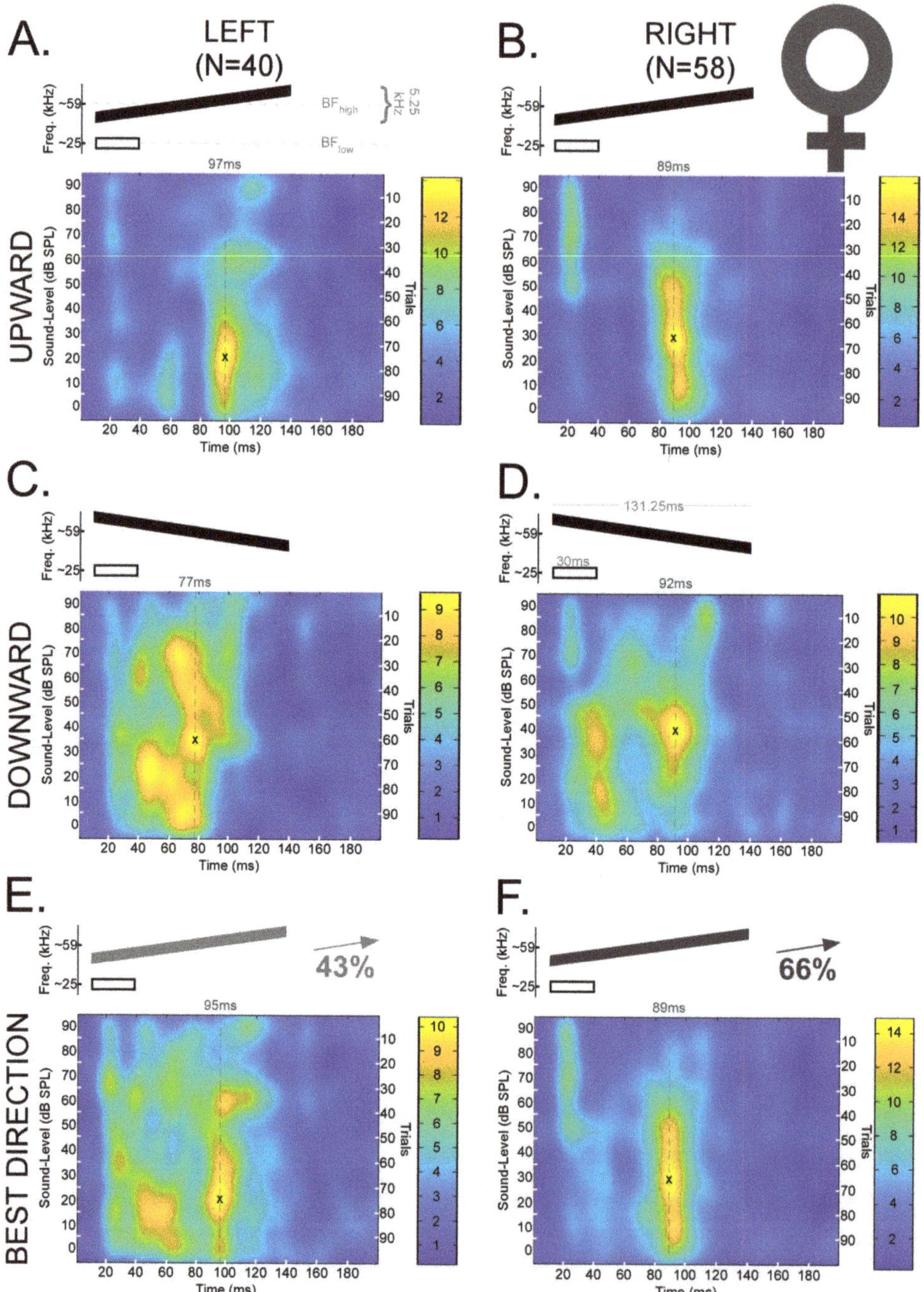

Figure 4. Mean spike density functions based on DSCF neural responses elicited by 100 trials of FMs with rates of 0.04 kHz/ms presented at 0–90 dB SPL and recorded from female mustached bats. (**A**) Top: Schematic of an upward FM (black) paired at the onset with a 30 ms CF (white) at BF_{low} (23–27 kHz). The upward FM had a bandwidth of 5.25 kHz, a

duration of 131.25 ms, and a central frequency equal to the BF_{high} (57.5–60 kHz) of the individual DSCF neuron under study. The FM and CF had an onset delay of 10 ms. Bottom left: Mean spike density function of DSCF neural responses to FMs with rates of 0.04 kHz/ms in the left hemispheres of female bats (N = 40). "X" marks the location of peak firing (97 ms, 20 dB SPL) in the spike density function, and the hatched line corresponds to the horizontal coordinate. Bottom right: Color bar encoding the firing rate intensity in the spike density function at left. (**B**) Top: Schematic of a CF and upward FM identical to those depicted in "A." Bottom left: Mean spike density function of DSCF neural responses to upward FMs with rates of 0.04 kHz/ms in the right hemispheres of female bats (N = 58). "X" marks the location of peak firing (89 ms, 30 dB SPL) in the spike density function, and the hatched line corresponds to the horizontal coordinate. Bottom right: Color bar encoding the firing rate intensity in the spike density function at left. (**C**) Top: Schematic of a downward FM (black) paired at the onset with a 30 ms CF at BF_{low} (white). All other parameters identical to "A." Bottom left: Mean spike density function of DSCF neural responses to downward FMs with rates of 0.04 kHz/ms in the left hemispheres of female bats (N = 40). "X" marks the location of peak firing (77 ms, 30–40 dB SPL) in the spike density function, and the hatched line corresponds to the horizontal coordinate. Bottom right: Color bar encoding the firing rate intensity in the spike density function at left. (**D**) Top: Schematic of a CF and downward FM identical to those depicted in "C." Bottom left: Mean spike density function of DSCF neural responses to downward FMs with rates of 0.04 kHz/ms in the right hemispheres of female bats (N = 58). "X" marks the location of peak firing (92 ms, 40 dB SPL) in the spike density function, and the hatched line corresponds to the horizontal coordinate. Bottom right: Color bar encoding the firing rate intensity in the spike density function at left. (**E**) Top: Schematic of an FM (gray or black at 43% opacity) paired at the onset with a 30 ms CF (white) at BF_{low} (23–27 kHz). Opacity denotes that the "best direction" for 43% of the neurons in this sample (17/40) was upward. Otherwise, stimuli are identical to those depicted in "A" and "C." Bottom left: Mean spike density function of DSCF neural responses to 0.04 kHz/ms FMs modulated in the "best directions" of each neuron in the left hemispheres of female bats. "X" marks the location of peak firing (95 ms, 20 dB SPL) in the spike density function, and the hatched line corresponds to the horizontal coordinate. Bottom right: Color bar encoding the firing rate intensity in the spike density function at left. (**F**) Top: Schematic of an FM (gray or black at 66% opacity) paired at the onset with a 30 ms CF (white) at BF_{low} (23–27 kHz). Opacity denotes that the "best direction" for 66% of the neurons in this sample (38/58) was upward. Otherwise, stimuli are identical to those depicted in "B" and "D." Bottom left: Mean spike density function of DSCF neural responses to 0.04 kHz/ms FMs modulated in the "best directions" of each neuron in the right hemispheres of female bats. "X" marks the location of peak firing (89 ms, 30 dB SPL) in the spike density function, and the hatched line corresponds to the horizontal coordinate. Bottom right: Color bar encoding the firing rate intensity in the spike density function at left. No responses occur in the last 50 ms, so they are omitted to provide greater detail.

There were highly sex dependent hemispheric differences in peak response latency. In males, there were no significant hemispheric differences in peak response latency in the upward (D (43,35) = 0.1495, p = n.s.), downward (D (43,35) = 0.1814, p = n.s.), or best (D (43,35) = 0.1761, p = n.s.) FM directions. Likewise, peak response latencies were similar between hemispheres in the downward (D (40,58) = 0.1250, p = n.s.) and best (D (40,58) = 0.2414, p = n.s.) FM directions in females. However, peak response latencies differed between hemispheres in the upward (D (40,58) = 0.3543, $p = 0.0037$; $|d| = 0.5221$, medium effect) FM direction in females, such that latencies of responses to upward FMs were shorter on the left (mean ± s.e.m.: 64.37 ms ± 7.45) than on the right (89.94 ms ± 6.59). Across sexes overall, latencies differed between hemispheres in the upward (D (83,93) = 0.2493, $p = 0.0068$, $|d| = 0.3579$, small-to-medium effect) but not downward (D (83,93) = 0.0963, p = n.s.) FM direction, such that responses to upward FMs had shorter latencies on the left (57.04 ms ± 4.96) than on the right (74.71 ms ± 5.48). There was a trend toward interhemispheric latency differences for the best FM direction across sexes (D (83,93) = 0.1929, $p = 0.0667$).

Further, latencies generally differed between sexes. Specifically, latencies of responses in the left hemisphere were shorter in males (48.11 ms ± 5.72) than in females (66.94 ms ± 7.05) in the downward (D (43,40) = 0.3366, $p = 0.0135$, $|d| = 0.4587$, small-to-medium effect) FM direction. There was a similar trend in the left hemisphere for the upward (D (43,40) = 0.2640, $p = 0.0925$) FM direction, indicating a tendency towards shorter latencies in males (50.22 ms ± 6.51) than in females (64.37 ms ± 7.45). Interestingly, there was neither a difference nor a trend for the best (D (43,40) = 0.2355, p = n.s.) FM direction between males (49.38 ms ± 5.74) and females (61.82 ms ± 6.60) in the left hemisphere. In the right hemisphere, there was a trend for the downward (D (43,40) = 0.2714, p = 0.0656) FM direction that likewise indicated a tendency towards shorter latencies in males (58.93 ms ± 8.34) than in females (68.84 ms ± 6.76). However, latencies of responses to upward FMs in the right hemisphere were substantially different between sexes (D (43,40) = 0.5650, $p = 7.42 \times 10^{-7}$, $|d|$ = 0.8214, large effect), again driven by shorter latencies in males (49.46 ms ± 8.06) than in females (89.94 ms ± 6.59). Likewise, responses to FMs modulated in the neuron's best (D (43,40) = 0.3414, p = 0.0090, $|d|$ = 0.4666, small-to-medium effect) direction differed in latency between males (55.05 ms ± 8.39) and females (78.38 ms ± 6.59) in the right hemisphere. Overall, latencies differed between sexes in the upward (D (78,98) = 0.4063, $p = 6.48 \times 10^{-7}$, $|d|$ = 0.6180, medium effect), downward (D (78,98) = 0.2700, p = 0.0027, $|d|$ = 0.3262, small-to-medium effect), and best (D (78,98) = 0.2889, p = 0.0011; $|d|$ = 0.4301, small-to-medium effect) FM directions. These differences were due to males (49.88 ms ± 5.06) having shorter latencies to upward FMs than females (79.50 ms ± 5.08), and males (52.96 ms ± 4.90) likewise had shorter latencies to downward FMs than females (68.06 ms ± 4.91).

There were highly sex dependent hemispheric differences in best amplitude (BA). In males, there were hemispheric differences in BA that were more marked in the upward (D (43,35) = 0.5136, $p = 3.83 \times 10^{-5}$; $|d|$ = 1.0159, large effect) and best (D (43,35) = 0.4100, p = 0.0020; $|d|$ = 0.7219, medium-to-large effect) FM direction than in the downward (D (43,35) = 0.3063, p = 0.0420; $|d|$ =0.5853, medium effect) direction. Specifically, for the upward FM direction in males, BAs were lower amongst left (38.51 dB SPL ± 3.10) than right (57.04 dB SPL ± 2.58) DSCF neurons. Likewise, BAs were lower amongst left (38.13 dB SPL ± 3.07) than right (50.38 dB SPL ± 3.70) DSCF neurons for the downward FM direction in males. In females, there were trends towards hemispheric differences for BAs in the upward (D (40,58) = 0.2586, p = 0.0698) and downward (D (40,58) = 0.2526, p = 0.0815) FM directions that were significant for the best (D (40,58) = 0.3069, p = 0.0177; $|d|$ = 0.1882, minute-to-small effect) direction. However, in the upward FM direction in females, this trend indicated a tendency for BAs to be greater amongst left (38.02 dB SPL ± 3.36) than right (33.24 dB SPL ± 2.24) DSCF neurons. Likewise, there was a tendency for BAs to be greater amongst left (42.92 dB SPL ± 2.85) than right (38.91 dB SPL ± 2.34) DSCF neurons in the downward FM direction in females. Largely due to these diametrically opposed tendencies between sexes, BAs showed no overall hemispheric differences in either the upward (D (83,93) = 0.1283, p = n.s.), downward (D (83,93) = 0.1347, p = n.s.), or best (D (83,93) = 0.1347, p = n.s.) FM directions.

BAs differed substantially between sexes. BAs were generally greater in males (43.62 dB SPL ± 2.19) than in females (40.55 dB SPL ± 1.81) in the downward (D (78,98) = 0.2449, p = 0.0088, $|d|$ = 0.1882, minute-to-small effect) FM direction. Similarly, BAs were greater in males (46.83 dB SPL ± 2.06) than in females (35.19 dB SPL ± 1.91) in the upward (D (78,98) = 0.2449, $p = 3.43 \times 10^{-4}$; $|d|$ = 0.5945, medium-to-large effect) FM direction. These same differences were reflected in the BAs for FMs modulated in the best (D (78,98) = 0.2658, p = 0.0033; $|d|$ = 0.3984, small-to-medium effect) direction of each neuron. Separating analyses by hemisphere provided greater detail to the sex-dependent hemispheric differences in BA described above. Specifically, BAs were similar between males (38.13 dB SPL ± 3.07) and females (42.92 dB SPL ± 2.85) in the downward FM direction in the left hemisphere (D (43,40) = 0.2686, p = n.s.). However, BAs were greater between males and females in the downward FM direction in the right hemisphere (D (35,58) = 0.4562, $p = 1.2876 \times 10^{-4}$; $|d|$ = 0.5899, medium-to-large effect). BAs were similar between males (38.51 dB SPL ± 3.10) and females (38.02 dB SPL ± 3.36) for the upward FM direction in the left hemisphere (D (43,40) = 0.1674, p = n.s.). Again, there was no such similarity in the right hemisphere for upward FMs (D (35,58) = 0.6616, $p = 3.0550 \times 10^{-9}$; $|d|$ = 1.4495, large-to-huge effect) where BAs were greater in males (50.38 dB SPL ± 3.70) than in females (38.91 dB SPL ± 2.34) for upward FMs. Following from these results, BAs were similar between sexes in the left (D (78,98) = 0.1186, p = n.s.) but not the right (D (78,98) = 0.5700, $p = 5.7264 \times 10^{-7}$, $|d|$ = 0.9413, large effect) hemisphere when FMs were modulated in the best direction for each neuron. These results underscore that, amongst DSCF neurons, BAs for linear FMs with rates of 0.04 kHz/ms differ between sexes, but these sex differences in BA are largely driven by the right hemisphere.

Figures 3 and 4 reveal latency shifts that coincide with decreases in sound level in the upward FM direction. On average, response latencies increased by nearly 50 ms as the sound level decreased in the upward FM directions in both hemispheres of males and females. However, this latency shift in response to upward FMs occurred at a greater sound level (around 20 dB SPL louder) on the right in males than on the left in males or in either hemisphere in females. This pattern is similar for the downward FM direction in males, though a prominent second response peak is also visible after 150 ms (i.e., a possible offset response) and at 50–70 dB SPL. Changes in sound level for downward FMs yielded a variety of response patterns in the male right hemisphere and bilaterally in females. Right-hemispheric responses to downward FMs in males did not shift in latency, despite decreases in sound level, and these responses largely ceased for sound levels <40 dB SPL. Unlike the other groups, the average peak response to downward FMs amongst right-hemispheric neurons in males occurred at the highest sound levels and near stimulus onset (20 ms or 10 ms post-stimulus onset). Left and right-hemispheric responses to downward FMs in females were characterized by a quasi-tonic firing pattern most prominent at lower sound levels ($\leq$40 dB SPL).

Previous research established that DSCF neurons are generally more responsive to upward than to downward FMs when the FMs were optimized for rate, bandwidth, central frequency, and BA [80]. In our sample, 101/176 (57%) DSCF neurons had greater responses to upward than to downward FMs when FM rates were all equal to 0.04 kHz/ms and the sound level changed by 10 dB SPL every 10 out of 100 trials. In males, maximal responses to FMs with rates of 0.04 kHz/ms were found in the upward direction in 51% (22/43) of left-hemispheric and 69% (24/35) of right-hemispheric neurons. In females, maximal responses to FMs with these same rates were found in the upward direction in 42.5% (17/40) of left-hemispheric and 66% (38/58) of right-hemispheric neurons. As stated above, previous research employed more optimal measures to assess the general directional preference of DSCF neurons and thus provides a better guide to this filter property. Nonetheless, these results show that a general upward FM directional preference in DSCF neurons is present even when assessed using FM stimuli not optimized for rate, BA, and other FM parameters.

Lastly, we selected 64 DSCF neurons with best-FM rates of 0.04 kHz/ms and measured their responses to 200 presentations of their best-FMs (optimized for rate, bandwidth, central frequency, and direction) at BA (Figure 5). This relatively small number of neurons (Left Male = 4; Left Female = 10; Right Male = 17; Right Female = 33) yielded no significant differences when compared across hemispheres and sexes via two-sample Kolmogorov–Smirnov tests. However, descriptive statistics reveal notable patterns across hemispheres and sexes. In males, latencies of responses to best-FMs with rates of 0.04 kHz/ms tended to be shorter on the left (14.75 ms $\pm$ 2.21) than on the right (31.68ms $\pm$ 6.08), and BAs tended to be greater on the right (74.71 dB SPL $\pm$ 3.11) than on the left (22.5 dB SPL $\pm$ 7.5). In females, response latencies were closer in time between hemispheres but still tended to be shorter on the left (22.00 ms $\pm$ 4.89) than on the right (31.38 ms $\pm$ 5.26). Further, in females, BAs were also closer in loudness between hemispheres, tending to be slightly greater on the left (44.00 dB SPL $\pm$ 5.42) than on the right (38.48 dB SPL $\pm$ 3.70).

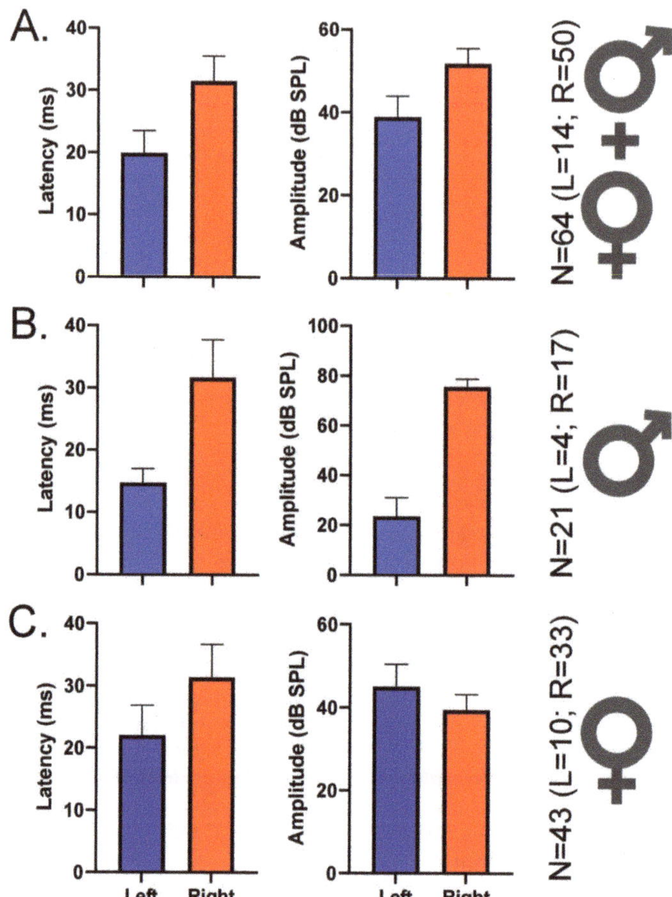

Figure 5. Mean peak response latencies and best amplitudes (BAs) of DSCF neurons selective for 0.04 kHz/ms. (**A**) Mean (± standard error of the mean) latencies (left column) and BAs (right column) of DSCF neuronal responses recorded from 64 neurons in male and female bats and elicited by 200 repetitions of their "best FMs" (optimized for BA, FM rate, FM bandwidth, FM central frequency, and FM direction). Recordings were from left (blue, N = 14) and right (red, N = 50). In this 64-neuron sample, 57.14% (8/14) of best FM directions for the left DSCF neurons and 64% (32/50) of best FM directions for the right DSCF neurons were upward. (**B**) Same depiction as "A" for responses recorded from 21 DSCF neurons (left, N = 4; right, N = 17) in male bats. In this 21-neuron sample, 50% (2/4) of best FM directions for the left DSCF neurons and 82.35% (14/17) of best FM directions for the right DSCF neurons were upward. (**C**) Same depiction as "A" for responses recorded from 43 DSCF neurons (left, N = 10; right, N = 33) in female bats. In this 43-neuron sample, 60% (6/10) of best FM directions for the left DSCF neurons and 54.54% (18/33) of best FM directions for the right DSCF neurons were upward. The lower number of left DSCF neurons across this sample reflects an earlier finding that, when presented with FM rates ranging from 0.04 to 4.0 kHz/ms, right DSCF neurons (especially in males) are far more likely to respond maximally to 0.04 kHz/ms than those on the left [67].

4. Discussion

Slow linear FMs (rate = 0.04 kHz/ms; bandwidth = 5.25 kHz; duration = 131.25 ms) when presented at sound levels ranging from 0 to 90 dB SPL (in 10 dB SPL attenuation steps) revealed sex-dependent hemispheric asymmetries in DSCF neural peak response latency and best amplitude (BA). Two results support our hypothesis that DSCF neural responses to slow linear FMs differ by hemisphere and sex. First, left DSCF neurons responded to slow, upward FMs with shorter latencies than right DSCF neurons, particularly in females. Second, BAs of right DSCF neurons responding to slow upward and downward FMs were greater than those of left DSCF neurons in males but not females. In other words, in males, DSCF neurons respond at their peak firing rates to quieter FM sounds on the left compared to the right. Furthermore, right DSCF neurons in males have higher minimum thresholds for slow upward and downward FMs than left DSCF neurons. With respect to sex differences, peak response latencies to both upward and downward FMs were generally shorter in males when compared to females. These results from 176 DSCF neurons are largely corroborated by responses from a subset of 64 DSCF neurons selective for FM rates of 0.04 kHz/ms elicited by 200 repetitions of optimized FM stimuli (i.e., best FM rate, bandwidth, center frequency, and direction) presented at BA.

Here, we reported results for the upward, downward, and best FM directions to provide greater detail and a wider scope. Though some results were more significant for downward FMs (e.g., shorter response latencies in males than in females for downward but not upward FMs), the majority of our findings were either significant for both FM directions or more significant for upward FMs. This aspect is critical to the interpretation of our results because DSCF neurons, in general, prefer (i.e., respond with greater magnitude to) upward FMs [80], and the data presented here reflects that same directional preference. Best FM results largely reflected those of the upward FM direction combined with some aspects of the downward direction.

The hemispheric differences in peak response latency primarily observed in female bats exist in the absence of any significant differences in BA. Hemispheric differences in temporal and spectral resolution, hypothesized to underlie speech and music processing in the human auditory cortex [28,29], operationally match those hypothesized to underlie social communication and echolocation in the DSCF area [50,67,84]. Though not direct evidence for this assertion, a longer peak response latency to a slowly modulated, long-duration FM signal could be elicited from a neuron tuned to a specific, narrow frequency band contained within the FM or a short segment of the FM itself, suggesting a relatively longer temporal integration window (i.e., low temporal resolution coinciding with high spectral resolution). Conversely, a shorter latency in response to a long, slow FM signal could be elicited from a neuron tuned to a broader range of frequencies contained within the FM, with earlier firing resulting from faster integration times (i.e., low spectral resolution but with a high temporal resolution, allowing the neuron to follow frequency changes across shorter time windows than those possible on the right).

Previous research suggests that interhemispheric temporal versus spectral processing differences in mustached bats [67], rodents [66], and humans [9] would either be more pronounced in males or similar between the sexes. A closer examination of previous sex-dependent asymmetry results in the DSCF area in the mustached bat, however, provides additional details. Left DSCF neurons in males are selective for FMs with faster rates and shorter durations than their right-hemispheric homologues or either hemisphere in females [67]. In females, whereas left DSCF neurons were selective for FMs with shorter durations than their right-hemispheric homologues, they had no significant hemispheric differences in FM rate selectivity. Despite these sex-dependent hemispheric differences in FM selectivity, left DSCF neurons responded to their best FMs (i.e., optimized for rate, bandwidth, central frequency, direction, and BA) nearly 10 ms faster than right DSCF neurons in both sexes. As reported here, peak response latencies were generally longer for females than males. These results, when contextualized by previous results, suggest that interhemispheric spectral versus temporal processing differences are present yet manifest differently within both sexes. Specifically, in males, the left-hemispheric advantage for temporal processing amongst DSCF neurons manifests as an enhanced ability to detect and quickly respond to short duration, rapidly changing FMs relative to those on the right. In females, on the other hand, a left-hemispheric temporal processing advantage results in earlier responses to FMs with slow-to-moderate rates relative to the right as opposed to selectivity for faster FM rates. Thus, hemispheric differences in peak response latency to best-FMs would appear similar in both sexes with potentially longer latencies in females overall, as reported previously [67].

The amplitude-related asymmetry observed amongst DSCF neurons in males, on the other hand, does not conform as elegantly to the theoretical framework of spectral versus temporal processing. Hemispheric differences in neural selectivity for amplitude (dB SPL) can be restated as hemispheric differences in acoustical energy or power (i.e., energy per unit time). In other words, right DSCF neurons in males required greater acoustic energy to respond to long-duration FMs with slow rates than did those on the left. This point is made even clearer when observing "latency shifts" within the recorded population of DSCF neurons. With the rare exception of auditory neurons that display "paradoxical latency shifts," neurons throughout the auditory system typically increase their latencies as the sound level (i.e., intensity or loudness) decreases [87]. Such latency shifts are observed in both hemispheres and sexes, especially in the upward FM direction. Our group analyses, however, demonstrate that the latency shifts that commonly accompany decreasing amplitude in auditory neurons are evident even after relatively slight amplitude decreases amongst right DSCF neurons in males. These high amplitude latency shifts likely explain why there are hemispheric differences in peak response latencies in males for the subset of DSCF neurons selective for 0.04 kHz/ms ($N = 21$ in males), to which we presented best FMs 200 times at BA, but not in the larger population ($N = 78$), where we presented FMs with rates of 0.04 kHz/ms across sound levels from 0 to 90 dB SPL.

The results above add new details to the previously reported hemispheric differences for processing social calls in the DSCF area of the mustached bat auditory cortex [50]. Comparisons between social calls and pulse–echo CF pairs revealed that left DSCF neurons responded equally well to both stimulus types, whereas right DSCF neurons were more responsive to pulse–echo CF pairs than to social calls. Subsequent comparisons between CF pairs and linear FMs revealed that right DSCF neurons only responded to CFs and relatively slow FM rates (<0.5 kHz/ms), left DSCF neurons responded to CFs and a wider range of FM rates, and this hemispheric difference appeared to be more pronounced in males than in females [67].

Most mustached bat social calls often contain FMs with rapid rates, so a lack of the right DSCF neural responses to many social calls is in accordance with their general lack of responses to rapid FMs. The robust responses of DSCF neurons in both hemispheres to slow FM rates (0.04 kHz/ms) of upward FMs also does not correspond to the gradually increasing Doppler shifts (<0.01 kHz/ms) of the CF in echoes returning from approaching targets (Mueller and Kanwal, unpublished). CF and quasi-CF calls emitted by mustached bats, however, do contain slow modulations near 0.04 kHz/ms in the echo-CF_2 range [81]. Further, CF and quasi-CF calls, particularly the QCFl call, are frequently emitted by males during their affiliative interactions between other males and females [85]. Our results of shorter peak response latencies and lower BAs for slow FMs on the left, therefore, support the left-hemispheric specialization for processing social calls [50] though sex differences in call processing need further investigation. We propose that studies of hemispheric specialization for social calls in mammals, typically restricted to tests at single sound levels, could benefit from testing wider amplitude ranges and FM rates present in conspecific social calls. Future investigations can also help determine whether the amplitude-related asymmetry observed here is a direct consequence of an asymmetry for spectral versus temporal processing or whether they are independent aspects of auditory processing asymmetries in highly social and vocal species, including humans.

One potential limitation to our study is that functional hemispheric asymmetries, in general, can be modulated over time by stress [88,89]. Such stress-related changes on functional hemispheric asymmetries have been found in classes as diverse as Chondrichthyes, Osteichthyes, amphibians, aves, and mammals (humans included). The influence of stress on functional hemispheric asymmetries can be attributed either to steroid hormones (e.g., androgens, estrogens, progestogens, glucocorticoids, and mineralocorticoids), hemispheric differences in emotional processing, or an interaction thereof. For instance, it has been suggested that the regulation of cortisol secretion during stress is regulated by right-hemispheric neural circuitry [90]. Further, the relationship between stress and functional hemispheric asymmetries may also depend on the developmental stage along with the type of stressor and its duration [91]. Experimental conditions such as those used in highly invasive techniques such as single-unit recording or even non-invasive techniques predicated on confined environments such as fMRI have the potential to induce stress. Certain functional hemispheric asymmetries in mustached bats reported above and previously [67] are highly sex-dependent, suggesting that such asymmetries are subject to changes in sex hormones. It is thus likely that other steroid hormones, such as those regulated by stress, would similarly impact the cortical asymmetries observed in mustached bats. The mustached bats used in our study were free to fly and socialize in a temperature and humidity-controlled environment that simulates the caves they naturally inhabit. Nonetheless, keeping these bats in a confined environment may have caused them stress and influenced their hemispheric differences in a multitude of ways. Thus, designing experiments to assess the influence of stress on hemispheric asymmetries for processing social calls in mustached bats represents a key future direction.

It is important to contextualize these and previous findings in mustached bats with hemispheric specializations reported in other species. One classic paper employed cladographic comparisons to identify vertebrate orders for which there is evidence supporting or negating lateralization of conspecific social call production and/or perception [57]. This extensive review details evidence from the literature supporting hemispheric specialization for conspecific social calls in species as diverse as fish, frogs, songbirds, parrots, eagles, horses, dogs, seals, rodents, and primates. Interestingly, based on the literature at the time of its publication, this review stated that "no general left-hemispheric dominance for the auditory perception of conspecific vocalization comparable to humans exists in bats ... ", apparently because a stereological study of Nissl-stained cells failed to find hemispheric differences between the left and right DSCF areas [92]. This conclusion was revised after two neurophysiological studies revealed functional hemispheric differences in the processing of social calls vs. echolocation signals [50] and linear FMs [67] in that same area.

Despite their exclusion from that cladographic analysis, mustached bats have much to offer the field of comparative cortical lateralization for social communication. First, the auditory cortices of mustached bats have evolved to process both their stereotypic biosonar signals during echolocation [68–71,93] and also to process their acoustically diverse social calls [76–80]. Thus, maps of the mustached bat auditory cortex reflect the stereotypic nature of their biosonar signals, making it one of the best-established auditory cortex maps amongst all mammals studied to date. Second, central and peripheral auditory structures evolved to be hypertrophic in mustached bats [83] and other high-duty cycle (i.e., CF-FM) chiroptera [94], making them relatively easy to probe and/or image. Third, the review described above advocates for neuroimaging (e.g., fMRI) approaches to studying hemispheric specialization for audiovocal communication in animals. Here, mustached bats have the advantage that the acoustic frequencies of MRI scanner noise largely fall below this species' range of hearing [95]. Thus, the results presented above and previous findings suggest that mustached bats have much to offer from their unique position amongst animal models of hemispheric specialization for auditory social communication.

Author Contributions: Conceptualization: S.D.W., D.L.P., G.A.K., and J.S.K.; Methodology: S.D.W., D.L.P., and J.S.K.; Software: S.D.W., D.L.P., and G.A.K.; Validation: S.D.W.; Formal Analysis: S.D.W. and D.L.P.; Investigation: S.D.W.; Resources: J.S.K.; Data Curation: S.D.W., G.A.K., J.S.K.; Writing—Original Draft Preparation: S.D.W.; Writing—Review and Editing: S.D.W., D.L.P., G.A.K., and J.S.K.; Visualization: S.D.W., D.L.P., and J.S.K.; Supervision: D.L.P., G.A.K., and J.S.K.; Project Administration: J.S.K.; Funding Acquisition: S.D.W. and J.S.K. All authors have read and agreed to the published version of the manuscript.

Funding: This work was supported in part by the National Institutes of Health Grants DC02054 and DC008822 (to J.S. Kanwal), DC75763 (to S.D. Washington), HD046388 (to V. Gallo), and an Erasmus Mundus Auditory Cognitive Exchange Program Fellowship (to S.D. Washington). This project was also supported in part by the NIH/NIMHD U54MD007597 grant and the NIH/HICHD P50HD105328 grant.

Data Availability Statement: The data presented in this study are available on request from the corresponding author.

Acknowledgments: The Ministry of Agriculture, Land and Marine Resources in Trinidad kindly permitted us to export mustached bats, and F. Muradali assisted with collection and exportation procedures. We thank Gerd Schüller and Leon Der for the design and fabrication of the stereotaxic recording setup. We also thank Tsang-Wei Tu, Annemie Van der Linden, Marleen Verhoye, and Paul Wang for supporting our work on this manuscript.

Conflicts of Interest: The authors declare no conflict of interest.

References

1. Wernicke, K. *Der Aphasische Symptomencomplex: Eine Psychologische Studie auf Anatomischer Basis Breslau*; Crohn, M., Ed.; Cohn Weigert: Breslau, Poland, 1874; Available online: https://wellcomecollection.org/works/dwv5w9rw (accessed on 15 February 2021).
2. Robson, H.; Sage, K.; Ralph, M.A.L. Wernicke's aphasia reflects a combination of acoustic-phonological and semantic control deficits: A case-series comparison of Wernicke's aphasia, semantic dementia and semantic aphasia. *Neuropsychologia* **2012**, *50*, 266–275. [CrossRef]
3. Belin, P.; Zilbovicius, M.; Crozier, S.; Thivard, L.; Fontaine, A.A.; Masure, M.-C.; Samson, Y. Lateralization of Speech and Auditory Temporal Processing. *J. Cogn. Neurosci.* **1998**, *10*, 536–540. [CrossRef]
4. Mäkelä, A.M.; Alku, P.; May, P.; Mäkinen, V.; Tiitinen, H. Left-hemispheric brain activity reflects formant transitions in speech sounds. *Neuro Rep.* **2005**, *16*, 549–553. [CrossRef]
5. Schwartz, J.; Tallal, P. Rate of acoustic change may underlie hemispheric specialization for speech perception. *Science* **1980**, *207*, 1380–1381. [CrossRef] [PubMed]
6. Flinker, A.; Doyle, W.K.; Mehta, A.D.; Devinsky, O.; Poeppel, D. Spectrotemporal modulation provides a unifying framework for auditory cortical asymmetries. *Nat. Hum. Behav.* **2019**, *3*, 393–405. [CrossRef] [PubMed]
7. Boemio, A.; Fromm, S.J.; Braun, A.R.; Poeppel, D. Hierarchical and asymmetric temporal sensitivity in human auditory cortices. *Nat. Neurosci.* **2005**, *8*, 389–395. [CrossRef] [PubMed]
8. Schönwiesner, M.; Rübsamen, R.; Von Cramon, D.Y. Hemispheric asymmetry for spectral and temporal processing in the human antero-lateral auditory belt cortex. *Eur. J. Neurosci.* **2005**, *22*, 1521–1528. [CrossRef] [PubMed]
9. Brown, C.P.; Fitch, R.H.; Tallal, P. Sex and Hemispheric Differences for Rapid Auditory Processing in Normal Adults. *Laterality Asymmetries Body Brain Cogn.* **1999**, *4*, 39–50. [CrossRef] [PubMed]
10. Divenyi, P.L.; Robinson, A.J. Nonlinguistic auditory capabilities in aphasia. *Brain Lang.* **1989**, *37*, 290–326. [CrossRef]
11. Merzenich, M.M.; Jenkins, W.M.; Johnston, P.; Schreiner, C.; Miller, S.L.; Tallal, P. Temporal Processing Deficits of Language-Learning Impaired Children Ameliorated by Training. *Science* **1996**, *271*, 77–81. [CrossRef]
12. Tallal, P.; Miller, S.; Fitch, R.H. Neurobiological Basis of Speech: A Case for the Preeminence of Temporal Processing. *Ann. N. Y. Acad. Sci.* **1993**, *682*, 27–47. [CrossRef] [PubMed]
13. Temple, E.; Poldrack, R.A.; Protopapas, A.; Nagarajan, S.; Salz, T.; Tallal, P.; Merzenich, M.M.; Gabrieli, J.D.E. Disruption of the neural response to rapid acoustic stimuli in dyslexia: Evidence from functional MRI. *Proc. Natl. Acad. Sci. USA* **2000**, *97*, 13907–13912. [CrossRef] [PubMed]
14. Robin, D.A.; Tranel, D.; Damasio, H. Auditory perception of temporal and spectral events in patients with focal left and right cerebral lesions. *Brain Lang.* **1990**, *39*, 539–555. [CrossRef]
15. Efron, R. Temporal Perception, Aphasia and D'ej'a Vu. *Brain* **1963**, *86*, 403–424. [CrossRef]
16. Jamison, H.L.; Watkins, K.E.; Bishop, D.; Matthews, P.M. Hemispheric Specialization for Processing Auditory Nonspeech Stimuli. *Cereb. Cortex* **2005**, *16*, 1266–1275. [CrossRef]
17. Zatorre, R.J.; Belin, P. Spectral and temporal processing in human auditory cortex. *Cereb. Cortex* **2001**, *11*, 946–953. [CrossRef]

18. Sidtis, J.J. On the nature of the cortical function underlying right hemisphere auditory perception. *Neuropsychologia* **1980**, *18*, 321–330. [CrossRef]
19. Sidtis, J.J. The complex tone test: Implications for the assessment of auditory laterality effects. *Neuropsychologia* **1981**, *19*, 103–112. [CrossRef]
20. Samson, S.; Zatorre, R.J. Melodic and harmonic discrimination following unilateral cerebral excision. *Brain Cogn.* **1988**, *7*, 348–360. [CrossRef]
21. Zatorre, R.J. Pitch perception of complex tones and human temporal-lobe function. *J. Acoust. Soc. Am.* **1988**, *84*, 566–572. [CrossRef]
22. Zatorre, R.J. Discrimination and recognition of tonal melodies after unilateral cerebral excisions. *Neuropsychologia* **1985**, *23*, 31–41. [CrossRef]
23. Samson, S.; Zatorre, R.J. Contribution of the right temporal lobe to musical timbre discrimination. *Neuropsychologia* **1994**, *32*, 231–240. [CrossRef]
24. Zatorre, R.J.; Evans, A.C.; Meyer, E. Neural mechanisms underlying melodic perception and memory for pitch. *J. Neurosci.* **1994**, *14*, 1908–1919. [CrossRef]
25. Milner, B. Laterality effects in audition. In *Interhemispheric Relations and Cerebral Dominance*; Mountcastle, V.B., Ed.; Johns Hopkins Press: Baltimore, MD, USA, 1962; pp. 177–195.
26. Robinson, T.; Fallside, F. A recurrent error propagation network speech recognition system. *Comput. Speech Lang.* **1991**, *5*, 259–274. [CrossRef]
27. Lakshminarayanan, K.; Ben Shalom, D.; van Wassenhove, V.; Orbelo, D.; Houde, J.; Poeppel, D. The effect of spectral manipulations on the identification of affective and linguistic prosody. *Brain Lang.* **2003**, *84*, 250–263. [CrossRef]
28. Zatorre, R.J.; Belin, P.; Penhune, V.B. Structure and function of auditory cortex: Music and speech. *Trends Cogn. Sci.* **2002**, *6*, 37–46. [CrossRef]
29. Poeppel, D. The analysis of speech in different temporal integration windows: Cerebral lateralization as 'asymmetric sampling in time'. *Speech Commun.* **2003**, *41*, 245–255. [CrossRef]
30. Joos, M. *Acoustic Phonetics*; Linguistic Society of America: Baltimore, MD, USA, 1948; 136p.
31. Schuller, B.; Batliner, A. *Computational Paralinguistics: Emotion, Affect and Personality in Speech and Language Processing*, 1st ed.; John Wiley and Sons, Ltd.: West Sussex, UK, 2014; p. 344.
32. Lansdell, H. Sex Differences in Hemispheric Asymmetries of the Human Brain. *Nat. Cell Biol.* **1964**, *203*, 550. [CrossRef] [PubMed]
33. McGlone, J. Sex Differences in the Cerebral Organization of Verbal Functions in Patients with Unilateral Brain Lesions. *Brain* **1977**, *100*, 775–793. [CrossRef] [PubMed]
34. Shaywitz, B.A.; Shaywltz, S.E.; Pugh, K.R.; Constable, R.T.; Skudlarski, P.; Fulbright, R.K.; Bronen, R.A.; Fletcher, J.M.; Shankweiler, D.P.; Katz, L.; et al. Sex differences in the functional organization of the brain for language. *Nat. Cell Biol.* **1995**, *373*, 607–609. [CrossRef]
35. Evans, T.M.; Flowers, D.L.; Napoliello, E.M.; Eden, G.F. Sex-specific gray matter volume differences in females with developmental dyslexia. *Brain Struct. Funct.* **2013**, *219*, 1041–1054. [CrossRef] [PubMed]
36. Liberman, A.; Mattingly, I. A specialization for speech perception. *Science* **1989**, *243*, 489–494. [CrossRef]
37. Ringo, J.L.; Doty, R.W.; Demeter, S.; Simard, P.Y. Time Is of the Essence: A Conjecture that Hemispheric Specialization Arises from Interhemispheric Conduction Delay. *Cereb. Cortex* **1994**, *4*, 331–343. [CrossRef]
38. Poremba, A.; Malloy, M.; Saunders, R.C.; Carson, R.E.; Herscovitch, P.; Mishkin, M. Species-specific calls evoke asymmetric activity in the monkey's temporal poles. *Nature* **2004**, *427*, 448–451. [CrossRef]
39. Heffner, H.E.; Heffner, R.S. Temporal lobe lesions and perception of species-specific vocalizations by Macaques. *Science* **1984**, *226*, 75–76. [CrossRef] [PubMed]
40. Heffner, E.H.; Heffner, R.S.; Hefner, H.E. Effect of unilateral and bilateral auditory cortex lesions on the discrimination of vocalizations by Japanese macaques. *J. Neurophysiol.* **1986**, *56*, 683–701. [CrossRef]
41. Hook, M.; Rogers, L. Lateralized use of the mouth in production of vocalizations by marmosets. *Neuropsychologia* **1998**, *36*, 1265–1273. [CrossRef]
42. Gannon, P.J.; Holloway, R.L.; Broadfield, D.C.; Braun, A.R. Asymmetry of Chimpanzee Planum Temporale: Humanlike Pattern of Wernicke's Brain Language Area Homolog. *Science* **1998**, *279*, 220–222. [CrossRef] [PubMed]
43. Poremba, A.; Mishkin, M. Exploring the extent and function of higher-order auditory cortex in rhesus monkeys. *Hear. Res.* **2007**, *229*, 14–23. [CrossRef]
44. Voss, H.U.; Tabelow, K.; Polzehl, J.; Tchernichovski, O.; Maul, K.K.; Salgado-Commissariat, D.; Ballon, D.; Helekar, S.A. Functional MRI of the zebra finch brain during song stimulation suggests a lateralized response topography. *Proc. Natl. Acad. Sci. USA* **2007**, *104*, 10667–10672. [CrossRef]
45. Poirier, C.; Boumans, T.; Verhoye, M.; Balthazart, J.; Van Der Linden, A. Own-Song Recognition in the Songbird Auditory Pathway: Selectivity and Lateralization. *J. Neurosci.* **2009**, *29*, 2252–2258. [CrossRef] [PubMed]
46. Nottebohm, F.; Arnold, A.P. Sexual dimorphism in vocal control areas of the songbird brain. *Science* **1976**, *194*, 211–213. [CrossRef] [PubMed]
47. DeVoogd, T.J.; Nottebohm, F. Sex differences in dendritic morphology of a song control nucleus in the canary: A quantitative Golgi study. *J. Comp. Neurol.* **1981**, *196*, 309–316. [CrossRef]

48. Phan, M.; Vicario, D.S. Hemispheric differences in processing of vocalizations depend on early experience. *Proc. Natl. Acad. Sci. USA* **2010**, *107*, 2301–2306. [CrossRef]
49. George, I.; Cousillas, H.; Richard, J.-P.; Hausberger, M. State-dependent hemispheric specialization in the songbird brain. *J. Comp. Neurol.* **2005**, *488*, 48–60. [CrossRef]
50. Kanwal, J.S. Right-left asymmetry in the cortical processing of sounds for social communication vs. navigation in mustached bats. *Eur. J. Neurosci.* **2011**, *35*, 257–270. [CrossRef]
51. Ehret, G. Left hemisphere advantage in the mouse brain for recognizing ultrasonic communication calls. *Nat. Cell Biol.* **1987**, *325*, 249–251. [CrossRef] [PubMed]
52. Levy, R.B.; Marquarding, T.; Reid, A.P.; Pun, C.M.; Renier, N.; Oviedo, H.V. Circuit asymmetries underlie functional lateralization in the mouse auditory cortex. *Nat. Commun.* **2019**, *10*, 2783. [CrossRef] [PubMed]
53. Wetzel, W.; Ohl, F.W.; Scheich, H. Global versus local processing of frequency-modulated tones in gerbils: An animal model of lateralized auditory cortex functions. *Proc. Natl. Acad. Sci. USA* **2008**, *105*, 6753–6758. [CrossRef]
54. Güntürkün, O.; Ströckens, F.; Ocklenburg, S. Brain Lateralization: A Comparative Perspective. *Physiol. Rev.* **2020**, *100*, 1019–1063. [CrossRef]
55. Tallal, P. Of bats and men. *J. Neurophysiol.* **2012**, *108*, 1545–1547. [CrossRef]
56. Andics, A.; Miklósi, Á. Neural processes of vocal social perception: Dog-human comparative fMRI studies. *Neurosci. Biobehav. Rev.* **2018**, *85*, 54–64. [CrossRef] [PubMed]
57. Ocklenburg, S.; Ströckens, F.; Güntürkün, O. Lateralisation of conspecific vocalisation in non-human vertebrates. *Laterality Asymmetries Body Brain Cogn.* **2013**, *18*, 1–31. [CrossRef] [PubMed]
58. Boye, M.; Güntürkün, O.; Vauclair, J. Right ear advantage for conspecific calls in adults and subadults, but not infants, California sea lions (*Zalophus californianus*): Hemispheric specialization for communication? *Eur. J. Neurosci.* **2005**, *21*, 1727–1732. [CrossRef] [PubMed]
59. Bauer, R.H. Lateralization of neural control for vocalization by the frog (*Rana pipiens*). *Psychobiology* **1993**, *21*, 243–248. [CrossRef]
60. Andics, A.; Gábor, A.; Gácsi, M.; Faragó, T.; Szabó, D.; Miklósi, Á. Neural mechanisms for lexical processing in dogs. *Science* **2016**, *353*, 1030–1032. [CrossRef]
61. Andics, A.; Gácsi, M.; Faragó, T.; Kis, A.; Miklósi, Á. Voice-Sensitive Regions in the Dog and Human Brain Are Revealed by Comparative fMRI. *Curr. Biol.* **2014**, *24*, 574–578. [CrossRef]
62. Siniscalchi, M.; D'Ingeo, S.; Quaranta, A. Lateralized emotional functioning in domestic animals. *Appl. Anim. Behav. Sci.* **2021**, *237*, 105282. [CrossRef]
63. Smith, A.V.; Proops, L.; Grounds, K.; Wathan, J.; Scott, S.K.; McComb, K. Domestic horses (*Equus caballus*) discriminate between negative and positive human nonverbal vocalisations. *Sci. Rep.* **2018**, *8*, 13052. [CrossRef]
64. Van Ruijssevelt, L.; Washington, S.D.; Hamaide, J.; Verhoye, M.; Keliris, G.A.; Van Der Linden, A. Song Processing in the Zebra Finch Auditory Forebrain Reflects Asymmetric Sensitivity to Temporal and Spectral Structure. *Front. Neurosci.* **2017**, *11*, 549. [CrossRef]
65. Pagliaro, A.H.; Arya, P.; Piristine, H.; Lord, J.S.; Gobes, S.M. Bilateral brain activity in auditory regions is necessary for successful vocal learning in songbirds. *Neurosci. Lett.* **2020**, *718*, 134730. [CrossRef]
66. Fitch, R.H.; Brown, C.P.; O'Connor, K.; Tallal, P. Functional lateralization for auditory temporal processing in male and female rats. *Behav. Neurosci.* **1993**, *107*, 844–850. [CrossRef] [PubMed]
67. Washington, S.D.; Kanwal, J.S. Sex-dependent hemispheric asymmetries for processing frequency-modulated sounds in the primary auditory cortex of the mustached bat. *J. Neurophysiol.* **2012**, *108*, 1548–1566. [CrossRef] [PubMed]
68. Suga, N.; Jen, P. Disproportionate tonotopic representation for processing CF-FM sonar signals in the mustache bat auditory cortex. *Science* **1976**, *194*, 542–544. [CrossRef] [PubMed]
69. Suga, N. The extent to which biosonar information is represented in the bat auditory cortex. In *Neurocomputing 2: Directions for Research*; Anderson, J.A., Pellionisz, A., Rosenfeld, E., Eds.; MIT Press: Cambridge, MA, USA, 1985; pp. 259–294.
70. O'Neill, W.; Suga, N. Target range-sensitive neurons in the auditory cortex of the mustache bat. *Science* **1979**, *203*, 69–73. [CrossRef]
71. Suga, N.; O'Neill, W. Neural axis representing target range in the auditory cortex of the mustache bat. *Science* **1979**, *206*, 351–353. [CrossRef] [PubMed]
72. Suga, N. Specialization of the auditory system for the processing of bio-sonar information in the frequency domain: Mustached bats. *Hear. Res.* **2018**, *361*, 1–22. [CrossRef]
73. Xiao, Z.; Suga, N. Reorganization of the cochleotopic map in the bat's auditory system by inhibition. *Proc. Natl. Acad. Sci. USA* **2002**, *99*, 15743–15748. [CrossRef]
74. Kanwal, J.S.; Fitzpatrick, D.C.; Suga, N. Facilitatory and inhibitory frequency tuning of combination-sensitive neurons in the primary auditory cortex of mustached bats. *J. Neurophysiol.* **1999**, *82*, 2327–2345. [CrossRef]
75. Fitzpatrick, D.C.; Kanwal, J.S.; Butman, J.; Suga, N. Combination-sensitive neurons in the primary auditory cortex of the mustached bat. *J. Neurosci.* **1993**, *13*, 931–940. [CrossRef] [PubMed]
76. Fitzpatrick, D.C.; Suga, N.; Olsen, J.F. Distribution of response types across entire hemispheres of the mustached bat's auditory cortex. *J. Comp. Neurol.* **1998**, *391*, 353–365. [CrossRef]

77. Washington, S.D.; Hamaide, J.; Jeurissen, B.; Van Steenkiste, G.; Huysmans, T.; Sijbers, J.; Deleye, S.; Kanwal, J.S.; De Groof, G.; Liang, S.; et al. A three-dimensional digital neurological atlas of the mustached bat (*Pteronotus parnellii*). *NeuroImage* **2018**, *183*, 300–313. [CrossRef] [PubMed]
78. Esser, K.-H.; Condon, C.J.; Suga, N.; Kanwal, J.S. Syntax processing by auditory cortical neurons in the FM-FM area of the mustached bat *Pteronotus parnellii*. *Proc. Natl. Acad. Sci. USA* **1997**, *94*, 14019–14024. [CrossRef]
79. Hauser, M.D.; Konishi, M. Processing Species-specific Calls by Combination-sensitive Neurons in an Echolocating Bat. In *The Design of Animal Communication*; The MIT Press: Cambridge, MA, USA, 2003; pp. 135–157.
80. Kanwal, J.S. A distributed cortical representation of social communication calls. In *Behavior and Neurodynamics for Auditory Communication*; Kanwal, J.S., Ehret, G., Eds.; Cambridge University Press: New York, NY, USA, 2006; pp. 156–188.
81. Medvedev, A.V.; Kanwal, J.S. Local Field Potentials and Spiking Activity in the Primary Auditory Cortex in Response to Social Calls. *J. Neurophysiol.* **2004**, *92*, 52–65. [CrossRef] [PubMed]
82. Washington, S.D.; Kanwal, J.S. DSCF Neurons Within the Primary Auditory Cortex of the Mustached Bat Process Frequency Modulations Present Within Social Calls. *J. Neurophysiol.* **2008**, *100*, 3285–3304. [CrossRef]
83. Kanwal, J.S.; Matsumura, S.; Ohlemiller, K.; Suga, N. Analysis of acoustic elements and syntax in communication sounds emitted by mustached bats. *J. Acoust. Soc. Am.* **1994**, *96*, 1229–1254. [CrossRef]
84. Washington, S.D.; Tillinghast, J.S. Conjugating time and frequency: Hemispheric specialization, acoustic uncertainty, and the mustached bat. *Front. Neurosci.* **2015**, *9*, 143. [CrossRef]
85. Clement, M.J.; Gupta, P.; Dietz, N.; Kanwal, J.S. Audiovocal Communication and Social Behavior in Mustached Bats. In *Behavior and Neurodynamics for Auditory Communication*; Kanwal, J.S., Ehret, G., Eds.; Cambridge University Press: New York, NY, USA, 2006; pp. 57–84.
86. Washington, S.D.; Kanwal, J.S. Linear FM Synthesis: Test Stimuli for Rapid Analysis of Auditory Neurodynamics. *Hoya Tech. Res. Bull.* **2004**, *2004120*, 2004.
87. Ma, X.; Suga, N. Corticofugal Modulation of the Paradoxical Latency Shifts of Inferior Collicular Neurons. *J. Neurophysiol.* **2008**, *100*, 1127–1134. [CrossRef]
88. Ocklenburg, S.; Korte, S.M.; Peterburs, J.; Wolf, O.T.; Güntürkün, O. Stress and laterality—The comparative perspective. *Physiol. Behav.* **2016**, *164*, 321–329. [CrossRef]
89. Cory-Slechta, D.A.; Weston, D.; Liu, S.; Allen, J.L. Brain Hemispheric Differences in the Neurochemical Effects of Lead, Prenatal Stress, and the Combination and Their Amelioration by Behavioral Experience. *Toxicol. Sci.* **2013**, *132*, 419–430. [CrossRef] [PubMed]
90. Wittling, W.; Pflüger, M. Neuroendocrine hemisphere asymmetries: Salivary cortisol secretion during lateralized viewing of emotion-related and neutral films. *Brain Cogn.* **1990**, *14*, 243–265. [CrossRef]
91. Brüne, M.; Nadolny, N.; Güntürkün, O.; Wolf, O.T. Stress induces a functional asymmetry in an emotional attention task. *Cogn. Emot.* **2013**, *27*, 558–566. [CrossRef] [PubMed]
92. Sherwood, C.C.; Raghanti, M.A.; Wenstrup, J.J. Is humanlike cytoarchitectural asymmetry present in another species with complex social vocalization? A stereologic analysis of mustached bat auditory cortex. *Brain Res.* **2005**, *1045*, 164–174. [CrossRef]
93. Suga, N.; O'Neill, W.; Manabe, T. Harmonic-sensitive neurons in the auditory cortex of the mustache bat. *Science* **1979**, *203*, 270–274. [CrossRef] [PubMed]
94. Hsiao, C.J.; Jen, P.H.-S.; Wu, C.-H. The cochlear size of bats and rodents derived from MRI images and histology. *Neuro Rep.* **2015**, *26*, 478–482. [CrossRef] [PubMed]
95. Kamada, K.; Pekar, J.J.; Kanwal, J.S. Anatomical and functional imaging of the auditory cortex in awake mustached bats using magnetic resonance technology. *Brain Res. Protoc.* **1999**, *4*, 351–359. [CrossRef]

Article

Lateralization of Auditory Processing of Silbo Gomero

Pamela Villar González, Onur Güntürkün and Sebastian Ocklenburg *

Institute of Cognitive Neuroscience, Biopsychology, Department of Psychology, Ruhr University Bochum, 44801 Bochum, Germany; Pamela.VillarGonzalez@ruhr-uni-bochum.de or pamela.villargonzalez@rub.de (P.V.G.); onur.guentuerkuen@rub.de (O.G.)

* Correspondence: sebastian.ocklenburg@rub.de; Tel.: +49-234-32-24323

Received: 22 April 2020; Accepted: 5 July 2020; Published: 17 July 2020

Abstract: Left-hemispheric language dominance is a well-known characteristic of the human language system. However, it has been shown that leftward language lateralization decreases dramatically when people communicate using whistles. Whistled languages present a transformation of a spoken language into whistles, facilitating communication over great distances. In order to investigate the laterality of Silbo Gomero, a form of whistled Spanish, we used a vocal and a whistled dichotic listening task in a sample of 75 healthy Spanish speakers. Both individuals that were able to whistle and to understand Silbo Gomero and a non-whistling control group showed a clear right-ear advantage for vocal dichotic listening. For whistled dichotic listening, the control group did not show any hemispheric asymmetries. In contrast, the whistlers' group showed a right-ear advantage for whistled stimuli. This right-ear advantage was, however, smaller compared to the right-ear advantage found for vocal dichotic listening. In line with a previous study on language lateralization of whistled Turkish, these findings suggest that whistled language processing is associated with a decrease in left and a relative increase in right hemispheric processing. This shows that bihemispheric processing of whistled language stimuli occurs independent of language.

Keywords: Silbo Gomero; whistle language; cerebral lateralization; brain asymmetry; dichotic listening task

---

1. Introduction

Both the left and the right hemispheres contribute to language processing, but they are relevant for different aspects of how language is processed. The auditory language comprehension model by Friederici [1] assumes that the left hemisphere is dominant for the processing of syntactic structures, semantic relations, grammatical and thematic relations, and information integration when spoken language is perceived. In contrast, the right hemisphere is dominant for the processing of prosody, intonational phrasing, and accentuation focus. This implies that if a language is processed that requires a greater amount of prosody processing to be understood correctly than spoken language, greater right-hemispheric activation should be expected.

Typically, processing spoken language activates a larger network of brain areas in the left than in the right hemisphere [2,3]. Overall, 96% of strong right-handers, 85% of ambidextrous individuals, and 83% of strong left-handers show left-hemispheric language dominance [4,5]. Left-hemispheric language dominance has been reported for both atonal [6,7] and tonal languages [8], and also for writing [9] as well as sign languages [10–12]. It has been suggested that this dominance of the left hemisphere is caused by superiority to assess fast temporal changes in auditory input, making the left hemisphere ideally suited to analyze voice onset times of different syllables [13–15].

One interesting way to investigate whether language lateralization indeed is based on these properties of the left hemisphere is to compare the processing of stimuli that require different degrees of fast temporal processing in order to be understood correctly. Here, it has been suggested that comparing vocal and whistled languages might be a meaningful approach to do so [16]. A whistled language is a system of communication based on whistling. Whistled articulated languages provide an unlimited number of messages [17], and a whistled sound has a higher pitch and intensity than one produced on the vocal cords, meaning a whistle is the most powerful sound that a human being can produce without any external tool [18]. Whistled languages utilize the vocabulary, grammar, syntax of the local speech, and even the phonology [19], with a reduction in phonemes [18,20], lending major importance to the context in the conversation. Except for isolated cases, whistled languages were, and still are, used for communication over long distances [19]. Depending on the atmospheric conditions (distance to the sea, air, presence or absence of mountains and valleys), a whistled message can be understood in distances over 3 km [19]. Today, 70 whistled languages are still in use [17].

One of the most common uses of whistle languages is communication among shepherds when they work in mountainous regions. This is also the main use of Silbo Gomero (meaning "whistle from La Gomera" in Spanish), a communication system based on Spanish still used today on the Canary Islands in Spain. While historically, there were a number of different whistled languages in use on the Canary Islands [18,20], Silbo Gomero is the only one still widely used today. Despite the arrival of newer communication technologies such as cell phones, Silbo Gomero is still taught in some schools on the Canary Islands. Younger pupils often learn it as part of their cultural heritage education, not so much for necessity. However, it still is very useful in natural contexts like trekking or in several parts of the islands where the telecommunication network is not powerful enough to ensure coverage.

Lateralization of whistled languages is still not well understood. The first and only study to investigate language lateralization in whistled languages was conducted in Turkey [16]. In this study, 31 proficient whistled Turkish speaking participants were tested with a dichotic listening task divided into two sections: hearing spoken Turkish syllables in the first one and whistled Turkish syllables in the second one. Dichotic listening is one of the most commonly used behavioral tasks to assess language lateralization [21,22]. Participants listen to pairs of syllables simultaneously on headphones. They have to indicate which syllable they understood best, and typically a right-ear/left-hemisphere advantage is observed for spoken syllables. In contrast to that, it was demonstrated that whistled Turkish is processed more bilaterally than spoken Turkish [16].

For Silbo Gomero, language lateralization has not been investigated yet. However, an fMRI study of Silbo Gomero has been performed, in which samples of spoken and whistled Spanish sentences and isolated words were presented to a group of five proficient whistlers (Silbadores) and a control group who were Spanish speakers but unfamiliar with Silbo Gomero [23]. The results indicate that the temporal regions of the left hemisphere that are usually associated with spoken-language function were also engaged during the processing of Silbo in experienced Silbadores. Both passive listening and active-monitoring tasks produced a common activation in the left superior posterior temporal gyrus. Activation of the right superior–midtemporal region was also evident across both the Silbo and Spanish speech conditions. Furthermore, activity increased in the right temporal lobe in response to non-linguistic pitch changes, tones, and complex sounds, but according to the authors, the same regions may also be associated with linguistic processing tasks. Group analysis indicated that the areas activated during both Spanish and Silbo processing in Silbadores differed significantly from those activated in non-whistlers. In particular, there was less ventral–anterior temporal activation during the processing of Silbo Gomero than during speech processing. The authors argued that this is due to there being less need to correctly identify specific phonological contrasts. Moreover, there was a stronger premotor activation for Silbo Gomero.

In the present study, we used the dichotic listening task to investigate language lateralization in Silbo Gomero. On the one hand, we wanted to test whether the results obtained in Turkish whistlers in Küşköy [16] could be replicated with Silbo Gomero in Tenerife and Gran Canaria. On the other hand, we wanted to test whether experience with the whistled language modulated the extent of left- and right-hemispheric contributions. To answer these questions, we tested 75 Spanish speakers separated into a non-whistling control group (CG) and an experimental group that was able to whistle and understand Silbo Gomero (WG) with vocal and whistle dichotic listening [16]. Based on the literature, we expected to find leftward lateralization in the spoken dichotic task, but a more bilateral pattern in the whistled dichotic listening. Moreover, within the WG, we assessed experience with Silbo Gomero by comparing individuals that were still learning Silbo Gomero with experienced whistlers. This was done since whistle experience and aptitude to learn a new language [24] could affect overall performance in the whistled dichotic listening task, as well as the lateralization pattern.

2. Materials and Methods

2.1. Participants

The sample consisted of 75 native Spanish speakers aged between 15 and 80 years. The cohort was separated into two groups according to their Silbo Gomero abilities. Participants in the control group (CG) were not able to whistle or understand Silbo Gomero (n = 25; aged between 22 and 80 years, mean age: 35.24 years, SD: 13.81; 12 women; 13 men). In contrast, participants in the whistlers' group (WG) were able to whistle and understand Silbo Gomero (n = 50; aged between 15 and 57 years, mean age: 38.27 years, SD: 10.30; 17 women and 33 men). There was no significant age difference between the WG and the CG ($t_{(67)} = -1.04$, $p = 0.30$).

Handedness was determined using the Edinburgh Handedness Inventory (EHI) [25]. The EHI is a ten-item questionnaire designed to assess handedness by self-report of the preferred hand for performing common activities such as writing and using utensils such as a toothbrush. Participants had five different answer options. They could indicate that they always used their right/left hand for a specific activity, mostly used their right/left hand for a specific activity, or used both hands equally for a specific activity. A laterality quotient (LQ) was calculated using the formula $LQ = [(R - L)/(R + L)] \times 100$. A score of 100 reflects consistent right-handedness, while a score of −100 reflects consistent left-handedness. In the CG, there were 3 left-handers and 22 right-handers. In the WG, there were 5 left-handers and 45 right-handers. There were no significant differences in the frequency of left-handedness between the two experimental groups ($p = 0.79$). Additionally, we compared the EHI LQs for the two groups. In the CG, the mean LQ was 51.62 (SD = 54.06; range: −100 to 100). In the WG, the mean LQ was 47.62 (SD = 40.92; range: −71.43 to 100). There was no significant difference in EHI LQ between the two groups ($t_{(73)} = 0.36$; $p = 0.72$). We did include left-handed participants on purpose in the sample. Left-handers represent a substantial portion of the human population (10.6%) and it has recently been argued that they need to be included in laterality studies, as they are an important part of the normal range of human diversity [26]. Thus, excluding them would give a skewed picture of the actual laterality patterns for Silbo Gomero.

In order to assess the effects of experience with Silbo Gomero on language lateralization, we further subdivided the WG into two groups. First, a learners group (LG) (aged between 15 and 57 years, mean age: 37.82 years, SD: 9.91; 11 women; 14 men) that included individuals that had been practicing Silbo Gomero for 3 years or less (mean time whistling: 1.16, SD: 0.47). Second, an advanced group (AG) (age between 18 and 57, mean age: 38.73 years, SD: 10.90; 6 women; 19 men), who had active experience with Silbo Gomero for more than 3 years (mean time whistling: 6.08, SD: 5.00). Participants had no history of any neurological or psychiatric diseases that could affect language perception or production. All participants had unimpaired hearing capabilities according to self-report. The local ethics committee of the psychological faculty at Ruhr-University Bochum approved the procedure. All participants gave written informed consent and were treated in accordance with the Declaration of Helsinki. For the one 15-year-old participant, parental informed consent was also obtained.

2.2. Language Skills

A questionnaire regarding language skills was handed to the participants. Subjects were asked to declare their first language: mother tongue(s), second spoken languages: the language and approximated level of competency (subjects who did not know their level according to the Common European Framework were given a subjective measure: low, medium or advanced). In addition, the subjects in the WG were asked to indicate the amount of time since they had learned to whistle. Subjects in the CG were asked whether they were aware of the existence of Silbo Gomero. Most of the participants in the WG learned Silbo Gomero in courses and the majority of the whistlers in the AG were active Silbo teachers at the time point at which the study was conducted. In general, participants presented a wide range of language skills including early bilingualism and second spoken languages. In the CG, 14 people were bilingual speakers (Spanish-Catalán, Spanish-Gallego, Spanish-Asturiano, and Spanish-German). The mean number of second spoken languages was 2.2 (SD: 1.04). In the LG, there were 3 bilingual speakers (Spanish-German). The mean number of second spoken languages in this group was 0.92 (SD: 0.76). In the AG, there were no bilingual speakers. In this group, the mean number of second spoken languages was 0.84 (SD: 0.62).

2.3. Dichotic Listening Task

Language lateralization was assessed using an auditory dichotic listening paradigm programmed and presented using Presentation® software (Neurobehavioral Systems, Inc., Albany, USA). A similar task has been used before to assess whistle language lateralization in Turkish whistle language speakers [16].

For the dichotic listening task, syllable pairs consisting of two out of five different consonant-vowel (CV) syllables (ba [ba], ca [ka], cha [tʃa], ga [ga], ya [ya]) were used as stimuli. The syllables were chosen according to the five groups of distinguishable consonants [18] for learners of Silbo Gomero. Overall, there were 25 different syllable pairs, five homonyms (ba/ba, ca/ca, cha/cha, ga/ga and ya/ya) and 20 heteronyms (ba/ca, ba/cha, ba/ga, ba/ya, ca/ba, ca/cha, ca/ga, ca/ya, cha/ba, cha/ca, cha/ga, cha/ya, ga/ba, ga/ca, ga/cha, ga/ya, ya/ba, ya/ca, ya/cha, ya/ga). The spoken stimuli were recorded by a native male Spanish speaker. The whistled stimuli were recorded by a proficient male whistler. Syllable onset within each syllable pair stimuli was set at the beginning of the sound file using Audacity® software (Trademark of Dominic Mazzoni, Pittsburgh, PA, USA). Stimuli had a mean duration of 300 ms for the spoken syllables, and 750 ms for the whistled syllables. The stimuli were presented via headphones (Beyerdynamic GmbH, Heilbronn, Germany) at 80 dB.

The keyboard was customized: five buttons were labeled with the presented syllables (ba, ca, cha, ga, ya). After the stimulus presentation, participants had to press one of five keys to indicate which of the syllables they had perceived more accurately. The inter-stimulus interval was fixed at two seconds.

The task was divided into two conditions: a vocal condition and a whistled condition. Each condition had one practice block that was not included in the final analysis and two experimental blocks. Independently from the preferred hand, participants started with right or left hand alternatively in a randomized fashion and changed the hand in the middle of each block. In the second test block, the headphones were reversed to avoid the possible confounding effects of slightly different noise levels coming from the left and the right headphone speaker (which should not exist). The practice block for both the vocal and the whistled condition consisted of 20 trials to get participants accustomed to the task and the tone of the syllables. Afterwards, the two experimental blocks for each condition were presented. Here, all of the possible combinations of the syllables were presented: 5 homonyms and 20 heteronyms twice for each ear (one for every ear-hand combination). Thus, the total number of trials was 100 for each condition (80 heteronym and 20 homonym trials)

2.4. Statistical Analysis

Statistics were performed using IBM SPSS Statistics for Windows, Version 21.0. (IBM Corp., Armonk, NY, USA). The dichotic laterality index (LI) of the participants was calculated as LI = [(RE − LE)/(RE + LE)] * 100, (RE = number of right ear responses, LE = number of left ear responses). This index varies between −100 and +100, with positive values indicating a right ear advantage (REA) and negative values indicating a left-ear advantage (LEA) [16]. Performance on the dichotic listening task was analyzed parametrically using ANOVAs. Neyman–Pearson correlation coefficients were determined in order to investigate possible relationships between the variables.

3. Results

3.1. Dichotic Listening

Table 1 shows the results of the dichotic listening task.

Table 1. Results of the dichotic listening task: number of correct right ear (RE) and left ear (LE) answers on heteronym trials for the control group (CG) and the whistlers' group (WG) for vocal and whistled dichotic listening, as well as errors and laterality indexes (LIs).

Condition	Variable	CG	WG
Vocal	RE	48.40 ± 1.99	46.66 ± 1.41
	LE	23.88 ± 1.92	26.36 ± 1.36
	Error	7.72 ± 1.11	6.98 ± 0.82
	LI	27.71 ± 4.04	22.01 ± 2.97
Whistle	RE	21.00 ± 1.50	29.52 ± 1.06
	LE	21.20 ± 1.06	24.10 ± 0.75
	Error	37.80 ± 1.46	26.38 ± 1.03
	LI	−0.48 ± 2.03	7.21 ± 2.46

3.2. Error Rates

In order to check whether participants in the WG showed better recognition of whistled syllables than participants in the CG, we compared error rates between the groups using a 2 × 2 repeated measures ANOVA with the within-subjects factor condition (VOCAL, WHISTLE) and the between-subjects factor group (WG, CG). The main effect of condition reached significance ($F_{(1,73)} = 690.19$; $p < 0.001$; partial $\eta^2 = 0.90$), indicating that overall, participants made more errors during whistled dichotic listening (32.09 ± 0.89) than during vocal dichotic listening (7.35 ± 0.67). Moreover, the main effect of group reached significance ($F_{(1,73)} = 23.14$; $p < 0.001$; partial $\eta^2 = 0.24$). This effect indicated that overall, participants in the WG made fewer errors (16.68 ± 0.73) than participants in the CG (22.76 ± 1.03). In addition, the interaction condition × group reached significance ($F_{(1,73)} = 32.16$; $p < 0.001$; partial $\eta^2 = 0.31$) and Bonferroni-corrected post hoc tests were used to further investigate this effect. The analysis revealed that there was no significant difference between CG and WG for vocal dichotic listening ($p = 0.58$). In contrast, there was a significant difference in whistled dichotic listening ($p < 0.001$). Here, the CG made substantially more errors (37.8 ± 1.46) than the WG (26.38 ± 1.03).

3.3. Laterality Index

In order to ensure comparability with a previous whistle language dichotic listening study in Turkish participants [16], we first compared dichotic listening LIs between the CG and WG, irrespective of experience (see Table 1). To this end, we used a 2 × 2 repeated measures ANOVA with the within-subjects factor condition (VOCAL, WHISTLE) and the between-subjects factor group (WG, CG). While there was no main effect of group ($F_{(1,73)} = 0.08$; $p = 0.77$), the main effect of condition reached significance ($F_{(1,73)} = 56.82$; $p < 0.001$; partial η^2=0.44), indicating a stronger REA for spoken syllables (LI = 24.86 ± 2.54) than for whistled syllables (LI = 3.37 ± 1.89). Moreover, the interaction condition × group reached significance ($F_{(1,73)} = 5.50$; $p < 0.05$; partial $\eta^2 = 0.07$) and Bonferroni-corrected post hoc tests were used to further investigate this effect. The analysis revealed that there was no significant difference between the CG and WG for vocal dichotic listening ($p = 0.27$). In contrast, there was a significant group difference for whistled dichotic listening ($p < 0.05$). Here, the CG showed a slight negative LI that was close to zero (−0.48 ± 2.03), indicating no lateralization in this group (one-sample *t*-test against zero, $p = 0.82$, no difference from zero). In contrast, the WG showed a positive LI (7.21 ± 2.46), indicating a significant REA (one-sample t-test against zero, $p = 0.01$).

3.4. Right-Ear Advantage

To test whether the percentages of individuals with a REA during vocal and whistled dichotic listening differed between CG and WG, we determined for each participant whether they showed a REA (positive LI) or a LEA (negative LI) during vocal and whistled dichotic listening. We then compared the numbers of left- and right-preferent individuals between the CG and WG using Mann–Whitney U-tests. For vocal dichotic listening, there was no significant difference between the CG and WG ($p = 0.37$). Here, participants from both groups were much more likely to show a REA than a LEA (CG: REA: 96%, LEA: 4%; WG: REA: 90%, LEA: 10%). However, the effect reached significance for whistled dichotic listening ($p < 0.01$). Here, participants in the CG showed a LEA more often than a REA (REA: 36%, LEA: 64%). In contrast, participants in the WG showed a REA more often than a LEA (REA: 68%, LEA: 32%).

3.5. Association between Whistled and Vocal Dichotic Listening

In order to investigate the association between whistled and vocal dichotic listening, we calculated Neyman–Pearson correlation coefficients between the LIs for vocal and whistled dichotic listening. In the CG, the effect failed to reach significance (r = −0.02; $p = 0.91$). In the WG, the correlation coefficient also failed to reach significance (r = 0.26; $p = 0.07$).

3.6. Association between Dichotic Listening and Handedness

In order to investigate the associations between whistled and vocal dichotic listening and handedness, we calculated Neyman–Pearson correlation coefficients between the LIs for vocal and whistled dichotic listening and EHI LQ. In both the CG and the WG, all effects failed to reach significance (all p's > 0.10).

3.7. The Effect of Experience with Silbo Gomero

In order to investigate whether the extent of experience with Silbo Gomero affected language lateralization, we re-analyzed the data from the WG by splitting it into a learner's groups (LG) and an advanced group (AG). We then compared dichotic listening LIs between LG and AG, using a 2 × 2 repeated measures ANOVA with the within-subjects factor condition (VOCAL, WHISTLE) and the between-subjects factor group (LG, AG). The main effect of condition reached significance ($F_{(1,48)} = 19.45$; $p < 0.001$; partial $\eta^2 = 0.29$), indicating a stronger right-ear advantage during spoken dichotic listening (LI = 22.01 ± 3.00) than during whistled dichotic listening (LI = 7.21 ± 2.48). All other effects failed to reach significance (all p's > 0.65).

In order to test whether the age of acquisition of Silbo Gomero had an impact on whistle language lateralization, we calculated Neyman–Pearson correlation coefficients between the age of acquisition on whistle language LI for both the LG and the AG. Both effects failed to reach significance (LG: r = 0.22; $p = 0.34$; AG: −0.36; $p = 0.104$). In addition, there were no significant correlations either between the LI and the number of years whistling (LG: r = 0.266; $p = 0.20$; AG: r = 0.091; $p = 0.67$).

4. Discussion

The aim of the present study was to investigate the brain lateralization of whistled Spanish. We hypothesized that contrary to spoken Spanish, Silbo Gomero is more bilaterally represented, as both the left and right hemispheres are needed to process whistled stimuli correctly. This hypothesis was confirmed by the data. For vocal dichotic listening, both groups showed a pronounced REA, replicating the main finding of a substantial body of evidence for this task [21,27–33]. Thus, both participants in the CG and participants in the WG on average showed leftward lateralization for the processing of vocal Spanish. As was to be expected, the two groups did not differ from each other in this condition, as participants in both groups were native Spanish speakers.

There was, however, a group difference in the whistled condition. Here, participants in the CG did not show any lateralization, indicating that they did not process the whistled syllables as language. In contrast, participants in the WG still showed a significant REA in this condition, which was, however, substantially decreased compared to vocal dichotic listening. This indicates that in the whistled condition, the right hemisphere likely played a more important role in stimulus processing than during vocal dichotic listening.

This reduction in the REA is in line with the main finding of a previous dichotic listening study in proficient Turkish whistlers [16]. Here, the authors reported that whistled language comprehension relies on symmetric hemispheric contributions, associated with a decrease in left and a relative increase in right hemispheric encoding mechanisms. While we did not find a completely symmetrical pattern for whistled Silbo Gomero, the LI of the WG in the whistle condition was substantially reduced compared to the vocal condition, indicating a similar principle to that observed in the Turkish study. Why whistled Silbo Gomero still elicits a slight REA in the present study might be explained by differences in cohort characteristics or differences between Turkish and Spanish. Interestingly, it has been shown that most participants show a LEA for musical stimuli [30,31]. It could be conceived that whistled languages present a system of communication with processing demands somewhere between that of languages and that of music, explaining the reduced REA in processing whistled stimuli.

Our findings are also in line with the only neuroimaging study on Silbo Gomero that has been conducted so far [23]. While it is difficult to directly compare the results of the two studies, as we did not perform fMRI scans, our results are largely in line with the overall findings of the previous work. Specifically, the authors showed that in proficient whistlers, left temporal brain areas commonly associated with language are also activated during the processing of Silbo Gomero. However, activation in the right temporal lobe also increased during whistle processing. The authors assumed that this is due to the need to process to non-linguistic pitch changes, tones, and complex sounds when listening to Silbo Gomero. This agrees with our finding of a reduced REA in the WG in the whistle condition. The idea that different cognitive processes are involved in whistle and vocal dichotic listening is also supported by the lack of significant correlation between spoken LI and whistled LI, which makes it likely that non-verbal processes are involved in understanding whistled dichotic listening.

Our finding that whistlers still show a REA for the processing of whistled dichotic listening is also in line with previous studies in non-verbal languages. Generally, it has been shown that independent of language modality, a left-lateralized pattern can be observed for both signed and spoken languages [34,35]. Similar results were also found for Morse code. Experienced Morse code operators show a significant REA, indicating left hemisphere lateralization, for the perception of dichotically presented Morse code letters [36].

We assumed that whistle experience could affect overall performance in the whistled dichotic listening task, as well as the lateralization pattern. However, both the direct statistical comparison of the two groups and the correlation analyses indicated that experience with Silbo Gomero did not significantly affect language lateralization. This indicates that the critical period in which individuals that learn Silbo Gomero switch from internally translating whistles to vocal language to natively understanding the whistle language might be outside the time range we tested.

Interestingly, our findings are also largely in line with a recent meta-analysis on language lateralization in bilinguals [37]. Here, it was shown that language lateralization differed between bilinguals who acquired both languages by 6 years of age and those who acquired the second language later. While the early bilinguals showed bihemispheric processing for both languages, the late bilinguals showed a left-hemispheric dominance for both languages. In our sample, both the LG and the AG acquired Silbo Gomero decidedly later in life than by 6 years of age. Thus, both groups could be considered late bilinguals, which would explain why they did not differ from each other. Since Silbo Gomero is a whistled communication system based on Spanish, it is somewhat unclear whether or not Spanish-speaking individuals able to communicate in Silbo Gomero could be considered truly bilingual or not. Nevertheless, these findings on language lateralization in bilinguals clearly suggest that for future studies on lateralization of Silbo Gomero processing, it would be meaningful to tests participants that acquired Silbo Gomero before their sixth birthday.

In addition to this meta-analysis, there is also an empirical study on language lateralization in bilinguals that might be of relevance for the understanding of the present results [38]. Here, the authors analyzed language lateralization assessed with the dichotic listening task for both the first and the second language in two groups of bilinguals. In the first group, both languages the bilingual participants spoke came from the same linguistic root. In the second group, the two languages the bilingual participants spoke came from different linguistic roots. Here, the authors found that when the second language came from a different linguistic root than the first language, the participants showed comparable brain lateralization for both languages. However, in the group were the two languages came from the same linguistic root, the second language showed a stronger REA than when the two languages came from different linguistic roots. As spoken Spanish and Silbo Gomero clearly have the same linguistic roots, this effect might explain why we found partly left-hemispheric processing for Silbo Gomero.

One effect of note we found was the above-chance recognition rates of the CG for whistled stimuli. It is, however, not unlikely that the participants in the CG correctly identified some syllables in the whistle condition despite having no knowledge of Silbo Gomero. For example, it was shown that native speakers of French and Spanish understood whistled vowels above chance, even if they did not speak any whistle language [39]. Moreover, the findings that the recognition rates of the CG for whistled stimuli were above-chance might also be related to phonetic symbolism. For example, it has been shown that participants are able to guess the meaning from word sounds of languages unknown to them based on the processing of phonetic symbolism [40]. Still, participants in the WG understood a significantly higher number of syllables than those in the CG in this condition, as evidenced by the analysis of the error rates.

Concerning methodological issues, a potential drawback of the present study was the impossibility to use the exact same syllables as in the whistled Turkish dichotic listening study [16], since the syllables "ba" and "pa" used in the Turkish experiment are not distinguishable in Silbo Gomero. As a result, the possible mechanisms involved in the two studies are very likely similar but possibly not identical. Moreover, due to the high relevance of context for understanding whistled languages, it is somewhat difficult to test a whistled language using just syllables as stimuli, especially CV syllables. The problem lies in the fact that the whistled language needs a context to be understood and some parts of the word or sentence are not intelligible for the receptor but are clarified thanks to the rest of the sentence. Furthermore, some syllables sound very similar to each other (like "ga" and "ya") and are thus potentially difficult to distinguish from one another without any other extra information. This explains the somewhat high error rate of the WG for whistled dichotic listening. In future studies, this issue could be remedied by using syllables VCV (vowel-consonant-vowel) that according to the comments of several participants in the WG would be easier to understand for them. Moreover, it was not optimal that potential hearing issues were assessed by self-report. Future dichotic listening studies on Silbo Gomero should include detailed audiometric testing prior to data collection. Another point that could be optimized in future studies is a stronger control of language background between the groups, specifically regarding bilingualism. In our study, a higher number of bilingual individuals were found in the CG than in the WG, which could potentially have affected results. Additionally, it needs to be mentioned is that handedness can be measured as both hand preference [25] and hand skill [41]. In our study, we only measured hand preference using the EHI. Future studies should also include a measure of hand skill, as the two variables can differ to some extent [42].

Our study has several interesting implications for future studies. For example, outside of the Carreiras et al., 2005, study, no neuroimaging studies have been conducted with Silbo Gomero speakers. Thus, using modern neuroimaging techniques to further unravel the brain networks involved in the understanding and production of whistled languages is a crucial next step. Moreover, using EEG to understand the electrophysiological correlates of Silbo Gomero would be a meaningful aim for future studies. Furthermore, a previous study has used transcranial electrical stimulation of the auditory cortex to modulate the REA in dichotic listening [43]. Similar study designs could help to disentangle differences between language lateralization for spoken and whistled languages. In addition, studies with people who suffered from damage in left hemisphere language areas could yield promising results for using whistled languages as a means of rehabilitation for communication impairments.

In conclusion, the processing of Silbo Gomero leads to a reduced REA compared to spoken Spanish. This is in line with previous findings for whistled Turkish, implying that processing of whistled languages occurs more bihemispherically, independently of which language is whistled. This shows that if left-hemispheric functions like fast temporal processing are less relevant for processing a specific form of language, leftward language lateralization decreases.

Author Contributions: Conceptualization, O.G., and S.O.; methodology, O.G., and S.O.; software, S.O. and P.V.G.; validation, O.G., P.V.G. and S.O.; formal analysis, S.O. and P.V.G.; investigation, P.V.G.; resources, P.V.G. and O.G.; data curation, S.O. and P.V.G.; writing—original draft preparation, P.V.G.; writing—review and editing, O.G., S.O., P.V.G.; visualization, P.V.G., and S.O.; supervision, S.O.; project administration, S.O.; funding acquisition, P.V.G. All authors have read and agreed to the published version of the manuscript.

Funding: This research was supported by a Promos grant from the DAAD and Ruhr University of Bochum and a prize from the RuhrScience Night. Infrastructure facilities were provided by Ruhr University Bochum (Bochum, Germany), University of Duisburg-Essen (Essen, Germany), and Instituto Andrés Bello, University of La Laguna (Tenerife, Canary Islands, Spain).

Acknowledgments: We would like to thank the following people and institutions: Instituto Andrés Bello, University of La Laguna, for facilitating infrastructure for the experiment. Asociación Yo Silbo, for provide materials, infrastructure, information, and helping to find volunteers in Tenerife and Gran Canaria. Asociación sociocultural ARUME, for providing volunteers and infrastructure. Kimberly Álvarez González, for the ofimatic support. To all the volunteers for participating in the study without any gratification further than helping science and support the cultural heritage of the Canary Islands.

Conflicts of Interest: The authors declare no conflict of interest. Also, the funders had no role in the design of the study; in the collection, analyses, or interpretation of data; in the writing of the manuscript, or in the decision to publish the results.

References

1. Friederici, A.D. The brain basis of language processing: From structure to function. *Physiol. Rev.* **2011**, *91*, 1357–1392. [CrossRef] [PubMed]
2. Ocklenburg, S.; Beste, C.; Arning, L.; Peterburs, J.; Güntürkün, O. The ontogenesis of language lateralization and its relation to handedness. *Neurosci. Biobehav. Rev.* **2014**, *43*, 191–198. [CrossRef] [PubMed]
3. Tzourio-Mazoyer, N.; Marie, D.; Zago, L.; Jobard, G.; Perchey, G.; Leroux, G.; Mellet, E.; Joliot, M.; Crivello, F.; Petit, L.; et al. Heschl's gyrification pattern is related to speech-listening hemispheric lateralization: FMRI investigation in 281 healthy volunteers. *Brain Struct. Funct.* **2015**, *220*, 1585–1599. [CrossRef] [PubMed]
4. Knecht, S.; Deppe, M.; Dräger, B.; Bobe, L.; Lohmann, H.; Ringelstein, E.; Henningsen, H. Language lateralization in healthy right-handers. *Brain* **2000**, *123 Pt 1*, 74–81. [CrossRef]
5. Knecht, S.; Dräger, B.; Deppe, M.; Bobe, L.; Lohmann, H.; Flöel, A.; Ringelstein, E.B.; Henningsen, H. Handedness and hemispheric language dominance in healthy humans. *Brain* **2000**, *123 Pt 12*, 2512–2518. [CrossRef]

6. Hickok, G.; Poeppel, D. The cortical organization of speech processing. *Nat. Rev. Neurosci.* **2007**, *8*, 393–402. [CrossRef] [PubMed]
7. Hugdahl, K. Lateralization of cognitive processes in the brain. *Acta Psychol. (Amst.)* **2000**, *105*, 211–235. [CrossRef]
8. Gu, F.; Zhang, C.; Hu, A.; Zhao, G. Left hemisphere lateralization for lexical and acoustic pitch processing in Cantonese speakers as revealed by mismatch negativity. *Neuroimage* **2013**, *83*, 637–645. [CrossRef] [PubMed]
9. Marsolek, C.J.; Deason, R.G. Hemispheric asymmetries in visual word-form processing: Progress, conflict, and evaluating theories. *Brain Lang.* **2007**, *103*, 304–307. [CrossRef]
10. Campbell, R.; MacSweeney, M.; Waters, D. Sign language and the brain: A review. *J. Deaf Stud. Deaf Educ.* **2008**, *13*, 3–20. [CrossRef]
11. Hickok, G.; Love-Geffen, T.; Klima, E.S. Role of the left hemisphere in sign language comprehension. *Brain Lang.* **2002**, *82*, 167–178. [CrossRef]
12. Levänen, S.; Uutela, K.; Salenius, S.; Hari, R. Cortical representation of sign language: Comparison of deaf signers and hearing non-signers. *Cereb. Cortex* **2001**, *11*, 506–512. [CrossRef] [PubMed]
13. Slevc, L.R.; Martin, R.C.; Hamilton, A.C.; Joanisse, M.F. Speech perception, rapid temporal processing, and the left hemisphere: A case study of unilateral pure word deafness. *Neuropsychologia* **2011**, *49*, 216–230. [CrossRef] [PubMed]
14. Zatorre, R.J.; Belin, P. Spectral and temporal processing in human auditory cortex. *Cereb. Cortex* **2001**, *11*, 946–953. [CrossRef] [PubMed]
15. Zatorre, R.J.; Belin, P.; Penhune, V.B. Structure and function of auditory cortex: Music and speech. *Trends Cogn. Sci. (Regul. Ed.)* **2002**, *6*, 37–46. [CrossRef]
16. Güntürkün, O.; Güntürkün, M.; Hahn, C. Whistled Turkish alters language asymmetries. *Curr. Biol.* **2015**, *25*, R706–R708. [CrossRef]
17. Meyer, J. Whistled languages. In *A Worldwide Inquiry on Human Whistled Speech*; Springer: Heidelberg, Germany; New York, NY, USA, 2015; ISBN 9783662458365.
18. Díaz Reyes, D. *El Lenguaje Silbado en la isla de El Hierro*; Cabildo: El Hierro, Spain, 2008; ISBN 8493514764.
19. Busnel, R.-G.; Classe, A. *Whistled Languages*; Springer: Berlin/Heidelberg, Germany, 1976; ISBN 9783642463372.
20. Trujillo, R. *El Silbo Gomero. Análisis Lingüístico*, 1st ed.; Interinsular Canaria: Santa Cruz de Tenerife, Spain, 1978; ISBN 8485543033.
21. Tervaniemi, M.; Hugdahl, K. Lateralization of auditory-cortex functions. *Brain Res. Brain Res. Rev.* **2003**, *43*, 231–246. [CrossRef]
22. Westerhausen, R.; Kompus, K. How to get a left-ear advantage: A technical review of assessing brain asymmetry with dichotic listening. *Scand. J. Psychol.* **2018**, *59*, 66–73. [CrossRef]
23. Carreiras, M.; Lopez, J.; Rivero, F.; Corina, D. Linguistic perception: Neural processing of a whistled language. *Nature* **2005**, *433*, 31–32. [CrossRef]
24. Reiterer, S.M.; Hu, X.; Erb, M.; Rota, G.; Nardo, D.; Grodd, W.; Winkler, S.; Ackermann, H. Individual differences in audio-vocal speech imitation aptitude in late bilinguals: Functional neuro-imaging and brain morphology. *Front. Psychol.* **2011**, *2*, 271. [CrossRef]
25. Oldfield, R.C. The assessment and analysis of handedness: The Edinburgh inventory. *Neuropsychologia* **1971**, *9*, 97–113. [CrossRef]
26. Willems, R.M.; van der Haegen, L.; Fisher, S.E.; Francks, C. On the other hand: Including left-handers in cognitive neuroscience and neurogenetics. *Nat. Rev. Neurosci.* **2014**, *15*, 193–201. [CrossRef] [PubMed]
27. Bradshaw, J.L.; Burden, V.; Nettleton, N.C. Dichotic and dichhaptic techniques. *Neuropsychologia* **1986**, *24*, 79–90. [CrossRef]
28. Hiscock, M.; Kinsbourne, M. Attention and the right-ear advantage: What is the connection? *Brain Cogn.* **2011**, *76*, 263–275. [CrossRef]
29. Hugdahl, K.; Westerhausen, R.; Alho, K.; Medvedev, S.; Laine, M.; Hämäläinen, H. Attention and cognitive control: Unfolding the dichotic listening story. *Scand. J. Psychol.* **2009**, *50*, 11–22. [CrossRef]
30. Kimura, D. Left-right differences in the perception of melodies. *Q. J. Exp. Psychol.* **1964**, *16*, 355–358. [CrossRef]
31. Kimura, D. From ear to brain. *Brain Cogn.* **2011**, *76*, 214–217. [CrossRef]
32. Westerhausen, R. A primer on dichotic listening as a paradigm for the assessment of hemispheric asymmetry. *Laterality* **2019**, *24*, 740–771. [CrossRef]

33. Westerhausen, R.; Hugdahl, K. The corpus callosum in dichotic listening studies of hemispheric asymmetry: A review of clinical and experimental evidence. *Neurosci. Biobehav. Rev.* **2008**, *32*, 1044–1054. [CrossRef]
34. Bellugi, U.; Poizner, H.; Klima, E.S. Language, modality and the brain. *Trends Neurosci.* **1989**, *12*, 380–388. [CrossRef]
35. Gordon, N. The neurology of sign language. *Brain Dev.* **2004**, *26*, 146–150. [CrossRef]
36. Papçun, G.; Krashen, S.; Terbeek, D. Is the Left Hemisphere Specialized for Speech, Language, or Something Else. *J. Acoust. Soc. Am.* **1972**, *51*, 79. [CrossRef]
37. Hull, R.; Vaid, J. Bilingual language lateralization: A meta-analytic tale of two hemispheres. *Neuropsychologia* **2007**, *45*, 1987–2008. [CrossRef] [PubMed]
38. D'Anselmo, A.; Reiterer, S.; Zuccarini, F.; Tommasi, L.; Brancucci, A. Hemispheric asymmetries in bilinguals: Tongue similarity affects lateralization of second language. *Neuropsychologia* **2013**, *51*, 1187–1194. [CrossRef] [PubMed]
39. Meyer, J.; Dentel, L.; Meunier, F. Categorization of Natural Whistled Vowels by Naïve Listeners of Different Language Background. *Front. Psychol.* **2017**, *8*, 25. [CrossRef] [PubMed]
40. D'Anselmo, A.; Prete, G.; Zdybek, P.; Tommasi, L.; Brancucci, A. Guessing Meaning From Word Sounds of Unfamiliar Languages: A Cross-Cultural Sound Symbolism Study. *Front. Psychol.* **2019**, *10*, 593. [CrossRef]
41. Brandler, W.M.; Morris, A.P.; Evans, D.M.; Scerri, T.S.; Kemp, J.P.; Timpson, N.J.; St Pourcain, B.; Smith, G.D.; Ring, S.M.; Stein, J.; et al. Common variants in left/right asymmetry genes and pathways are associated with relative hand skill. *PLoS Genet.* **2013**, *9*, e1003751. [CrossRef]
42. Papadatou-Pastou, M.; Ntolka, E.; Schmitz, J.; Martin, M.; Munafò, M.R.; Ocklenburg, S.; Paracchini, S. Human handedness: A meta-analysis. *Psychol. Bull.* **2020**, *146*, 481–524. [CrossRef]
43. Prete, G.; D'Anselmo, A.; Tommasi, L.; Brancucci, A. Modulation of the dichotic right ear advantage during bilateral but not unilateral transcranial random noise stimulation. *Brain Cogn.* **2018**, *123*, 81–88. [CrossRef]

Article

A Simulation on Relation between Power Distribution of Low-Frequency Field Potentials and Conducting Direction of Rhythm Generator Flowing through 3D Asymmetrical Brain Tissue

Hao Cheng [1,2], Manling Ge [1,2], Abdelkader Nasreddine Belkacem [3], Xiaoxuan Fu [1,2,4,5], Chong Xie [1,2], Zibo Song [1,2], Shenghua Chen [1,2,*] and Chao Chen [6,*]

1 State Key Laboratory of Reliability and Intelligence of Electrical Equipment, Hebei University of Technology, Tianjin 300130, China; 202031403008@stu.hebut.edu.cn (H.C.); gemanling@hebut.edu.cn (M.G.); 201711401003@stu.hebut.edu.cn (X.F.); 201931403038@stu.hebut.edu.cn (C.X.); 201931403037@stu.hebut.edu.cn (Z.S.)
2 Key Laboratory of Electromagnetic Field and Electrical Apparatus Reliability of Hebei Province, Hebei University of Technology, Tianjin 300130, China
3 Department of Computer and Network Engineering, College of Information Technology, United Arab Emirates University, Al Ain 15551, United Arab Emerites; belkacem@uaeu.ac.ae
4 Department of Neuroscience, Medical University of South Carolina, Charleston, SC 29425, USA
5 Athinoula A. Martinos Center for Biomedical Imaging, Department of Radiology, Massachusetts General Hospital/Harvard Medical School, Charlestown, MA 02115, USA
6 Key Laboratory of Complex System Control Theory and Application, Tianjin University of Technology, Tianjin 300384, China
* Correspondence: chenshenghua@hebut.edu.cn (S.C.); chao_chen@email.tjut.edu.cn (C.C.)

Citation: Cheng, H.; Ge, M.; Belkacem, A.N.; Fu, X.; Xie, C.; Song, Z.; Chen, S.; Chen, C. A Simulation on Relation between Power Distribution of Low-Frequency Field Potentials and Conducting Direction of Rhythm Generator Flowing through 3D Asymmetrical Brain Tissue. *Symmetry* **2021**, *13*, 900. https://doi.org/10.3390/sym13050900

Academic Editors: Sebastian Ocklenburg and Onur Güntürkün

Received: 28 April 2021
Accepted: 14 May 2021
Published: 19 May 2021

Abstract: Although the power of low-frequency oscillatory field potentials (FP) has been extensively applied previously, few studies have investigated the influence of conducting direction of deep-brain rhythm generator on the power distribution of low-frequency oscillatory FPs on the head surface. To address this issue, a simulation was designed based on the principle of electroencephalogram (EEG) generation of equivalent dipole current in deep brain, where a single oscillatory dipole current represented the rhythm generator, the dipole moment for the rhythm generator's conducting direction (which was orthogonal and rotating every 30 degrees and at pointing to or parallel to the frontal lobe surface) and the (an)isotropic conduction medium for the 3D (a)symmetrical brain tissue. Both the power above average (significant power value, SP value) and its space (SP area) of low-frequency oscillatory FPs were employed to respectively evaluate the strength and the space of the influence. The computation was conducted using the finite element method (FEM) and Hilbert transform. The finding was that either the SP value or the SP area could be reduced or extended, depending on the conducting direction of deep-brain rhythm generator flowing in the (an)isotropic medium, suggesting that the 3D (a)symmetrical brain tissue could decay or strengthen the spatial spread of a rhythm generator conducting in a different direction.

Keywords: finite element method; electrical field potential; dipole moment; power; EEG

1. Introduction

Theta oscillations (4–8 Hz), which originate in deep brain cortex region, are associated with cognition and memory [1]. They can be measured not only by conducting a deep brain electroencephalogram (EEG) in vivo (local field potentials, LFPs) but also by the oscillatory field potentials (FPs) on either the frontal or temporal lobe surface via a scalp EEG [2,3]. Power fluctuation is the fundamental parameter to evaluate theta rhythms, which can reveal important information about a neural network, e.g., the extent of synchronous neurons in a local assembly. The power spectral density may be dependent on a reference scheme at frequency bands less than 100 Hz [4,5]. In addition, power could be related to various factors such as age, long-term synaptic modification, brain structure, network state and pathology. Thus, power was generally used to investigate the scale of synchronized neurons in cognition and memory, human and animal behaviors, and even functional connectivity of rhythmic brain activity [4,6–10]. Recent evidence indicates that neural disinhibition would vary the frequency dependent LFP states such as burst, suppression and continuous. Increasing power can be observed at lower frequencies (less than 20 Hz), whereas decreasing power can be observed at higher frequencies (more than 20 Hz) in the hippocampus [11].

A consensus has not been achieved regarding the spatial spread of LFPs or FPs in the cortical medium [12–16]. The traveling theta oscillations in deep brain is an important recent observation [17], which implies that the measurement of scalp EEG rhythm at the theta frequency band considers not only some factors studied traditionally, such as the amplitude of rhythmic source current, capacitive extracellular medium, electrical conductivity and position of a rhythm generator, but also other factor such as the conducting direction of a rhythm generator. However, only some reports have investigated the latter factor [14]. Therefore, by considering the frontal lobe as an example, we attempt to map the relation between the FP power distribution on the frontal lobe surface and the conducting direction of the low-frequency rhythm generator in deep brain, based on the theory of equivalent dipole current that pertains to the generation of EEG (forward problem) [18,19].

Here, a quasi-real head surface was reconstructed from 256 T1-weighted MRI slices based on the concept of inverse engineering. The electrical conductivity of the brain tissue was described by the 3D (a)symmetrical tensor and inclusion of (an)isotropy, and the brain rhythm generator was depicted as a quasi-static dipole current in deep brain. The activity of the latter could be representative of a sine oscillation function at a low frequency (here, 6 Hz) and the moment of which could simulate the conducting direction of rhythmic source current (here, the orthogonal conducting directions, pointing to or parallel to the frontal lobe surface). Thus, a distribution of FPs evoked by the dipole current at a time point could be estimated by FEM on a quasi-real head surface by changing some simulation conditions, e.g., conductivity tensor and (an)isotropy, the position of a single dipole current and dipole moment, such as pointing to or parallel to the frontal lobe surface. During a certain time period, a simulated rhythm could be obtained based on the time series of oscillatory FPs. The instantaneous power of simulated rhythms was estimated by Hilbert transform and displayed by FPs when considering the amplitude of a single dipole current. Then, the SP area estimated by global statistics was used to study the influence of a single dipole current, such as its moment (e.g., rotating every 30°) and position (e.g., three positions inside the frontal lobe) on the SP area, by comparatively studying the anisotropic and isotropic medium with 3D asymmetrical conduction tensor and 3D symmetrical conduction tensor [20]. The flow diagram of research is shown in Figure 1.

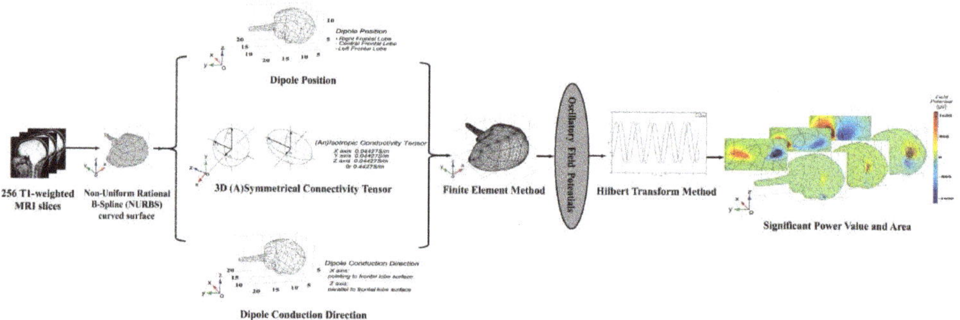

Figure 1. Flow diagram of study.

This study presents the combined effects of the low-frequency rhythm generator's conducting direction and position as well as (a)symmetrical brain tissue conductivity in mapping the FP power distribution on the head surface. Therefore, it may be helpful to further understand the generative mechanism of spontaneous low-frequency brain oscillations at the system level.

2. Methods

The power distribution was analyzed by changing some factors, i.e., the conductivity tensor and inclusion of (an)isotropy, (a)symmetrical conductivity tensor in 3D, position and conducting direction of a single dipole current on a homogeneous single-layer quasi-real head model.

2.1. Model Building

2.1.1. Reconstructed 3D Quasi-Real Head Model

Initially, a numerical model of the cortical surface was reconstructed using Simpleware (Simpleware Ltd. Corp. Exeter, UK) and then digitized using Geomagic Studio (Raindrop Ltd. Corp. Morrisville, USA), based on 256 T1-weighted MRI slices (structural MRI slices). The process required two steps. the first step was the reconstruction of a 3D quasi-real head surface model, wherein 256 sMRI images (DICOM format and 256 $\times$ 256 pixel matrix) were imported into scan image processing (Scan IP) module of Simpleware. In order to strip the cortical tissue from the brain tissues rapidly, the slice images were trimmed and segmented according to the gray value of the image using the interactive threshold function to segment the target area. The target area (the cerebral cortex) was removed. Surface smoothing and model configuration were employed to make the surface model more realistic, and the scalp surface model was exported in the STL format. Then, the DISCRETIZED model of head was rebuilt. The STL-format surface model was imported into Geomagic Studio. It is worth mentioning that there were still some unusable points left and the holes were missing. For the editing of unusable points, Select Outliers function and Reduce Noise function were used to remove the noise points generated by the scan moving and the noise points outside the target area. To edit the holes, they were either filled directly or a large hole diameter was dug and then filled. If the filling effect was still unsatisfactory, they were filled using the Create Point Cover function to reconstruct the area point space to repair the hole. After editing and filtering to optimize the tissue data, the holes and other drawbacks in the model were filled and repaired. The traits of the model were distinguished and extracted to build a high quality Non-Uniform Rational B-Spline (NURBS) curved surface and to generate the cortical space model (also known as reverse engineering model, R-E model). The flow chart of reconstructing the 3D quasi-real head model is shown in Figure 2.

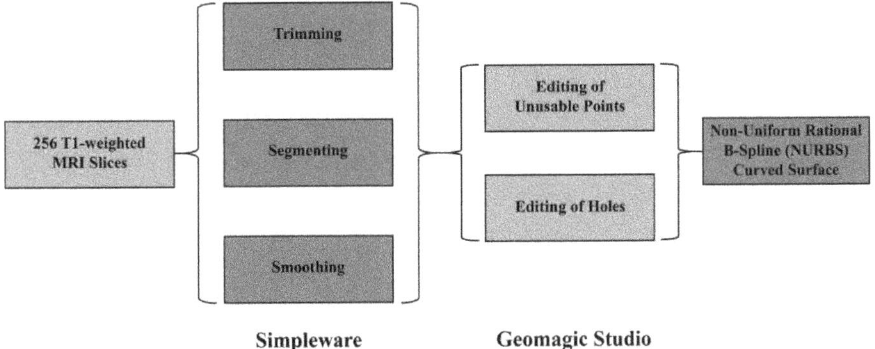

Figure 2. Flow chart of reconstructing NUBS model.

To account for the 3D elements in the finite element method, the NUBS model was layered and the key points were selected by the equal angle or equal distance method, and the nodes were connected between the layers to form a split element. The NUBS model was imported into the COMSOL Multiphysics (COMSOL Inc. Stockholm, SWE) software platform for splitting, and 35,273 elements were automatically generated under the user-controlled element/standard mode, as shown in Figure 3a. Since each element was divided into a plane on the surface of the model, the larger the curvature of the boundary of the model, the smaller the element was divided, and the more the number, the denser the unit element. Among all the elements, 3600 elements were on the frontal lobe. The isopotential lines are shown in Figure 3b, the zero-potential surface was the FP reference as shown in Figure 3c, and the alternative dipole positions are shown in Figure 3d.

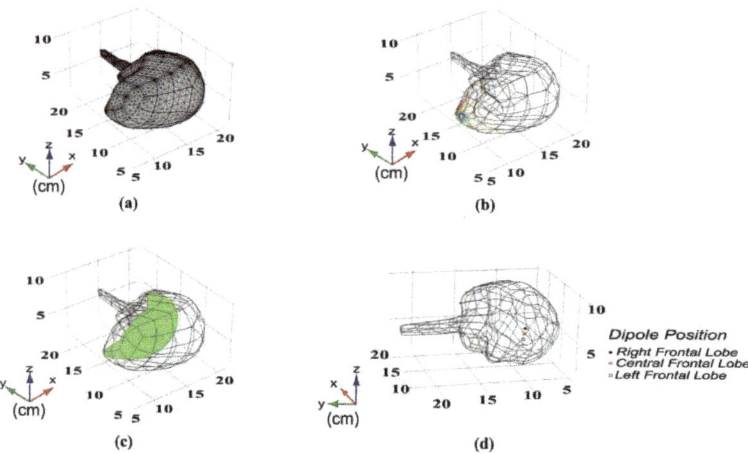

Figure 3. FP computation on the quasi-real head model surface. (**a**) Finite element method; (**b**) contour lines when the dipole current was located inside the central frontal lobe, x, y, z: 8.2, 10.9, 5.8; (**c**) zero-potential surface (green part); (**d**) schematic of alternative positions of a single dipole current (1 cm displacement along the Z axis). The dipole current amplitude was 0.1 nA.

2.1.2. Position and Conducting Direction of a Single Dipole Current

To study the influence of the position and conducting direction of a single dipole current inside the frontal lobe, the dipole direction was assumed to be the Z axis for simulating the conducting direction of dipole current parallel to the frontal lobe surface and the -X axis for the conducting direction of the dipole current pointing to the frontal lobe surface. At each dipole moment, there were three positions (denoted as right frontal lobe, central frontal lobe and left frontal lobe) with a displacement of 1 cm. Along the Z axis, the dipole position was assumed to be in the normal direction in the X–Y plane, and could be changed from the right part and middle part to the left part, the coordinates (cm) of which were (x, y, z: 8.2,10.9,5.8) (right frontal lobe), (x, y, z: 8.2,10.9,6.8) (central frontal lobe), and (x, y, z: 8.2,10.9,7.8) (left frontal lobe), respectively, as shown in Figure 3d. In addition, along the -X axis, the dipole moment could simulate the conducting direction of a single rhythm generator pointing to the frontal lobe surface, where, the dipole position (x, y, z: 8.2,10.9,6.8) was localized inside the central frontal lobe. Thus, the near frontal lobe (x, y, z: 7.2,10.9,6.8) represented carrying the dipole current near to the lobe surface, whereas the far frontal lobe (x, y, z: 9.2,10.9,6.8) represented carrying the dipole current far away from the lobe surface.

2.1.3. Symmetrical Conductivity and Asymmetrical Conductivity

The mean electrical conductivity was considered in this study [21]. Ideally, the conducting medium must be isotropic with the 3D symmetrical conductivity tensor of $\sigma_x = \sigma_y = \sigma_z = 0.14\,\mathrm{S/m}$. However, the conducting medium was anisotropic in a real brain with the 3D asymmetrical conductivity tensor of $\sigma_x = \sigma_y = 0.04427\,\mathrm{S/m}$, $\sigma_z = 0.4427\,\mathrm{S/m}$, i.e., $\sigma_x : \sigma_z = 1:10$. Here, the influence of anisotropic conductivity was compared with that of isotropic conductivity, where the influence of a 3D asymmetrical conductivity tensor was comparatively studied with that of a 3D symmetrical conductivity tensor. The isotropic conductivity medium tensor is shown in Formula (1), and the anisotropic conductivity medium tensor is shown in Formula (2).

$$\sigma = \begin{bmatrix} \sigma_x & \sigma_{xy} & \sigma_{xz} \\ \sigma_{yx} & \sigma_y & \sigma_{yz} \\ \sigma_{zx} & \sigma_{zy} & \sigma_z \end{bmatrix} = \begin{bmatrix} 0.14 & 0 & 0 \\ 0 & 0.14 & 0 \\ 0 & 0 & 0.14 \end{bmatrix} \quad (1)$$

$$\sigma = \begin{bmatrix} \sigma_x & \sigma_{xy} & \sigma_{xz} \\ \sigma_{yx} & \sigma_y & \sigma_{yz} \\ \sigma_{zx} & \sigma_{zy} & \sigma_z \end{bmatrix} = \begin{bmatrix} 0.04427 & 0 & 0 \\ 0 & 0.04427 & 0 \\ 0 & 0 & 0.4427 \end{bmatrix} \quad (2)$$

2.2. Calculation Derivation

2.2.1. FP Derivations of Low-Frequency Simulated Rhythms

A generalized expression of the forward problem, i.e., the theory of the equivalent dipole current with respect to the generation of EEG, could be helpful to facilitate mapping [18,19]. Given the position and moment of a static equivalent dipole current generator and the geometry and electrical conductivity (σ) profile of the volume conductor (Ω, i.e., model of the head), the electrical FPs (ψ) can be considered as EEG and could be expressed by Poisson's equation and the Neumann boundary condition. This is a very authoritative method since it was released in 1978 and is still in use today [18] on the head surface (S).

$$\nabla\cdot(\sigma\cdot\nabla\psi) = -\sum_{\Omega} J_s \quad (in\ \Omega) \quad (3)$$

$$\sigma(\nabla\psi)\cdot n = 0 \quad (on\ S) \quad (4)$$

Here, n is the normal direction of the boundary and J_s is the electric current density of a conductor.

The rhythms were simulated by oscillatory electrical FPs on a 3D quasi-real head surface model (single-layer homogeneous medium), which was obtained via FEM using COMSOL Multiphysics. The activity of the dipole current was a sine function with an amplitude of 0.1 nA, a frequency of 6 Hz and an initial phase of 0 radian when considering a time window of 1 s. The dipole current can be considered as the rhythm generator and the electrical field evoked by a single dipole current was quasi-static. Additionally, the electrical FP could be calculated by FEM at each time point at each generated element. A time course of 1 s with respect to the dipole current was split into 128 equal time points. The oscillating electrical FP was considered as the simulated rhythm on each element during this 1 s period.

2.2.2. FP Power Derivations of Low-Frequency Simulated Rhythms

The low-frequency oscillatory FP is a continuous signal, which is denoted as $FP(t)$. $FP(t)$ was Hilbert transformed (hilbert, MATLAB) to get $\widetilde{FP}(t)$ and can be expressed as follows.

$$\widetilde{FP}(t) = \frac{1}{\pi} p.v. \int_{-\infty}^{+\infty} \frac{A(\tau)e^{j\phi_{FP}(\tau)}}{t-\tau} d\tau \quad (5)$$

Here, $A(\tau)$ is the instantaneous amplitude of $FP(t)$, $\phi_{FP}(\tau)$ is the instantaneous phase of $FP(t)$, and $p.v.$ is the Cauchy principal value.

For a continuous signal, its instantaneous power value is equal to the square of the signal modulus after Hilbert transformation of the signal. This is to say the instantaneous power was a sum of the square of the real part plus the square of the imaginary part of the oscillatory FPs., The power was then averaged over time.

2.2.3. Value and Area of Significant Power

Power values greater than the mean power (significant power, denoted as, i.e., SP), were a concern to study the influence of the position strength and conducting direction of a rhythm generator on the significant power distribution. In addition, the SP area (%) was defined as the ratio of the number of elements with an SP value to the total number of elements needed to study the influence space. Moreover, the SP area was displayed by a spatial sum of a warm color tone plus a cold color tone on a standard colormap of FPs corresponding to a rhythm generator at an amplitude of 0.1 nA. According to the partial volume effect [22], the larger the SP area, the smaller the significant FP power value. This suggests that a greater space influence will mean a weaker influence of strength, and vice versa.

The subsequent statistical significance of the FP power was determined between anisotropic medium and isotropic medium, where the dipole moment was parallel to and pointing to the frontal lobe surface and the adjacent displacement distance of a single dipole current. The student's unpaired t-test was used to determine the statistical significance.

2.3. Validation Influence of Conducting Direction

The influence of the conducting direction of the rhythm generator on the SP area was validated by rotating the conduction direction of a single dipole current at every 30° of the rotation under two conditions, i.e., pointing to or parallel to the lobe surface, when the dipole current was at the central inside frontal lobe. The dipole moment was rotated (i) departing from the -X axis via the Y axis and arriving at the opposite site (X axis); (ii) departing from the Z axis via the Y axis and arriving at the opposite site (-Z axis).

3. Results

There were 3600 elements present from left to right on the frontal lobe surface; thus, the power values of these elements were considered during the analysis.

3.1. Distribution of Significant Power at Dipole Moment Pointing to Frontal Lobe Surface

The distribution of FP power when a dipole current was flowing directly to the frontal surface (dipole moment at the -X axis) is shown in Figure 4. From the horizontal perspective, i.e., isotropic medium (left panel) compared to anisotropic medium (right panel), the SP area was greatly reduced in the anisotropic medium relative to that in the isotropic medium ($p < 0.001$), suggesting that the 3D asymmetrical conductivity tensor of brain tissue could strengthen an ongoing dipole current in deep brain, thereby resulting in the SP area to reduce on the frontal lobe surface. From the vertical perspective, because of the dipole current position, the partial volume effect could be observable, i.e., the smaller the SP area, the greater the SP values, suggesting the opposite effects indicated by the SP area and the SP value. At a distance of 1 cm, the maximum power values differed by more than a factor of two, as indicated by the maximum value using the color bars.

3.2. Distribution of Significant Power at Dipole Moment Parallel to Frontal Lobe Surface

The FP power distribution when a dipole current was flowing parallel to the frontal lobe surface (dipole moment at the Z axis) is shown in Figure 5. From the horizontal perspective, the SP area considerably increased in the anisotropic medium (right panel) compared to that in the isotropic medium (left panel) ($p < 0.001$), suggesting that the 3D asymmetrical conductivity tensor of brain tissue could weaken an ongoing dipole current in deep brain. This would y result in the SP area increasing on the frontal lobe surface, which is the reverse of the effect shown in Figure 4. However, from the vertical perspective, the partial volume effect is similar to that shown in Figure 4, i.e., the smaller the SP area, the greater the SP values.

3.3. Validation of Influence of Dipole Moment on Area of Significant Power

In Figure 6, the SP area was shown by rotating the dipole moment after every 30° in two directions, pointing to and parallel to the frontal surface. When considering the condition of the dipole moment flowing along the frontal lobe, an increased SP area was observed on the frontal lobe surface in the anisotropic medium relative to the isotropic medium, implying that the 3D asymmetrical conductivity of brain tissue could decay the ongoing dipole current. In reverse, a decreased SP area was observed when considering the condition of the dipole moment flowing directly to the frontal lobe. This implies that the 3D asymmetrical conductivity of brain tissue could enhance the ongoing dipole current, except under some specific conditions, such as the Y axis (parallel to the frontal lobe surface) and X axis (pointing to deep brain), as shown in Figure 6a.

Taken together, as long as the dipole moment is flowing along the frontal lobe, the 3D asymmetrical conductivity of brain tissue could weaken the ongoing dipole current, leading to an increase in SP area which was observed on the frontal lobe surface in the anisotropic medium.

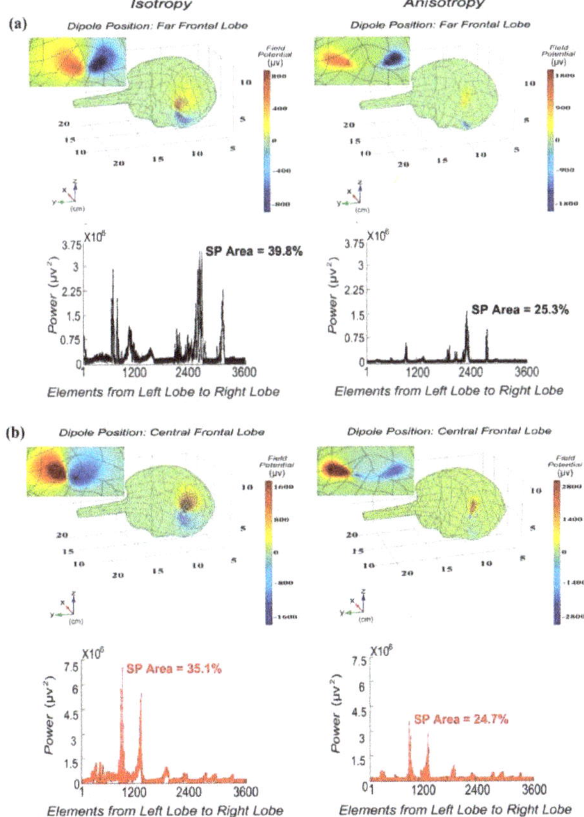

Figure 4. *Cont.*

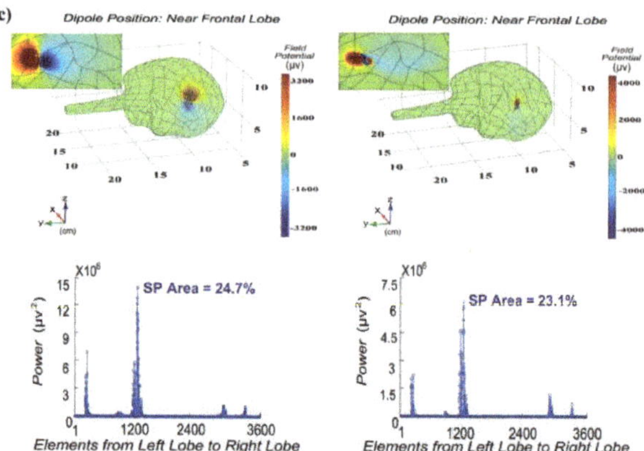

Figure 4. Influence of dipole current position on distribution of FP power, when a dipole current is flowing directly to the frontal lobe surface and is conducted in an isotropic medium (left panel) or an anisotropic medium (right panel). (**a**) Located far from the frontal lobe; (**b**) located inside the central frontal lobe; (**c**) located near to the frontal lobe. Each sub-picture contains an FP distribution figure with partially enlarged picture.

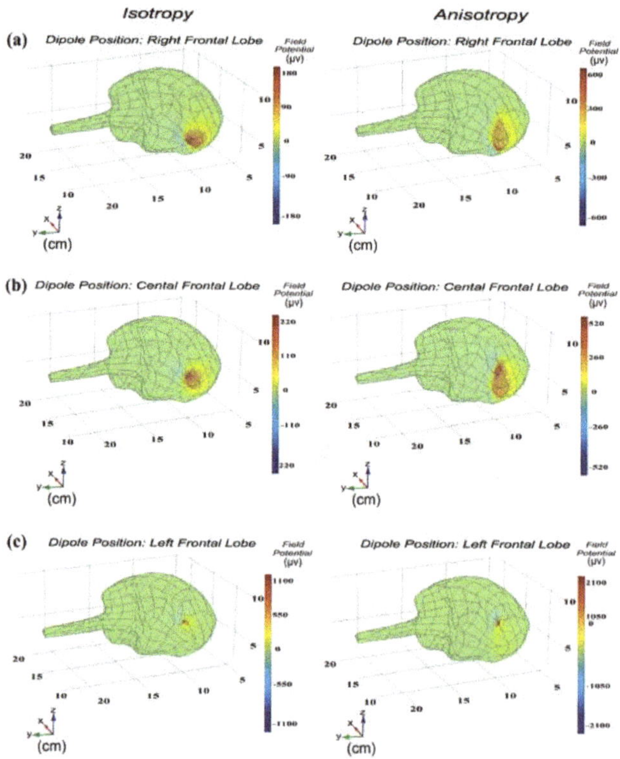

Figure 5. Influence of dipole current's position on distribution of FP power, when a dipole current is flowing parallel to the frontal lobe surface and is conducted in an isotropic medium (left panel) or an anisotropic medium (right panel). (**a**) Located inside the right frontal lobe; (**b**) located inside the central frontal lobe; (**c**) located inside the left frontal lobe.

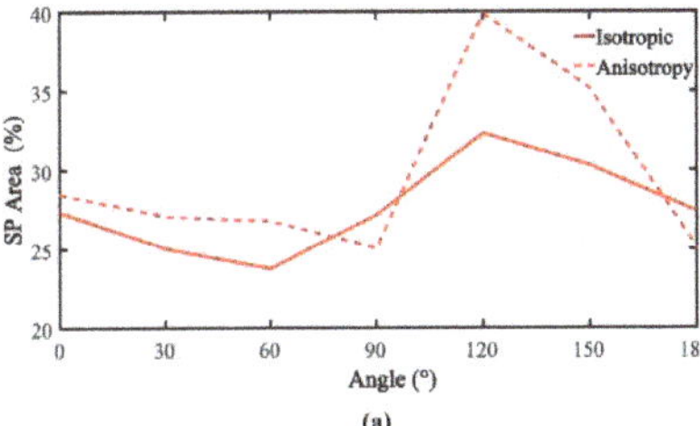

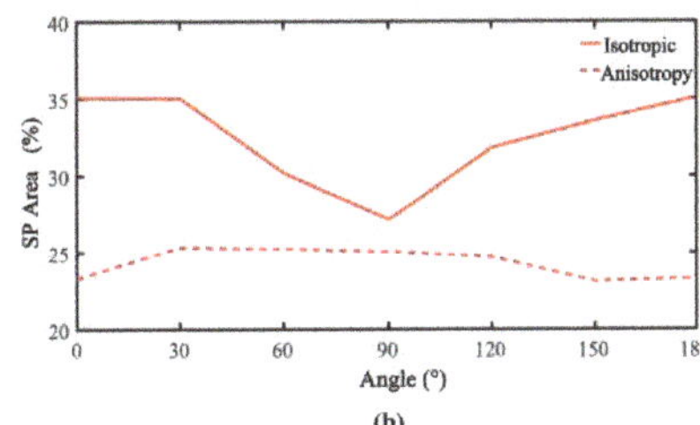

Figure 6. Influence of dipole moment on the SP area. (**a**) Direct to the frontal lobe surface, departing from -X axis and arriving at backward via Y axis (90 degree); (**b**) along the frontal lobe surface, departing from Z axis and arriving at backward via Y axis (90 degree).

4. Discussion

4.1. Power

As shown in Figures 4 and 5, the distribution of significant FP power could be reversed in the (an)isotropic medium in the orthogonal directions of the rhythm generator in deep brain. This implies a joint consequence of the conducting direction of rhythmic source current and the complex conductivity of brain tissue. This consequence can be further explained by the solution of FPs on a spherical surface in the forward problem, i.e., the FPs evoked by the dipole current were inversely proportional to the square of the distance to the rhythm generator. The closer the distance between an element and the rhythm generator, the stronger the FP rhythms and the greater the significant FP power values [18,23].

The distance between the dipole source current and the lobe surface was approximately equal when the dipole current flowed along the frontal lobe; thus, distance was a minor factor. However, the disturbance in anisotropic conductivity was a significant factor that resulted in more SP area in the 3D electrical field relative to the balance in isotropic conductivity. Consequently, the SP area in the anisotropic medium was greater than that in the isotropic medium, as shown in Figures 4 and 5.

Moreover, in Figure 6a, in some specific conditions such as the Y axis (parallel to the frontal lobe surface) the effect of anisotropy of brain tissue isn't an exception, because its effect is convergent to that as shown in Figure 6b.

In previously conducted neural measurement studies, power reflected the strength of brain rhythms, which were related to many factors such as brain regions, scale of synchronous neurons, brain function, pathology and physiology [2,7,10,24]. Our theoretical work further suggests that FP power was dependent on the combined factors that were difficult to measure at a system level, including the position of a rhythm generator, brain tissue conductivity and even the conducting direction of a rhythm generator in deep brain.

4.2. Anisotropy

A formation of grey matter and white matter is a basis of the conductivity of brain tissue. In the human brain, anisotropy is related to the fiber architecture of cortex and laminae and the gross anatomical regions related to cortical gyri and local curvature. The anisotropy is a consequence of a difference in conductivity of different tissues and matter types in the brain. Anisotropy could be measured using various technologies such as echo planar imaging and diffusion tensor imaging. Previous work has shown anisotropy could be associated with brain physiology and pathology [25,26], with a 10-fold relationship at the Z axis tensor where only one variable factor was required to produce an anisotropy. Our work chose this setting by which the influence of asymmetric conductivity on the propagation of deep-brain rhythm sources could be found significantly compared to symmetric conductivity. In a real system, either the anisotropy or the propagation direction of a deep-brain rhythm generator could be far more complicated than this anisotropic setting and the representation of conducting direction in this study. There is currently no clear conclusion on the parameters that would model the real brain anisotropy and the propagation pathway of a low-frequency brain rhythm generator. From the perspective of overall brain physiology, the anisotropic model is more realistic, and isotropic model is ideal. However, by comparing the conduction of isotropic media, it can also provide a reference for the conduction effect of isotropic local brain tissues.

5. Conclusions

On the frontal lobe surface, the space distribution of significant EEG rhythm power (SP area) in a low-frequency frequency band was investigated via simulation. The SP area was considerably affected by many factors. The anisotropy of brain tissue with 3D asymmetrical conduction tensor could limit or enlarge the available recording area of significant power, depending on the conducting direction of the rhythm generator in deep brain. Considering the partial volume effect, the narrower the SP area, the greater the power values. This result implies that the accurate and cautious placement of EEG electrodes is very important during measurement, and that traditional analysis on mean power must be conducted under very strict conditions such as potential localization and possible spread pathway of the rhythm generator in deep brain. Therefore, this study may be helpful to researchers and practitioners involved with the measurement and analysis of spontaneous low-frequency EEG rhythms at a system level.

To the best of our knowledge, the simulation described in this study is the first to report the 3D asymmetrical conduction characteristic of anisotropic conductivity tensor. Future investigations should focus on linking the shapes of gray and white matter, anisotropic tissue and other related factors to study their influence on low-frequency rhythms on the head surface.

Author Contributions: Conceptualization, H.C. and M.G.; methodology, H.C., M.G. and S.C.; software, H.C. and X.F.; validation, H.C and C.X.; formal analysis, H.C. and Z.S.; resources, M.G. and S.C.; data curation, H.C. and X.F.; writing—original draft preparation, H.C. and M.G.; writing—review and editing, H.C., M.G., A.N.B. and C.C.; visualization, H.C. and X.F.; supervision, M.G., S.C. and C.C.; project administration, H.C. and M.G.; funding acquisition, M.G., S.C. and C.C. All authors have read and agreed to the published version of the manuscript.

Funding: The work is partly financially supported by Grants sponsored by Hebei Province (E2019202019/ZD2021025). Fu Xiaoxuan is supported by Medical University of South Carolina as a research assistant (Program Number: P-1-00715) and Chen Chao is supported by in part by National Natural Science Foundation of China (61806146) and National Key Research & Development Program of China (2018YFC1314500).

Institutional Review Board Statement: The study was conducted according to the guidelines of the Declaration of Helsinki, and approved by the Biomedical Ethics Committee of Hebei University of Technology (protocol code: HEBUThMEC2021015 and date of approval: January 2021).

Informed Consent Statement: Informed consent was obtained from all subjects involved in the study.

Data Availability Statement: Not applicable.

Conflicts of Interest: The authors declare no conflict of interest.

References

1. McDonnell, J.; Murray, N.P.; Ahn, S.; Clemens, S.; Everhart, E.; Mizelle, J.C. Examination and Comparison of Theta Band Connectivity in Left- and Right-Hand Dominant Individuals throughout a Motor Skill Acquisition. *Symmetry* **2021**, *13*, 728. [CrossRef]
2. Buzsáki, G. *Rhythms of the Brain*; Oxford University Press: Oxford, UK, 2009.
3. Herweg, N.A.; Solomon, E.A.; Kahana, M.J. Theta Oscillations in Human Memory. *Trends Cogn. Sci.* **2020**, *24*, 208–277. [CrossRef]
4. Fu, X.; Wang, Y.; Ge, M.; Wang, D.; Gao, R.; Wang, L.; Guo, J.; Liu, H. Negative Effects of Interictal Spikes on Theta Rhythm in Human Temporal Lobe Epilepsy. *Epilepsy Behav.* **2018**, *87*, 207–212. [CrossRef]
5. Shirhatti, V.; Borthakur, A.; Ray, S. Effect of Reference Scheme on Power and Phase of the Local Field Potential. *Neural Comput.* **2016**, *28*, 882–913. [CrossRef]

6. Alarcon, G.; Binnie, C.D.; Elwes, R.D.C.; Polkey, C.E. Power Spectrum and Intracranial EEG Patterns at Seizure Onset in Partial Epilepsy. *Electroencephalogr. Clin. Neurophysiol.* **1995**, *94*, 326–337. [CrossRef]
7. Chauvière, L.; Rafrafi, N.; Thinus-Blanc, C.; Bartolomei, F.; Esclapez, M.; Bernard, C. Early Deficits in Spatial Memory and Theta Rhythm in Experimental Temporal Lobe Epilepsy. *J. Neurosci.* **2009**, *29*, 5402–5410. [CrossRef]
8. Ge, M.; Wang, D.; Dong, G.; Guo, B.; Gao, R.; Sun, W.; Zhang, J.; Liu, H. Transient Impact of Spike on Theta Rhythm in Temporal Lobe Epilepsy. *Exp. Neurol.* **2013**, *250*, 136–142. [CrossRef] [PubMed]
9. Lindén, H.; Pettersen, K.H.; Einevoll, G.T. Intrinsic Dendritic Filtering Gives Low-pass Power Spectra of Local Field Potentials. *J. Comput. Neuroence* **2010**, *29*, 423–444. [CrossRef] [PubMed]
10. Winson, J. Loss of Hippocampal Theta Rhythm Results in Spatial Memory Deficit in the Rat. *Science* **1978**, *201*, 160–163. [CrossRef]
11. Miriam, G.; Markus, B.; Tobias, B. Frequency- and State-dependent Effects of Hippocampal Neural Disinhibition on Hippocampal Local Field Potential Oscillations in Anesthetized Rats. *Hippocampus* **2020**, *30*, 1021–1043.
12. Agrita, D.; Supratim, R. Spatial Spread of Local Field Potential is Band-pass in the Primary Visual Cortex. *J. Neurophysiol.* **2016**, *116*, 1986–1999.
13. Bédard, C.; Rodrigues, S.; Roy, N.; Contreras, D.; Destexhe, A. Evidence for Frequency-dependent Extracellular Impedance from the Transfer Function between Extracellular and Intracellular Potentials. *J. Comput. Neurosci.* **2010**, *29*, 389–403. [CrossRef]
14. Ge, M.; Fu, X.; Zhang, J.; Chen, S.; Chen, Y.; Gao, R.; Zhang, H. The Influences of Tissue Anisotropy and Source Activity on Power and Phase Stability of Low-frequency EEG Rhythms: A Mathematical Observation of the Forward Problem Model. *Biomed. Phys. Eng. Express* **2016**, 2. [CrossRef]
15. Łęski, S.; Lindén, H.; Tetzlaff, T.; Pettersen, K.H.; Einevoll, G.T. Frequency Dependence of Signal Power and Spatial Reach of the Local Field Potential. *PLOS Comput. Biol.* **2013**, *9*, e1003137. [CrossRef]
16. Logothetis, N.K.; Kayser, C.; Oeltermann, A. In vivo Measurement of Cortical Impedance Spectrum in Monkeys: Implications for Signal Propagation. *Neuron* **2007**, *55*, 809–823. [CrossRef]
17. Lubenov, E.V.; Siapas, A.G. Hippocampal Theta Oscillations are Travelling Waves. *Nature* **2009**, *459*, 534–539. [CrossRef]
18. Brody, D.A.; Terry, F.H.; Ideker, R.E. Eccentric Dipole in a Spherical Medium: Generalized Expression for Surface Potentials. *IEEE Trans. Bio. Med. Eng.* **1973**, *20*, 141–143. [CrossRef] [PubMed]
19. Torres, F. Electroencephalography: Basic Principles, Clinical Applications and Related Fields. *Arch. Neurol.* **1983**, *40*, 191–192. [CrossRef]
20. Vecchio, A.; De Pascalis, V. EEG Resting Asymmetries and Frequency Oscillations in Approach/Avoidance Personality Traits: A Systematic Review. *Symmetry* **2020**, *12*, 1712. [CrossRef]
21. Nicholson, P.W. Specific Impedance of Cerebral White Matter. *Exp. Neurol.* **1965**, *13*, 386–401. [CrossRef]
22. Soret, M.; Bacharach, S.L.; Buvat, I. Partial-volume Effect in PET Tumor Imaging. *J. Nucl. Med.* **2007**, *48*, 932–945. [CrossRef]
23. De Munck, J.C.; Wolters, C.; Clerc, M. EEG and MEG: Forward Modeling. In *Handbook of Neural Activity Measurement*; Brette, R., Destexhe, A., Eds.; Cambridge University Press: Cambridge, UK, 2012; pp. 192–256.
24. Ding, N.; Simon, J.Z. Power and Phase Properties of Oscillatory Neural Responses in the Presence of Background Activity. *J. Comput. Neuroence* **2013**, *34*, 337–343. [CrossRef]
25. Satzer, D.; Lanctin, D.; Eberly, L.E.; Abosch, A. Variation in Deep Brain Stimulation Electrode Impedance over Years Following Electrode Implantation. *Stereotact. Funct. Neurosurg.* **2014**, *92*, 94–102. [CrossRef] [PubMed]
26. Wolters, C.H.; Anwander, A.; Tricoche, X.; Weinstein, D.; Koch, M.A.; Macleod, R.S. Influence of Tissue Conductivity Anisotropy on EEG/MEG Field and Return Current Computation in a Realistic Head Model: A Simulation and Visualization Study Using High-resolution Finite Element Modeling. *NeuroImage* **2006**, *30*, 813–826. [CrossRef]

MDPI
St. Alban-Anlage 66
4052 Basel
Switzerland
Tel. +41 61 683 77 34
Fax +41 61 302 89 18
www.mdpi.com

Symmetry Editorial Office
E-mail: symmetry@mdpi.com
www.mdpi.com/journal/symmetry